Principles of Interferometric and Polarimetric Radiometry

IEEE Press
445 Hoes Lane
Piscataway, NJ 08854

Principles of Interferometric and Polarimetric Radiometry

Ignasi Corbella
Universitat Politècnica de Catalunya
Barcelona
Spain

IEEE*Press*
WILEY

Published by John Wiley & Sons, Inc., Hoboken, New Jersey.
Published simultaneously in Canada.

Library of Congress Cataloging-in-Publication Data is applied for:

Hardback ISBN 9781394255108

Cover Design: Wiley
Cover Image: © ESA – P. Carrill

Set in 9.5/12.5pt STIXTwoText by Straive, Chennai, India

To Roser, Pau and Oriol. And also to my parents Leopoldo and Isabel

Contents

Foreword

To my knowledge, this is the first book ever published dedicated to the topic of aperture synthesis (or interferometry) for the remote sensing of the Earth. Much of its contents can, however, be applied to other types of microwave radiometers and applications. The author, Prof. Ignasi Corbella of the Polytechnic University of Catalunya (UPC), Barcelona (Spain), is, in my humble opinion, the engineer, among all I know, who best understands aperture synthesis for Earth Observation, a field that started in the 1980s. Readers of this book will have the privilege to learn directly from him.

In fact, I started learning microwave theory having Ignasi as teacher at UPC back in 1983, and he was also the supervisor of my master's thesis in 1986, and of my PhD in 1996. Therefore, it is an honor to write this foreword to his book some 41 years after I attended his lessons at UPC. From 1993 till 1999, he was involved in several studies and experiments to develop the Microwave Imaging Radiometer with Aperture Synthesis (MIRAS) within contracts with the European Space Agency (ESA) I was responsible for. Since 2000 till today, Prof. Corbella has been part of the team supporting, in many aspects, ESA's Soil Moisture and Ocean Salinity (SMOS) mission, which carries MIRAS as its only payload. It is the SMOS mission, in which I am involved from ESA side, which has allowed us to enjoy this very long professional link and friendship.

The development of MIRAS is a story truly worth telling. It all started with the work of David LeVine (NASA-GSFC) and Carl Swift (University of Massachusetts at Amherst) in the late 1980s on the Electronically Scanning Thinned-Array Radiometer (ESTAR). ESTAR was a one-dimensional interferometer presenting a new technological solution to deploy a very large L-band antenna in space, to make observations in the 1400–1427 MHz protected band, to provide global maps of soil moisture and ocean salinity. Inspired by this research, and following some work carried out by the French Space Agency (CNES), in 1992, ESA decided to initiate the development of a two-dimensional interferometer: MIRAS.

During the first 10 years of MIRAS, Prof. Corbella was involved in the theoretical aspects, the calibration, and the image processing of the instrument. In principle, although aperture synthesis had never been applied to remote sensing, it was a matter of translating the knowledge accumulated in radio-astronomy to the field of Earth Observation. Given this, we were all quite confident that the new type of radiometer should work very finely.

Radio-astronomy was (and is) based on a theorem that was first formulated by the Dutch physicist Pieter Hendrik van Cittert in 1934, and four years later proved in a simpler way by another Dutch physicist, Frits Zernike. Zernike was a prominent scientist who was awarded the Nobel Prize in 1953 for having invented the Phase-Contrast Microscope. The so-called Van Cittert-Zernike theorem, or in short, the VCZ theorem, was the theoretical basis of radio-astronomy, and in turn, of ESA's initiative to embark on two-dimensional aperture synthesis for remote sensing.

Nobody would have anticipated that such solid theoretical grounds were going to be shaken by the development of MIRAS, to the point of proving them completely inadequate for the application of interferometry to Earth Observation.

Based on the so-far successful breadboarding of MIRAS subsystems and the scientific need to monitor soil moisture and ocean salinity, ESA's Earth Science Advisory Committee recommended the implementation of the SMOS mission on April 27, 1999, as second Earth Explorer Opportunity Mission.

By 2002, the breadboarding activities (MIRAS Demonstrator Pilot Project – MDPP) were running in parallel with the Phase B of the SMOS project. The prototype built within the MDPP consisted of three linear arrays of four receivers each which could be mechanically moved along three horizontal tracks 120° equi-spaced from each other. The MDPP also included the optical harness, the correlator, and the internal calibration subsystems.

In December 2002, the MDPP demonstrator was brought inside the Electro Magnetic Compatibility (EMC) chamber of INTA (Instituto Nacional de Técnica Aeroespacial, Madrid) for the very first end-to-end test. The imaging algorithm, based on the VCZ theorem, was predicting high correlation values due to the high emissivity of the microwave absorber material (at room temperature) of the walls of the EMC chamber. To our surprise, the correlations were two orders of magnitude (!) smaller than those predicted by the VCZ theorem. After carefully assessing the breadboard, and realizing that the hardware was working well, we had to accept that the value of the correlations was real. As mentioned, SMOS was already in Phase B …

The UPC team, led by Prof. Corbella, was responsible for processing the data from the INTA test. Therefore, we all packed the hardware and went on Christmas Holidays with the heavy weight of the insignificant correlation values on our shoulders…

It was during the Christmas of 2022 that Ignasi set himself into resolving the mystery. He found a paper on noise waves and passive linear multiports of 1991 which gave the expression of the correlation of the outgoing noise waves of a passive microwave circuit. Such problem resembled the configuration of the MDPP breadboard inside the EMC chamber. The article referred to a theorem stated by a Dutch engineer, Hendrik Bosma, in his PhD thesis of 1967. According to Bosma's theorem, the correlation of the noise waves coming out of a passive microwave circuit in thermal equilibrium with its terminations, as the MDPP breadboard inside the EMC chamber, had to null.

Based on these findings, Prof. Corbella carefully reformulated the basic measurement of interferometry, the correlation between the output signals of two receivers forming a baseline, considering the presence of the neighboring receivers as well as the physical temperature of the receivers and the target. By February 2003, he had come with a new equation, more general than the CVZ theorem, which could explain the correlations in the INTA chamber as well as those measured by radio-telescopes (for large antennas and spacings, in wavelength units, the new formulation reduced to the CVZ theorem). Later the same year, Prof. Corbella had derived the polarimetric version of his equation. The new equation, which I very deservedly coined as the *Corbella Equation*, was first published in TGARS, August 2004.

The fact that the VCZ theorem could not explain all interferometric experiments, like the test of the MDPP breadboard in the EMC chamber (by two orders of magnitude), was further evidenced when the opposite test was carried out: the imaging of the Cold Sky at the Dwingeloo radio-observatory facility in The Netherlands. In the Dwingeloo experiment, four receivers were manually moved along three arms to get an image of the Cold Sky. While the VCZ theorem was predicting now very low correlations based on the low brightness temperature of the Cosmic Microwave Background Radiation (CMBR), around 2.7 K, the expected values according to the Corbella equation were two orders of magnitude larger in this case. The measurements were perfectly in accordance with Corbella's predictions and, once again, about two orders of magnitude away from the VCZ theorem-predicted values.

The Corbella Equation was then adopted to process the data of the SMOS mission instead of the VCZ theorem. The calibration approach had to be adapted, and, fortunately, the Flat Target Response of the instrument could be obtained by using the CMBR and the subsequent Flat Target Transformation could be devised to remove the $-T_r$ term Corbella had introduced into the VCZ equation. With the data processing and calibration following his equation, Ignasi had brought the SMOS project back on track half way through Phase B.

It is worth noting that it took some time for some engineers to accept and understand the profound implications of the "Corbella Equation." After all, the VCZ theorem, established by renowned (Nobel Prize awarded) scientists, had been working perfectly well in radio-astronomy for decades. It was only a matter of making new experiments in new conditions (as taking the MDDP breadboard inside the INTA EMC chamber), realizing of other scientists' research (as Bosma's PhD thesis), and having a clear and humble mind to put the pieces of the puzzle together. This ultimately culminated in a scenario where *David beat Goliath* with a new more general formulation of aperture synthesis.

This book starts with a tedious Chapter 1 on probability concepts and stochastic processes, Hilbert transform, and analytic signals. However, the usefulness of its contents will be much appreciated throughout the rest of the book, as they will be frequently called upon to make progress in the formulation of the different flavors of interferometers.

Chapter 2 introduces the basic concepts of radiometry. Prof. Corbella has made an effort to explain interferometry and all its equations in a way that collapses to the case of a conventional total power radiometer. In this respect, it makes interferometry easier to digest for engineers knowledgeable in conventional radiometry.

The Corbella Equation is derived in Chapter 3. This chapter can be considered central and including the nucleus of the aperture synthesis theory applied to remote sensing. An engineer confronted with the design of this type of radiometers should absorb these contents thoroughly. This chapter introduces also the tools necessary to derive the sensitivity of an interferometer, and the subtleties which occurr when digitizing the output signals.

The discussion on sensitivity continues more deeply in Chapter 4, which focuses on the image processing. The basic theory on how to retrieve the brightness temperature from the visibility function are explained in great detail. Key concepts such as the reciprocal grids, the apodization window, the G-matrix, the aliasing, and the noise floor are found here.

The implementation of the interferometry theory into practical instruments is covered by Chapter 5. The nomenclature of this chapter is dense, but precise and consistent with the previous chapters. The reader is invited to understand and carefully realize the subtle differences between the different instrument architectures, including the effects of digitalization of the signals and the different ways complex correlations can be obtained (whether using two or four multipliers, with or without digitalization).

Finally, Chapter 6 gives the reader an overview on the characterization and calibration of an interferometric radiometer, based on the Corbella Equation. The key parameters which have to be found to correct the raw visibilities into a calibrated set are described, together with the methods to retrieve their value.

For every new topic in engineering and science, there is always a "first reference book," a "bible." I believe this book is the "bible" on polarimetric aperture synthesis for remote sensing. This book will be tremendously useful for engineers working with this type of instruments, as well as a valuable reference for a university course on interferometry for Earth Observation.

I can see a future for aperture synthesis in remote sensing, as demonstrated by the SMOS mission and, in great part, thanks to Prof. Corbella's contributions. Interferometry combines well with formation flying as a way to deploy extremely large apertures. When the techniques to master formation flying will be sufficiently developed, interferometry will spread further. I do hope graduated students as well as practicing engineers will find in this book a great support for decades to come.

Oegstgeest 26 June, 2024
(14 years, 7 months, and 24 days from SMOS launch, still up and running)

Manuel Martín-Neira
Senior Microwave Radiometer Engineer
European Space Agency

About the Author

Ignasi Corbella was born in Barcelona, Spain, in 1955. He received the telecommunication engineering degree and doctoral degree in telecommunication engineering from the Universitat Politècnica de Catalunya (UPC), Barcelona, Spain, in 1977 and 1983, respectively.

In 1976, he was with the Microwave Laboratory, School of Telecommunication Engineering, UPC, as a research assistant, where he worked on passive microwave integrated-circuit design and characterization. In 1979, he was with Thomson-CSF, Paris, France, working on microwave oscillators' design. In 1982, he was Assistant Professor, in 1986, Associate Professor, and, in 1993, Full Professor with UPC, where he is currently teaching basic microwaves and antennas at the undergraduate level and graduate courses on nonlinear microwave circuits. Since 1993, he has been actively participating as a researcher with the European Space Agency (ESA) SMOS mission in the frame of several contracts, directly with ESA, with the payload prime contractor EADS-Casa Espacio (Spain) or with the operational processor prime contractor Deimos Engenharia (Portugal). His expertise includes, among others, fundamentals of interferometric aperture synthesis radiometry, radiometer calibration, image reconstruction, radiometer hardware specification, and payload characterization.

From 1993 to 1997, he was Academic Director of the School of Telecommunications Engineering. From 2001 to 2003, he was Director of the Department of Signal Theory and Communications, UPC. From 1998 to 1999, he was Guest Researcher at the NOAA/Environmental Technology Laboratory (Boulder-Colorado), developing methods for total-power radiometer calibration and data analysis. In 2015, he was a visiting scholar at the University of Colorado at Boulder, working on emission models from layered media. From 2004 to 2010, he was a member of the SMOS Science Advisory Group (SAG) and since 2010 of the SMOS Quality Working Group (QWG).

Dr. Corbella was the General Chairman of the IEEE 2007 International Geoscience and Remote Sensing Symposium (IGARSS), Barcelona.

He also was the Scientific Coordinator of a Dictionary of Telecommunication terms in Catalan language, with more than 4000 entries, published in March 2007.

Preface

Invention, it must be humbly admitted, does not consist in creating out of void, but out of chaos. (Mary Shelley, 1831)

This book compiles the expertise the author has gathered during his 30-year-long work on Microwave Interferometric and Polarimetric Radiometry, particularly in relation to the "Soil Moisture and Ocean Salinity" (SMOS) mission of the European Space Agency. The SMOS satellite was put into orbit in November 2009 with the objective of acquiring global measurements of the two important geophysical parameters mentioned in its full name. The satellite includes a single payload named MIRAS (Microwave Imaging Radiometer with Aperture Synthesis) that is the first ever microwave interferometric radiometer designed for observing the Earth surface from space.

Except in some very few cases where text previously written by the author is reused, practically all writing is completely new, not extracted from any existing source. Many concepts and methods were indeed published in open literature or internal reports, often in collaboration with other authors (see acknowledgments) but, for the present work, the author has made an important effort to rewrite all the material from scratch in order to present the information with consistent terminology and complemented with detailed procedures and concepts not sufficiently and/or rigorously explained in the original papers. In consequence, even for the more established themes, for example in Chapters 1 and 2, the reader may find some explanations or points of view different from those found in published treatises. This feature will be probably useful to complement different visions of similar concepts.

The book starts with a review of the fundamental concepts that are used throughout the rest of the volume. They range from random variables and stochastic processes to microwave circuits and antennas. This first chapter is essential to set the terminology and symbols and is frequently referenced in all others. Even though the reader may be familiar with these themes in general, the author recommends to start exploring this chapter, at least in a "fast read" mode, before going to others more specific.

The second chapter is dedicated to the quite established topic of Microwave Radiometry, focused especially on instrumental aspects. It is included as a kind of introduction of the main subject of the book, interferometric and polarimetric radiometry, in order to present the fundamental concepts, formulate the needed hypotheses, and derive the theorems that will be extended later to more complex systems. It should be helpful in facilitating the readers the study of the rest of the book, even if they have a sound knowledge of microwave radiometric systems in general.

Interferometric and polarimetric radiometry is presented in the third chapter. This is perhaps the core of the book as it presents the fundamentals of the interferometric technique in the microwave region of the spectrum. Optical systems are simply not covered, although they may have important similitudes with what is here presented. Unavoidably, the formulation is biased to its application in Earth observation, as this is the background of the author. Nevertheless, most of the results are perfectly applicable to Radioastronomy since the fundamental concept is exactly the same. It may happen that experts in this field find some material in this book interesting for their own research.

Combination of multiple baselines to form an image is the subject of the fourth chapter. Here, the content is even more specific than that of the previous one in terms of application to large instruments dedicated to Earth observation from space or aircraft. As a matter of fact, much of the material relies on algorithms or procedures actually implemented in the MIRAS instrument. This chapter should be useful to people designing similar systems for related applications.

Chapter 5 deals with specific instrument techniques. Consistent with a top-down vision, the general theory developed in Chapters 2–4 is here materialized into dedicated technologies. Many of the items included are already implemented in existing instruments or intended for their use in future ones. Digital processing, essential in today's systems, has been intentionally relegated to this chapter as a convenient way to treat measured data. Up to this point the world in this book is analog.

Finally, calibration and characterization are covered in Chapter 6. These are essential aspects of any radiometer, either total power or interferometric, maybe less conceptual, more oriented to techniques and procedures, and mathematically simpler. Commonly used methods are here described and particularly those utilized in the MIRAS instrument. This chapter has been written with the aim of helping engineers designing an instrument of the same or similar kind.

An appendix is also included with some specific and supposedly well-known concepts and definitions, such as Fourier transform (analog and discrete), special functions, spherical coordinates, solid angles, and others. Surely, in many cases the advanced reader will not need to consult it, although it can serve as a quick reference guide.

The author hopes that this book will help engineers and scientists facing the design of any kind of Microwave Interferometric and/or Polarimetric Radiometer, especially for application in Earth observation.

Barcelona
September 12, 2024

Ignasi Corbella

Acknowledgments

Above all, I would like to express my most sincere gratitude to Dr. Manuel Martin-Neira, Senior engineer at the European Space Agency, for his careful revision of the manuscript. His opportune comments always hit in some weakness of the original writing and, once endorsed, enhanced the quality of the final version. I am also thankful to him for having written the excellent Foreword and for the generous words he expresses in it about my work and person.

A substantial amount of information in this book can be found scattered in articles and conference proceedings published by myself, always in collaboration with other colleagues. Three persons deserve to be mentioned in this regard: the already mentioned Dr. Manuel Martin-Neira, Dr. Francesc Torres and Dr. Adriano Camps, both professors at the "Universitat Politècnica de Catalunya." This book would not have been possible without the exceptional job they have carried out during many years of fruitful cooperation. I am indebted to all three.

Ignasi Corbella

1

Signals, Receivers, and Antennas

A radiometer, as a measuring system, processes input information in the form of *signals* and converts them to usable magnitudes proportional to observational parameters. Since all signals encountered in radiometry are of random nature, it is of paramount importance to establish a solid background of knowledge on stochastic processes. With the objective in mind of establishing notation and in order to have concepts and definitions at hand, the most relevant of them are collected in this chapter. Section 1.1 is an introduction to random variables, a fundamental tool on which random signals, developed in the section 1.2, are based. The reader is assumed to be familiar with basic random variables and processes, so only the definitions and methods needed for this book are included. More complete treatments can be found in classical books on statistics and its application to signal theory [1–5].

Receivers and *antennas* are the two basic building blocks of any radiometer, so mastering the main concepts about these two components is fundamental to perceive the details of radiometric operation. Both of them are also used in Telecommunications or Radar and are subjects of basic university courses. Consequently, there is a considerable amount of literature about them. This chapter presents their usage in microwave radiometry, aimed at gathering all important definitions and relations used later in this book. To know more, the interested reader may wish to complement the information with basic textbooks, as for example [6–10] among others. In any case, some familiarity with microwave circuits, including noise, and antenna theory is assumed.

1.1 Random Variables, Real and Complex

A real random variable is rigorously defined in, for example, [1]. It is fundamentally described by the "probability density function," which is the probability that the variable takes a given value within a differential interval. For two random

Principles of Interferometric and Polarimetric Radiometry, First Edition. Ignasi Corbella.

variables, the "joint density function" is the same concept applied to a differential area in a cartesian plane defined by all possible values of both. Two real random variables are said to be *independent* if they describe completely unrelated events. In this case, their joint density function is equal to the product of both individual density functions. Not surprisingly, if two variables are independent, arbitrary functions of each one of them are also independent.

A complex random variable is defined by the joint density function of its real and imaginary parts. Most of the definitions and equations provided in this section apply to complex variables, although they are also valid for real ones just by setting the imaginary part equal to zero. To avoid confusion, *real variables are represented by upright roman typeface* ($\mathrm{x, y, z}\ldots$) *while complex variables use italic shape* ($x, y, z \ldots$). Complex conjugation is denoted by a superscript asterisk.

1.1.1 Definitions

The "expected value" or "mean" of a real random variable x is

$$\overline{\mathrm{x}} \triangleq \int_{-\infty}^{\infty} \mathrm{x} f(\mathrm{x}) d\mathrm{x} \tag{1.1}$$

where $f(\mathrm{x})$ is the probability density function. In [1], it is demonstrated that this operation is linear so, for any two random variables x and y, $\overline{\alpha \mathrm{x} + \beta \mathrm{y}} = \alpha \overline{\mathrm{x}} + \beta \overline{\mathrm{y}}$ where α and β are arbitrary constants.

Given a *complex* random variable $z = \mathrm{x} + j\mathrm{y}$, the following definitions apply:

$$\begin{aligned}
\text{Mean:}\quad & \overline{z} = \overline{\mathrm{x}} + j\overline{\mathrm{y}} \\
\text{Variance:}\quad & \sigma_z^2 = \overline{|z - \overline{z}|^2} = \overline{|z|^2} - |\overline{z}|^2 \\
& \sigma_z^2 = \sigma_\mathrm{x}^2 + \sigma_\mathrm{y}^2 \\
\text{Standard deviation:}\quad & \sigma_z \quad \text{(square root of variance)} \\
\text{Autocorrelation:}\quad & \overline{|z|^2} = \overline{zz^*}
\end{aligned} \tag{1.2}$$

The standard deviation is *always* real and positive and it has the same units as the random variable itself. Conjugating a random variable does not change the standard deviation. The following property is immediate from the definition:

$$\text{if } z = \alpha x + \delta \text{ then } \sigma_z = |\alpha| \sigma_x \tag{1.3}$$

where α and δ are complex constants, and x is a complex random variable, not to be confused with x, the real part of z. The additive constant δ does not change the standard deviation.

The generalization of variance and autocorrelation for two arbitrary complex random variables z and w is

$$\begin{aligned}
\text{Covariance:}\quad & C_{zw} = \overline{(z - \overline{z})(w - \overline{w})^*} = \overline{zw^*} - \overline{z}\,\overline{w}^* \\
\text{Cross-correlation:}\quad & \overline{zw^*}
\end{aligned} \tag{1.4}$$

The covariance is a complex number satisfying the Hermitian property: $C_{wz} = C^*_{zw}$. By convention, if z and w represent both the same variable, the covariance becomes the variance: $C_{zz} \equiv \sigma_z^2$.

1.1.1.1 Correlated, Uncorrelated, and Independent Random Variables

Variables z and w are said to be "uncorrelated" if their covariance vanishes, which is equivalent to

$$\overline{zw^*} = \overline{z}\,\overline{w}^* \quad \text{(uncorrelated variables)} \tag{1.5}$$

This definition does not necessarily imply that the real and imaginary parts of one of the variables are uncorrelated with those of the other, as it is readily deduced by expanding the product of two complex numbers. But the opposite is true: if the real and imaginary parts are uncorrelated, then the complex variables also. In this case, the two following identities hold

$$\begin{aligned} \overline{zw} &= \overline{z}\,\overline{w} \\ \overline{zw^*} &= \overline{z}\,\overline{w}^* \end{aligned} \quad \text{(uncorrelated real and imaginary parts)} \tag{1.6}$$

Complex variables are independent only if the corresponding real and imaginary parts are independent. As expected, independent variables are always uncorrelated (see [1] for a proof) so the above equations hold also for them. The converse is only true in some cases, as for example in *real* variables with Gaussian statistics. In this particular situation uncorrelated and independent are equivalent concepts. In general, it cannot be assumed that individual functions of two uncorrelated variables are also uncorrelated.

A property similar to (1.3) is:

$$\left.\begin{aligned} z &= \alpha x + \delta \\ w &= \beta y + \gamma \end{aligned}\right\} \Rightarrow C_{zw} = \alpha\beta^* C_{xy} \tag{1.7}$$

where all Greek letters are complex constants and x and y complex random variables. Furthermore, the covariance of the two linearly related variables z and x above is

$$C_{zx} = \alpha\sigma_x^2 = \frac{\sigma_z^2}{\alpha^*} = \frac{\alpha}{|\alpha|}\sigma_z\sigma_x \tag{1.8}$$

where (1.3) has been used to write the last equality. These variables are said to be "correlated," which in many textbooks is quantified by the so-called "correlation coefficient"

$$r_{zx} = \frac{C_{zx}}{\sigma_z\sigma_x} \tag{1.9}$$

with an absolute value equal to one in this case. In [1] it is demonstrated the inverse statement: if the absolute value of the correlation coefficient is 1, then the variables are linearly related. For uncorrelated variables this coefficient is null and in general its amplitude is bounded between zero and one. The correlation coefficient is no longer used in this book.

As expected, if two variables are correlated with respect to a third one they are correlated to each other. Making y equal to x in (1.7)

$$\left.\begin{aligned} z &= \alpha x + \delta \\ w &= \beta x + \gamma \end{aligned}\right\} \Rightarrow C_{zw} = \alpha\beta^* \sigma_x^2 = \frac{\alpha\beta^*}{|\alpha||\beta|}\sigma_z\sigma_w \tag{1.10}$$

and therefore the correlation coefficient has unit amplitude.

1.1.2 Operations

The sum of N complex random variables

$$s = \sum_{i=1}^{N} z_i \tag{1.11}$$

has variance

$$\sigma_s^2 = \overline{|s - \overline{s}|^2} = \sum_{i=1}^{N}\sum_{j=1}^{N} \overline{(z_i - \overline{z}_i)(z_j - \overline{z}_j)^*} \tag{1.12}$$

The double sum can be split into N terms with equal indices ($i = j$) and the rest, in turn grouped by mutually conjugated terms. Then

$$\sigma_s^2 = \sum_{i=1}^{N} \sigma_{z_i}^2 + 2\sum_{i=1}^{N-1}\sum_{j=i+1}^{N} \Re e[C_{z_i z_j}] \tag{1.13}$$

If variables are uncorrelated (or, more specifically, if the real part of the covariance vanishes), then the second term vanishes and the variance of the sum becomes equal to the sum of variances.

$$\sigma_s^2 = \sum_{i=1}^{N} \sigma_{z_i}^2 \quad \text{Uncorrelated variables} \tag{1.14}$$

If furthermore all variables have equal standard deviation ($\sigma_{z_i} = \sigma_z \forall i$), then $\sigma_s = \sqrt{N}\sigma_z$. On the other hand, if all variables are replicas of the same one ($z_i = z \forall i$), then $s = Nz$ and the standard deviation is $\sigma_s = N\sigma_z$. In this case, the variables are correlated. These two important results are highlighted here below for clarity

$$\begin{aligned} &\text{if } \sigma_{z_i} = \sigma_z \forall i \Rightarrow \sigma_s = \sqrt{N}\sigma_z \quad (z_i \text{ uncorrelated}) \\ &\text{if } z_i = z \forall i \quad \Rightarrow \sigma_s = N\sigma_z \quad (z_i \text{ correlated}) \end{aligned} \tag{1.15}$$

The *average* of N random variables is their sum (1.11) divided by the number of them N. Using (1.3) makes it apparent that the standard deviation is also that of the sum divided by N

$$m = \frac{1}{N}\sum_{i=1}^{N} z_i \Rightarrow \sigma_m = \frac{\sigma_s}{N} \tag{1.16}$$

where σ_s is the square root of the variance (1.13). If the real part of the covariance between any two random variables is null and all variables have the same standard deviation $\sigma_{z_i} = \sigma_z \forall i$, then

$$\sigma_m = \frac{\sigma_z}{\sqrt{N}} \quad \text{(uncorrelated with equal standard deviation)} \tag{1.17}$$

which means that averaging N uncorrelated random variables reduces the standard deviation of a single one (assumed the same for all) by a factor of $\sqrt{N}$. If they are correlated, the standard deviation is unchanged, as is clearly deduced from the second line of (1.15). Obviously, due to the linearity of the mean operator, the mean of m is, in any case, equal to the average of the mean values $\overline{z_i}$

$$\overline{m} = \frac{1}{N}\sum_{i=1}^{N}\overline{z_i} \tag{1.18}$$

which, in case of equal mean variables $\overline{z_i} = \overline{z} \forall i$ reduces to $\overline{m} = \overline{z}$. In other words, *averaging preserves the mean.*

1.1.2.1 Product

Consider now the product of two random variables $p = zw$. The variance of p, according to the definition (1.2) is

$$\sigma_p^2 = \overline{pp^*} - |\overline{p}|^2 = \overline{zw\,z^*w^*} - |\overline{zw}|^2 \tag{1.19}$$

In general this expression cannot be easily evaluated. However, if z and w are *independent*, then (1.6) allows separating the second term into the product of two individual means squared. Furthermore, in this case, the two real products zz^* and ww^* are independent so, again using (1.6), the first term is also separable,

$$\sigma_p^2 = \overline{|z|^2}\;\overline{|w|^2} - |\overline{z}|^2|\overline{w}|^2 \quad (z \text{ and } w \text{ independent}) \tag{1.20}$$

Using now (1.2), this variance can be written in terms of the corresponding individual ones,

$$\sigma_p^2 = |\overline{z}|^2\sigma_w^2 + |\overline{w}|^2\sigma_z^2 + \sigma_z^2\sigma_w^2 \tag{1.21}$$

Note that, if one of the two variables is a constant so it has zero standard deviation, this equation is consistent with (1.3). In general, for zero mean variables

$$\sigma_p = \sigma_z\sigma_w \quad (z \text{ and } w \text{ zero mean and independent}) \tag{1.22}$$

which means that the standard deviation of the product is equal to the product of standard deviations. Otherwise, it is convenient to rewrite (1.21) as

$$\sigma_p^2 = \left(|\overline{z}|^2 + \frac{\sigma_z^2}{2}\right)\sigma_w^2 + \left(|\overline{w}|^2 + \frac{\sigma_w^2}{2}\right)\sigma_z^2 \tag{1.23}$$

which, if the two variables involved have small standard deviation with respect to their means, can be approximated to

$$\sigma_p^2 \approx |\overline{z}|^2 \sigma_w^2 + |\overline{w}|^2 \sigma_z^2 \tag{1.24}$$

so, in this case, the variance of the product is equal to the sum of variances of the individual variables weighted by the mean of the other variable.

1.1.3 Normal Random Variables

A complex random variable $z = x + jy$ is said to be *normal* if its real and imaginary parts are *jointly* normal, meaning that their joint density function is Gaussian:

$$f(x, y) = \frac{1}{2\pi\sqrt{\|C\|}} e^{-\frac{1}{2}\Delta^t C^{-1} \Delta} \tag{1.25}$$

where $\|C\|$ its determinant of the "covariance matrix"

$$C = \begin{bmatrix} \sigma_x^2 & C_{xy} \\ C_{xy} & \sigma_y^2 \end{bmatrix} \tag{1.26}$$

Δ the column vector

$$\Delta = \begin{bmatrix} x - \overline{x} \\ y - \overline{y} \end{bmatrix} \tag{1.27}$$

and Δ^t its transpose. The definition (1.25) is extended to N jointly normal *complex* random variables with only changing the 2π factor by $(2\pi)^N$ and extending correspondingly the matrices. In this case, they are conveniently partitioned as

$$C = \begin{bmatrix} C_{rr} & C_{ri} \\ C_{ir} & C_{ii} \end{bmatrix} \; ; \; \Delta = \begin{bmatrix} \Delta_r \\ \Delta_i \end{bmatrix} \tag{1.28}$$

where the subscripts *r* and *i* refer to the real and imaginary parts of the variables, respectively.

Having four *zero-mean* jointly normal variables z_1 to z_4, it can be shown that they satisfy the following important property

$$\overline{z_1 z_2 z_3 z_4} = \overline{z_1 z_2}\;\overline{z_3 z_4} + \overline{z_1 z_3}\;\overline{z_2 z_4} + \overline{z_1 z_4}\;\overline{z_2 z_3} \tag{1.29}$$

Its proof is out of the scope of this book, but the interested reader may consult for example [2], in which it is termed as "moment theorem for normal random variables." It is demonstrated for *real* random variables using the so-called "characteristic function," defined as the Fourier transform of the probability density function. Besides the formal proof of [11], it is quite easy to infer that it also holds for *complex* variables: linear combinations of jointly normal variables are known to be also jointly normal (see example 7–12 of [1]). Consequently, if x_i

and y_i ($i = 1$ to 4) are jointly normal variables, the four random variables defined by their sum $x_i + y_i$ are also jointly normal, so they satisfy (1.29). On the other hand, both x_i and y_i individually satisfy the same equation, so it can be concluded that, in general, *if* (1.29) *holds for* x_i *and* y_i, *it holds also for their sum* $x_i + y_i$. Now, since multiplying by j has no effect on the four-term product, it follows that (1.29) also applies for jy_i, so it must hold for $z_i = x_i + jy_i$.

1.1.3.1 Circularity

Of particular interest is the case of "circular" complex jointly normal variables. They are defined such that they have *zero-mean* and covariance matrix (1.28) satisfying

$$C_{rr} = C_{ii} \ ; \ C_{ri} = -C_{ir} \tag{1.30}$$

For example, a single complex zero-mean normal random variable $z = x + jy$ is circular if

$$\sigma_x = \sigma_y = \sigma \ ; \ \overline{xy} = 0 \tag{1.31}$$

and (1.25) reduces to

$$f(x, y) = \frac{1}{2\pi\sigma^2} e^{-\frac{1}{2\sigma^2}(x^2+y^2)} \tag{1.32}$$

Since this function depends only on the amplitude of z, constant probability contours are circles in the complex plane, hence the name of circular. It is interesting to highlight that (1.31) implies that the variable z is uncorrelated with its conjugate z^* by virtue of the definition (1.5). The reader is invited to check this nonintuitive result.

Similarly, two zero-mean jointly normal complex random variables $z_1 = x_1 + jy_1$ and $z_2 = x_2 + jy_2$ are circular if they satisfy the following conditions derived from (1.30):

$$\begin{aligned} &\sigma_{x_1} = \sigma_{y_1} = \sigma_1 \qquad \sigma_{x_2} = \sigma_{y_2} = \sigma_2 \\ &\overline{x_1 x_2} = \overline{y_1 y_2} \qquad \overline{x_1 y_2} = -\overline{y_1 x_2} \end{aligned} \tag{1.33}$$

The corresponding joint density function is not provided here but it is opportune to mention that these conditions imply that the variances and cross-correlation of the complex variables are

$$\begin{aligned} &\sigma_{z_1}^2 = 2\sigma_1^2 \ ; \ \sigma_{z_2}^2 = 2\sigma_2^2 \\ &\overline{z_1 z_2^*} = 2\,\overline{x_1 x_2} + j2\,\overline{y_1 x_2} \end{aligned} \tag{1.34}$$

Furthermore, the application of (1.29) to the product of four variables with two of them conjugated results in

$$\overline{z_1 z_2 z_3^* z_4^*} = \overline{z_1 z_3^*}\ \overline{z_2 z_4^*} + \overline{z_1 z_4^*}\ \overline{z_2 z_3^*} \tag{1.35}$$

where the first term $\overline{z_1 z_2}\ \overline{z_3^* z_4^*}$ vanishes due to the circularity condition (1.33). It may be worth to remind that all four variables must have zero mean.

1.1.4 The Arc Sine Law

The so-called "arc sine law" is an elegant relation often used to simplify digital correlation measurements. Given two real, zero-mean, jointly normal variables, this law relates their cross correlation with that of the following associated variables

$$\check{\mathrm{x}} = \mathrm{sgn}\ \mathrm{x}\ ;\ \check{\mathrm{y}} = \mathrm{sgn}\ \mathrm{y} \tag{1.36}$$

where the signum function is defined in (A.30). These new variables contain only the sign information so they both have zero mean and unit variance. Their cross-correlation is related to that of the original variables through this law [see (1.42) below].

An easy method to derive it is by using the classical "Price's theorem." If p is an arbitrary function of x and y

$$\mathrm{p} = \mathrm{g}(\mathrm{x}, \mathrm{y}) \tag{1.37}$$

this theorem states (see for example [1]) that

$$\frac{\partial^n \overline{\mathrm{p}}}{\partial C_{\mathrm{xy}}^n} = \int_{-\infty}^{\infty} \int_{-\infty}^{\infty} \frac{\partial^{2n} \mathrm{g}(\mathrm{x}, \mathrm{y})}{\partial \mathrm{x}^n \partial \mathrm{y}^n} f(\mathrm{x}, \mathrm{y}) d\mathrm{x} d\mathrm{y} \tag{1.38}$$

with n arbitrary, $f(\mathrm{x}, \mathrm{y})$ the joint density function (1.25) and C_{xy} the covariance of x and y. This theorem does not require that these variables have zero mean.

Application of this theorem to compute the cross correlation of $\check{\mathrm{x}}$ and $\check{\mathrm{y}}$ is straightforward. Equation (1.37) is now written as

$$\mathrm{p} = \check{\mathrm{x}}\check{\mathrm{y}} = \mathrm{sgn}\ \mathrm{x}\ \ \mathrm{sgn}\ \mathrm{y} \tag{1.39}$$

and (1.38) is used with $n = 1$. Taking into account that the derivative of the signum function is equal to twice the Dirac delta function,

$$\frac{\partial \overline{\mathrm{p}}}{\partial C_{\mathrm{xy}}} = \int_{-\infty}^{\infty} \int_{-\infty}^{\infty} 4\delta(\mathrm{x})\delta(\mathrm{y}) f(\mathrm{x}, \mathrm{y}) d\mathrm{x} d\mathrm{y} = 4 f(0, 0) \tag{1.40}$$

and substituting (1.25) knowing that $\overline{\mathrm{x}} = \overline{\mathrm{y}} = 0$,

$$\frac{\partial \overline{\mathrm{p}}}{\partial C_{\mathrm{xy}}} = \frac{2}{\pi \sqrt{\sigma_{\mathrm{x}}^2 \sigma_{\mathrm{y}}^2 - C_{\mathrm{xy}}^2}} \tag{1.41}$$

which is easily integrated to yield

$$\overline{\check{\mathrm{x}}\check{\mathrm{y}}} = \frac{2}{\pi} \arcsin \frac{\overline{\mathrm{x}\mathrm{y}}}{\sqrt{\overline{\mathrm{x}^2}\ \overline{\mathrm{y}^2}}} \tag{1.42}$$

in which $\overline{p}$ has been substituted by $\overline{\check{x}\check{y}}$ and the definitions of covariance (1.4) and standard deviation (1.2) have been used. Remind that x and y are two real, zero-mean, jointly normal variables and $\check{x}$ and $\check{y}$ are related to them through (1.36).

The above equation is the arc sine law. A generalization for transitions at a nonzero threshold (i.e. $\text{sgn}\,(x-\delta)$) can be found in [12].

1.1.4.1 A Useful Approximation

Related to the usage of this law in retrieving correlation measurements, an approximation is needed in section 3.4.5 to handle non infinite integration times. The demonstration is provided here for clarity. Consider two real random variables related through a sinus function

$$z = \sin w \tag{1.43}$$

In general the mean of z differs from the sinus of the mean of w ($\overline{z} \neq \sin \overline{w}$). Putting $w = \overline{w} + \delta$, it can be written as

$$\overline{z} = \sin \overline{w}\, \overline{\cos \delta} + \cos \overline{w}\, \overline{\sin \delta} \tag{1.44}$$

which for small δ, using Taylor linear and quadratic expansions for the sinus and cosinus, respectively, is approximated as

$$\overline{z} \approx \left(1 - \frac{\sigma_w^2}{2}\right) \sin \overline{w} \tag{1.45}$$

where $\sigma_w^2 = \overline{\delta^2}$ is the variance of w.

1.2 Stochastic Processes

A stochastic process is a mathematical model to describe the random fluctuations of a physical quantity and/or its measurement with time. The concept is fundamental for understanding many physical phenomena, particularly in this book to interpret radiometric measurements, as for example the output voltage of a total power radiometer or the correlation between output signals in an interferometric radiometer. In what follows the terms "stochastic process," "random process" or "random signal" will be used indistinguishably. And often the adjective "stochastic" or "random" will be omitted to simplify writing.

Signals in the real world are real. However, in order to produce more elegant results and simplify formulations, it is often useful to use a complex representation of real signals. This is a common practice in engineering and is profusely used in this book. For this reason all properties and definitions provided all along this book assume complex processes unless mentioned otherwise. Following the same convention as for random variables in section 1.1, real signals are represented by upright roman typeface [$\mathrm{x}(t), \mathrm{y}(t), \mathrm{z}(t) \ldots$] and complex signals use italic shape [$x(t), y(t), z(t) \ldots$] except, exceptionally, if denoted by Greek letters. Variable t denotes time (in seconds) and f frequency (in Hertz).

1.2.1 Stationarity

To systematize and simplify the study of physical observations, it is fundamental to understand the concept of "stationarity" applied to a stochastic process. Later a mathematical definition will be given, but for now just an intuitive one is needed. Figure 1.1 shows two sketches of stochastic processes, both representing for example the output voltage of a radiometer. The plot on figure 1.1(a) is stationary and the one on figure 1.1(b) is not. Without any rigorous definition, the reader can understand why this adjective is adequate to describe each of them.

Since nothing remains forever stable in time, all physical phenomena are described by nonstationary processes. For example, a radiometric measurement evolves as the target is moved in front of the antenna and experiences changes as a function of the receiver's physical temperature and other effects. Needless to say, nonstationarity is difficult to handle mathematically so, in practice, a convenient time scale is defined for which the process is assumed stationary; and then, after characterized, its (stable) parameters are allowed to vary ("slowly") with time. For example, the signal on figure 1.1(b) can be interpreted as a *locally stationary* process with slowly varying average and amplitude evolving with time. Moreover, a nonstationary process can be viewed as the result of applying a function of time to a stationary one, so fundamental studies can be safely limited to this last

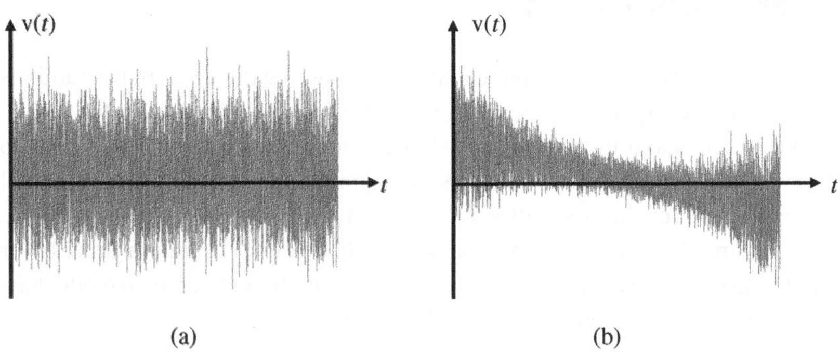

Figure 1.1 Qualitative examples of stochastic processes. (a) Stationary. (b) Non stationary.

case. With this approach in mind, stationarity will always be implicitly assumed, understanding that the results apply only to a finite time scale and may change by long-term time evolution.

Given a stationary process characterized by a real signal x(t), its time average is a *constant* defined as

$$\langle x(t) \rangle = \lim_{T\to\infty} \frac{1}{T} \int_T x(t)dt \tag{1.46}$$

The "instantaneous power" is in turn a stochastic process defined by the square of the signal $x^2(t)$. The term "power" is just a convenient name, but the physical meaning depends on that of the original signal x(t). In this context, power does not necessarily have units of Watt. In radiometry only "finite mean power" signals are encountered, meaning that the time average of the instantaneous power is a constant

$$\left\langle x^2(t) \right\rangle = \lim_{T\to\infty} \frac{1}{T} \int_T x^2(t)dt \tag{1.47}$$

A set of samples of any process at different epochs is a random variable. This is an axiom based on the fundamental definition of a random variable, always linked to the outcome of a certain experiment (see for example section 4.1 of [1]). The two above definitions are thus interpreted as the corresponding *mean* (1.1) and *autocorrelation* (1.2) of the random variable formed by all possible values of x(t). Therefore, all definitions and equations defined for random variables in section 1.1 are directly applicable to samples of stochastic processes. To be precise though, there is a mathematical subtlety called "ergodicity," which is the property that a process must satisfy for the above statement to be true. In practice one needs not care about it since *stationary* processes describing physical experiments are ergodic.

1.2.2 Correlation and Power

Given a complex stationary random process $z(t) = x(t) + jy(t)$, the following definitions apply:

$$\begin{aligned}
\text{Mean:}\quad & \langle z(t)\rangle = \langle x(t)\rangle + j\,\langle y(t)\rangle \\
\text{Variance:}\quad & \sigma_z^2 = \left\langle |z(t) - \langle z(t)\rangle|^2 \right\rangle \\
& \quad = \left\langle |z(t)|^2 \right\rangle - |\langle z(t)\rangle|^2 \\
& \quad = \sigma_x^2 + \sigma_y^2 \\
\text{Standard deviation:}\quad & \sigma_z \quad \text{(square root of variance)} \\
\text{Autocorrelation function:}\quad & R_z(\tau) = \langle z(t) z^*(t-\tau)\rangle \\
\text{Mean power (or power):}\quad & R_z(0) = \left\langle |z(t)|^2 \right\rangle = \sigma_z^2 + |\langle z(t)\rangle|^2 \\
& \quad = R_x(0) + R_y(0)
\end{aligned} \tag{1.48}$$

The absolute value of the mean squared $|\langle z(t)\rangle|^2$ is sometimes called "DC power" and the variance "AC power"[1] so the mean power is the sum of AC plus DC powers. A zero mean process has no DC power. For most processes used all along this book the mean power, and hence the autocorrelation function, has units of Watt (see section 1.5.1 for details). The standard deviation has always the same units as the signal itself.

Subtracting the mean to a process implies subtracting the DC power to the autocorrelation function:

$$\text{if } z'(t) = z(t) - \langle z(t)\rangle \Rightarrow R_{z'}(\tau) = R_z(\tau) - |\langle z(t)\rangle|^2 \tag{1.49}$$

which, particularized for $\tau = 0$ becomes the variance of $z(t)$: $R_{z'}(0) = \sigma_z^2$.

For stationary processes, both mean and variance are constants and the autocorrelation function depends only on time difference τ. This statement is in fact the mathematical definition of (wide sense) stationarity [1]. It implies that a random variable defined as the average of N samples of $z(t)$ has a mean equal to that of the process

$$m = \frac{1}{N}\sum_{i=1}^{N} z(t_i) \Rightarrow \overline{m} = \langle z(t)\rangle \tag{1.50}$$

Similarly, the mean of a random variable defined by averaging the amplitude squared of a process is equal to the mean power

$$\mathrm{p} = \frac{1}{N}\sum_{i=1}^{N} |z(t_i)|^2 \Rightarrow \overline{\mathrm{p}} = \langle |z(t)|^2\rangle \tag{1.51}$$

Given a second complex process $w(t) = \mathrm{u}(t) + j\mathrm{v}(t)$, the following additional definitions apply

$$\begin{aligned} \text{Cross-correlation function:} \quad & R_{zw}(\tau) = \langle z(t)w^*(t-\tau)\rangle \\ \text{Cross-correlation:} \quad & R_{zw}(0) = \langle z(t)w^*(t)\rangle \\ \text{Covariance:} \quad & C_{zw} = R_{zw}(0) - \langle z(t)\rangle \langle w(t)\rangle^* \end{aligned} \tag{1.52}$$

The two processes $z(t)$ and $w(t)$ are uncorrelated if samples of them at any time delay are uncorrelated. Using (1.5), the cross-correlation function becomes then a constant,

$$R_{zw}(\tau) = \langle z(t)\rangle \langle w(t)\rangle^* \quad \text{(uncorrelated processes)} \tag{1.53}$$

which vanishes if one of the two variables has zero-mean.

As a generalization of (1.49), the cross-correlation function of the incremental signals that result of subtracting the mean $z'(t) = z(t) - \langle z(t)\rangle$ and $w'(t) = w(t) - \langle w(t)\rangle$ is

$$R_{z'w'}(\tau) = R_{zw}(\tau) - \langle z(t)\rangle \langle w(t)\rangle^* \tag{1.54}$$

1 In Electrical Engineering "DC" means direct current and "AC" alternating current.

from which it follows that the covariance is $C_{zw} = R_{z'w'}(0)$. In some texts, $R_{z'w'}(\tau)$ is named "covariance function" with the symbol $C_{zw}(\tau)$, but this definition is not used in this book. The cross-correlation function of the "primed" variables will be used instead. Note that for uncorrelated processes $R_{z'w'}(\tau) = 0$.

If both $z(t)$ and $w(t)$ represent the same signal, the cross-correlation function becomes the autocorrelation function $R_{zz}(\tau) \equiv R_z(\tau)$, so no double subscript is used, in this case. Similarly, the covariance reduces to the variance $C_{zz} \equiv \sigma_z^2$.

Sometimes, it is useful to normalize the correlation functions with respect to their value at the origin. The following naming convention is used

$$r_{zw}(\tau) \equiv \frac{R_{zw}(\tau)}{R_{zw}(0)} \; ; \; r_z(\tau) \equiv \frac{R_z(\tau)}{R_z(0)} \tag{1.55}$$

As a direct consequence of the definition (1.52), swapping the subscripts of the cross-correlation function is equivalent to changing the sign of the time delay and conjugate,

$$R_{wz}(\tau) = R_{zw}^*(-\tau) \Rightarrow R_z(\tau) = R_z^*(-\tau) \tag{1.56}$$

which implies that $R_z(0)$ is always real, consistent with the definition of mean power in (1.48). Moreover, the amplitude of the autocorrelation function is an even function of τ and the phase an odd function

$$|R_z(\tau)| = |R_z(-\tau)| \; ; \; \arg[R_z(\tau)] = -\arg[R_z(-\tau)] \tag{1.57}$$

For real signals both auto- and cross-correlation functions are real, so (1.56) is valid without the conjugate operation. The autocorrelation function of a real process is an even function of τ.

The following identity, deduced directly from the definition (1.52), relates the cross-correlation function of two complex signals to those of their real and imaginary parts: if $z(t) = \mathrm{x}(t) + j\mathrm{y}(t)$ and $w(t) = \mathrm{u}(t) + j\mathrm{v}(t)$, then

$$R_{zw}(\tau) = [R_{\mathrm{xu}}(\tau) + R_{\mathrm{yv}}(\tau)] + j[R_{\mathrm{yu}}(\tau) - R_{\mathrm{xv}}(\tau)] \tag{1.58}$$

which, for $w = z$ (single signal) reduces to

$$R_z(\tau) = [R_{\mathrm{x}}(\tau) + R_{\mathrm{y}}(\tau)] + j[R_{\mathrm{yx}}(\tau) - R_{\mathrm{xy}}(\tau)] \tag{1.59}$$

1.2.2.1 Sum and Product of Signals

Having two signals $z(t)$ and $w(t)$, the autocorrelation function of a linear combination of them $s(t) = \alpha\, z(t) + \beta\, w(t)$, with α and β complex constants, is readily deduced from the definition (1.48)

$$R_s(\tau) = |\alpha|^2 R_z(\tau) + |\beta|^2 R_w(\tau) + \alpha\beta^* R_{zw}(\tau) + \beta\alpha^* R_{wz}(\tau) \tag{1.60}$$

valid for either complex or real processes.

The autocorrelation function of the complex product $p(t) = z(t)w^*(t)$ is

$$R_p(\tau) = \langle z(t)w^*(t)z^*(t-\tau)w(t-\tau)\rangle \tag{1.61}$$

which, in general, cannot be easily expanded. However, if $z(t)$ and $w(t)$ are independent processes, so they are the products $z(t)z^*(t-\tau)$ and $w(t)w^*(t-\tau)$. Since independent variables are also uncorrelated, from (1.5)

$$R_p(\tau) = R_z(\tau)R_w^*(\tau) \quad \text{independent processes} \tag{1.62}$$

applicable also to real signals. Other situations are analyzed here below.

1.2.3 Jointly Normal Processes

In the particular case where the signals at a given time and their delayed versions are *circular jointly normal* complex random variables, property (1.35) applies. Identifying $z_1 = z(t)$, $z_2 = w(t-\tau)$, $z_3 = z(t-\tau)$ and $z_4 = w(t)$, (1.61) reduces to

$$R_p(\tau) = |\langle p(t)\rangle|^2 + R_z(\tau)R_w^*(\tau) \tag{1.63}$$

where $\langle p(t)\rangle = \langle z(t)w^*(t)\rangle$. The reader may easily check that this result is compatible with (1.62). In general, particularizing the above expression for $\tau = 0$ results in

$$\sigma_p = \sigma_z\sigma_w \tag{1.64}$$

which is a very elegant result. Unlike (1.22), the above result *does not* require the signals to be uncorrelated. Note that the standard deviation of the complex product of jointly normal circular processes does not depend on the cross-correlation between the signals.

1.2.3.1 Real Signals

A slightly different conclusion is found for *real signals*. Defining the product $\mathrm{p}(t) = \mathrm{z}(t)\mathrm{w}(t)$, the autocorrelation function is

$$R_\mathrm{p}(\tau) = \langle \mathrm{z}(t)\mathrm{w}(t)\mathrm{z}(t-\tau)\mathrm{w}(t-\tau)\rangle \tag{1.65}$$

and assuming that signal samples and their delayed versions are zero-mean jointly normal random variables, property (1.29) can be used by identifying $z_1 = \mathrm{z}(t)$, $z_2 = \mathrm{w}(t)$, $z_3 = \mathrm{z}(t-\tau)$ and $z_4 = \mathrm{w}(t-\tau)$. The above expression then reduces to

$$R_\mathrm{p}(\tau) = \langle \mathrm{p}(t)\rangle^2 + R_\mathrm{z}(\tau)R_\mathrm{w}(\tau) + R_\mathrm{zw}(\tau)R_\mathrm{wz}(\tau) \tag{1.66}$$

where $\langle \mathrm{p}(t)\rangle = \langle \mathrm{z}(t)\mathrm{w}(t)\rangle$. The reader may easily check that for uncorrelated (zero mean) signals this reduces to (1.62) as expected. Particularization for $\tau = 0$, knowing that both processes have zero mean, yields

$$\sigma_\mathrm{p}^2 = \sigma_\mathrm{z}^2\sigma_\mathrm{w}^2 + \langle \mathrm{z}(t)\mathrm{w}(t)\rangle^2 \tag{1.67}$$

which, compared to (1.64), includes an additive term that depends on the cross-correlation of the signals. Only if z(t) and w(t) are uncorrelated, $\langle z(t)w(t)\rangle$ vanishes and this equation becomes formally the same as (1.64). Maybe it is worth to remind that (1.66) and (1.67) apply to the product of two zero-mean real jointly normal processes while (1.63) and (1.64) apply to the product of two complex circular jointly normal processes, one of them conjugated.

1.2.3.2 Check

It is interesting to check that (1.63) and (1.66) [and hence (1.64) and (1.67)] are consistent with each other. Separating the complex signals into their real and imaginary parts, $z(t) = \mathrm{x}(t) + j\mathrm{y}(t)$ and $w(t) = \mathrm{u}(t) + j\mathrm{v}(t)$, the complex product $p(t) = z(t)w^*(t)$ becomes

$$p(t) = \mathrm{p}_r(t) + j\mathrm{p}_i(t) = \left[\mathrm{x}(t)\mathrm{u}(t) + \mathrm{y}(t)\mathrm{v}(t)\right] + j\left[\mathrm{y}(t)\mathrm{u}(t) - \mathrm{x}(t)\mathrm{v}(t)\right] \tag{1.68}$$

where $\mathrm{p}_r(t)$ and $\mathrm{p}_i(t)$ are the real and imaginary parts of $p(t)$. Using (1.59)

$$R_p(\tau) = [R_{\mathrm{p}_r}(\tau) + R_{\mathrm{p}_i}(\tau)] + j[R_{\mathrm{p}_i\mathrm{p}_r}(\tau) - R_{\mathrm{p}_r\mathrm{p}_i}(\tau)] \tag{1.69}$$

On the other hand, applying also (1.59) to $R_z(\tau)$ and $R_w(\tau)$ taking into account the circularity condition (1.33) [see also (1.71) below], (1.63) becomes

$$R_p(\tau) = |\langle p(t)\rangle|^2 + 4[R_{\mathrm{x}}(\tau)R_{\mathrm{u}}(\tau) + R_{\mathrm{yx}}(\tau)R_{\mathrm{vu}}(\tau)] + j4[R_{\mathrm{yx}}(\tau)R_{\mathrm{u}}(\tau) - R_{\mathrm{x}}(\tau)R_{\mathrm{vu}}(\tau)] \tag{1.70}$$

This equation is identical to (1.69) when the correlation functions of p_r and p_i are expanded using (1.60) and (1.66) and the circularity conditions (1.33) for $z(t)$ and $w(t)$ are considered. For completion, these conditions are written here below in terms of the correlation functions.

$$\begin{aligned} &R_{\mathrm{x}}(\tau) = R_{\mathrm{y}}(\tau)\ ;\ R_{\mathrm{u}}(\tau) = R_{\mathrm{v}}(\tau) \\ &R_{\mathrm{xy}}(\tau) = -R_{\mathrm{yx}}(\tau)\ ;\ R_{\mathrm{uv}}(\tau) = -R_{\mathrm{vu}}(\tau) \\ &R_{\mathrm{xu}}(\tau) = R_{\mathrm{yv}}(\tau) \Rightarrow \langle \mathrm{x}(t)\mathrm{u}(t)\rangle = \langle \mathrm{y}(t)\mathrm{v}(t)\rangle = 1/2\ \langle \mathrm{p}_r(t)\rangle \\ &R_{\mathrm{xv}}(\tau) = -R_{\mathrm{yu}}(\tau) \Rightarrow \langle \mathrm{x}(t)\mathrm{v}(t)\rangle = -\langle \mathrm{y}(t)\mathrm{u}(t)\rangle = -1/2\ \langle \mathrm{p}_i(t)\rangle \end{aligned} \tag{1.71}$$

Using these identities, after some (tedious) algebra, the following results are obtained

$$R_{\mathrm{p}_r}(\tau) = \langle \mathrm{p}_r(t)\rangle^2 + 2R_{\mathrm{x}}(\tau)R_{\mathrm{u}}(\tau) + 2R_{\mathrm{yx}}(\tau)R_{\mathrm{vu}}(\tau) + 2R_{\mathrm{xu}}(\tau)R_{\mathrm{ux}}(\tau) - 2R_{\mathrm{yu}}(\tau)R_{\mathrm{uy}}(\tau) \tag{1.72}$$

$$R_{\mathrm{p}_i}(\tau) = \langle \mathrm{p}_i(t)\rangle^2 + 2R_{\mathrm{x}}(\tau)R_{\mathrm{u}}(\tau) + 2R_{\mathrm{yx}}(\tau)R_{\mathrm{vu}}(\tau) - 2R_{\mathrm{xu}}(\tau)R_{\mathrm{ux}}(\tau) + 2R_{\mathrm{yu}}(\tau)R_{\mathrm{uy}}(\tau) \tag{1.73}$$

$$R_{\mathrm{p}_i\mathrm{p}_r}(\tau) = 2R_{\mathrm{yx}}(\tau)R_{\mathrm{u}}(\tau) - 2R_{\mathrm{x}}(\tau)R_{\mathrm{vu}}(\tau) - R_{\mathrm{xv}}(\tau)R_{\mathrm{ux}}(\tau)$$
$$+ R_{\mathrm{xu}}(\tau)R_{\mathrm{uy}}(\tau) - R_{\mathrm{yv}}(\tau)R_{\mathrm{vx}}(\tau) + R_{\mathrm{yu}}(\tau)R_{\mathrm{vy}}(\tau) \quad (1.74)$$

$$R_{\mathrm{p}_r\mathrm{p}_i}(\tau) = -2R_{\mathrm{yx}}(\tau)R_{\mathrm{u}}(\tau) + 2R_{\mathrm{x}}(\tau)R_{\mathrm{vu}}(\tau) - R_{\mathrm{xv}}(\tau)R_{\mathrm{ux}}(\tau)$$
$$+ R_{\mathrm{xu}}(\tau)R_{\mathrm{uy}}(\tau) - R_{\mathrm{yv}}(\tau)R_{\mathrm{vx}}(\tau) + R_{\mathrm{yu}}(\tau)R_{\mathrm{vy}}(\tau) \quad (1.75)$$

and (1.70) follows immediately by substituting these four equations into (1.69). Furthermore, the variances of the real and imaginary parts of $p(t)$ are obtained by particularizing (1.72) and (1.73) for $\tau = 0$,

$$\begin{aligned} \sigma^2_{\mathrm{p}_r} &= 2\sigma^2_{\mathrm{x}}\sigma^2_{\mathrm{u}} + \frac{1}{2}\langle \mathrm{p}_r(t)\rangle^2 - \frac{1}{2}\langle \mathrm{p}_i(t)\rangle^2 \\ \sigma^2_{\mathrm{p}_i} &= 2\sigma^2_{\mathrm{x}}\sigma^2_{\mathrm{u}} - \frac{1}{2}\langle \mathrm{p}_r(t)\rangle^2 + \frac{1}{2}\langle \mathrm{p}_i(t)\rangle^2 \end{aligned} \quad (1.76)$$

and their sum $\sigma^2_p = \sigma^2_{\mathrm{p}_r} + \sigma^2_{\mathrm{p}_i}$ is consistent with (1.64), taking into account that $\sigma_{\mathrm{x}} = \sigma_{\mathrm{y}}$ and $\sigma_{\mathrm{u}} = \sigma_{\mathrm{v}}$ due to the circularity conditions (1.71). Note that these two expressions, like (1.67), include additive terms that depend on the mean values of the real and imaginary parts. However, they are canceled out when the variance of the complex product is formed.

It is finally interesting to check the consistency of (1.76) with the variances of the different components of (1.68). For example, the real part can be written as $\mathrm{p}_r(t) = \mathrm{p}_A(t) + \mathrm{p}_B(t)$ with

$$\mathrm{p}_A(t) = \mathrm{x}(t)\mathrm{u}(t) \text{ and } \mathrm{p}_B(t) = \mathrm{y}(t)\mathrm{v}(t) \quad (1.77)$$

Using (1.67) together with the circularity conditions (1.71)

$$\sigma^2_{\mathrm{p}_A} = \sigma^2_{\mathrm{p}_B} = \sigma^2_{\mathrm{x}}\sigma^2_{\mathrm{u}} + \frac{1}{4}\langle \mathrm{p}_r(t)\rangle^2 \quad (1.78)$$

which allows to write the first equation (1.76) as

$$\sigma^2_{\mathrm{p}_r} = \sigma^2_{\mathrm{p}_A} + \sigma^2_{\mathrm{p}_B} - \frac{1}{2}\langle \mathrm{p}_i(t)\rangle^2 \quad (1.79)$$

On the other hand, the general equation (1.13) states that the standard deviation of the sum is equal to the sum of standard deviations plus twice the covariance between the summands. By simple identification, this one becomes

$$C_{\mathrm{p}_A\mathrm{p}_B} = -\frac{1}{4}\langle \mathrm{p}_i(t)\rangle^2 \quad (1.80)$$

which can be easily confirmed from the definition of covariance (1.4), the property (1.29) and the circularity conditions (1.71). The reader is invited to check it. A similar result is derived for the imaginary part. In this case, the covariance is minus one-fourth of the mean of the real part of the correlation squared.

1.2.4 Spectral Densities

The "power spectral density" (or "power spectrum" or simply "spectrum") of a stochastic process $z(t)$ is the Fourier transform of its autocorrelation function

$$S_z(f) = \mathcal{F}[R_z(\tau)] = \int_{-\infty}^{\infty} R_z(\tau)e^{-j2\pi f\tau}d\tau \tag{1.81}$$

If the autocorrelation function has units of W, then the power spectral density has units of W/Hz. From (1.56) and the conjugation property of the Fourier transform (Table A.1), it follows that the power spectral density $S_z(f)$ *is always real and positive*. Using (1.54) the spectral density can also be separated into a contribution from the mean and the rest

$$S_z(f) = S_{z'}(f) + |\langle z(t)\rangle|^2\delta(f) \tag{1.82}$$

where $z'(t) = z(t) - \langle z(t)\rangle$ is the signal with the mean removed and $\delta(f)$ is the Dirac delta function (A.35). The mean of a random process is always seen as a delta at the origin in the power spectral density.

Equation (1.81) implies the corresponding inverse

$$R_z(\tau) = \mathcal{F}^{-1}[S_z(f)] = \int_{-\infty}^{\infty} S_z(f)e^{j2\pi f\tau}df \tag{1.83}$$

The mean power defined in (1.48) is then

$$\text{mean power: } R_z(0) = \left\langle |z(t)|^2 \right\rangle = \int_{-\infty}^{\infty} S_z(f)df \tag{1.84}$$

which justifies the name of power spectral density for $S_z(f)$. As indicated, the integrals above must include both positive and negative frequencies.

For a real signal $z(t)$, $R_z(\tau)$ is also real and then $S_z(f)$ is an even function of frequency. In this case, using (A.45) the autocorrelation function can be expressed as

$$\begin{aligned} R_z(\tau) &= 2\Re e\left[\mathcal{F}^{-1}[S_z(f)u(f)]\right] \\ &= 2\int_0^{\infty} S_z(f)\cos(2\pi f\tau)df \quad \text{(real signals)} \end{aligned} \tag{1.85}$$

where $u(f)$ is the unit step function (A.31) and $S_z(0)$ includes a factor of $1/2$ for consistency. Particularizing for $\tau = 0$

$$\text{mean power: } R_z(0) = \left\langle z^2(t) \right\rangle = 2\int_0^{\infty} S_z(f)df \quad \text{(real signals)} \tag{1.86}$$

concluding that the total power of a signal is twice the integrated spectral density in the positive part of the spectrum.

Using a similar definition for two processes z and w, their "cross-power spectral density" (or "cross-power spectrum") is the Fourier transform of the cross-correlation function

$$S_{zw}(f) = \mathcal{F}\,[R_{zw}(\tau)] = \int_{-\infty}^{\infty} R_{zw}(\tau)e^{-j2\pi f\tau}d\tau \tag{1.87}$$

with units of W/Hz if the cross-correlation function is in W. In general $S_{zw}(f)$ is complex and satisfies the following property derived from (1.56) along with the properties of conjugation and sign change (Table A.1)

$$S_{wz}(f) = S^*_{zw}(f) \tag{1.88}$$

Note that for $z = w$, the above property reduces to $S_z(f) = S^*_z(f)$ consistent with the fact that $S_z(f)$ is always real. Using (1.54), the contribution of the mean of both signals is seen as an impulse at $f = 0$,

$$S_{zw}(f) = S_{z'w'}(f) + \langle z(t)\rangle \langle w(t)\rangle^* \delta(f) \tag{1.89}$$

where the primed variables are those resulting of subtracting their respective means. If the signals $z(t)$ and $w(t)$ are uncorrelated, (1.53) implies that

$$S_{zw}(f) = \langle z(t)\rangle \langle w(t)\rangle^* \delta(f) \quad \text{(uncorrelated processes)} \tag{1.90}$$

which is equivalent to putting $S_{z'w'}(f) = 0$.

The cross-correlation function is the inverse of (1.87)

$$R_{zw}(\tau) = \mathcal{F}^{-1}[S_{zw}(f)] = \int_{-\infty}^{\infty} S_{zw}(f)e^{j2\pi f\tau}df \tag{1.91}$$

and in consequence,

$$R_{zw}(0) = \langle z(t)w^*(t)\rangle = \int_{-\infty}^{\infty} S_{zw}(f)df \tag{1.92}$$

If both signals are real, their cross-correlation function is also real so $S_{\text{zw}}(f)$ has hermitic symmetry with respect to the origin $S_{\text{zw}}(-f) = S^*_{\text{zw}}(f)$. In this case, the integrals can also be limited to the positive side of the spectrum using (A.45)

$$\begin{aligned} R_{\text{zw}}(\tau) &= 2\Re e\left[\mathcal{F}^{-1}[S_{\text{zw}}(f)\text{u}(f)]\right] \\ &= 2\int_{0}^{\infty} \Re e[S_{\text{zw}}(f)e^{j2\pi f\tau}]df \quad \text{(real signals)} \end{aligned} \tag{1.93}$$

which, particularized for $\tau = 0$ results in

$$R_{\text{zw}}(0) = \langle \text{z}(t)\text{w}(t)\rangle = 2\int_{0}^{\infty} \Re e[S_{\text{zw}}(f)]df \quad \text{(real signals)} \tag{1.94}$$

compatible with (1.86) when z = w.

1.2.5 Linear Systems

If a signal $z(t)$ is passed through a linear system, the output is

$$w(t) = \int_{-\infty}^{\infty} h(t')z(t-t')dt' = h(t) * z(t) \tag{1.95}$$

where $h(t)$ is the system's impulse response and $*$ is the convolution operator (A.40). The Fourier transform of the impulse response is the frequency response:

$$H(f) = \mathcal{F}\,[h(t)] = \int_{-\infty}^{\infty} h(t)e^{-j2\pi ft}dt \tag{1.96}$$

Due to the linearity property pointed out in section 1.1.1, the mean of $w(t)$ is the same integral (1.95) but applied to the mean of $z(t)$, which is constant. Using (1.96),

$$\langle w(t)\rangle = H(0)\,\langle z(t)\rangle \tag{1.97}$$

which is an expected result: the respective means of input and output signals are related to each other by the frequency response at DC ($f = 0$).

Since the integral (1.95) operates in the variable t', a multiplication by a function of time can be introduced inside the integral. The autocorrelation function (1.48) of the output signal is then

$$\begin{aligned} R_w(\tau) = \langle w(t)w^*(t-\tau)\rangle &= \textstyle\int_{-\infty}^{\infty} h(t')\,\langle z(t-t')w^*(t-\tau)\rangle\, dt' \\ &= h(\tau) * R_{zw}(\tau) \end{aligned} \tag{1.98}$$

where the mean and integral operators have been interchanged, and the definition of cross-correlation function (1.52) has been used. Proceeding in the same way,

$$\begin{aligned} R_{zw}(\tau) = \langle z(t)w^*(t-\tau)\rangle &= \textstyle\int_{-\infty}^{\infty} h^*(t')\,\langle z(t)z^*(t-t'-\tau)\rangle\, dt' \\ &= h^*(-\tau) * R_z(\tau) \end{aligned} \tag{1.99}$$

where the change of variables $t'' = -t'$ has been applied to the last integral. Combining the above two equations knowing that the convolution is associative (A.41), it follows that

$$R_w(\tau) = h(\tau) * h^*(-\tau) * R_z(\tau) \tag{1.100}$$

Convolution in time domain is equivalent to multiplication in the frequency domain (see Table A.1), so the above expression is straightforwardly converted to the frequency domain as the product of three terms. Furthermore, combining the properties of conjugation and sign change, it follows that $\mathcal{F}\,[h^*(-t)] = H^*(f)$, therefore

$$S_w(f) = |H(f)|^2 S_z(f) \tag{1.101}$$

meaning that the power spectral densities of input and output signals are related to each other by the amplitude squared of the frequency response. Similarly (1.99) converts to

$$S_{zw}(f) = H^*(f)S_z(f) \quad \Rightarrow \quad S_{wz}(f) = H(f)S_z(f) \tag{1.102}$$

where the last expression results from applying (1.88) taking into account that $S_z(f)$ is always real.

The methodology used to derive these equations has been borrowed from that provided in chapter 10 of [1]. In the same reference, the reader can find details on how (1.101) is generalized to the case of having two linear systems applied to two different signals.

$$\begin{aligned} w_1(t) &= h_1(t) * z_1(t) \\ w_2(t) &= h_2(t) * z_2(t) \end{aligned} \tag{1.103}$$

In this case, the cross-power spectral density of the two output signals is

$$S_{w_1w_2}(f) = S_{z_1z_2}(f)H_1(f)H_2^*(f) \tag{1.104}$$

where $H_1(f)$ and $H_2(f)$ are the frequency responses of the two systems. Note that this equation reduces to (1.101) for a single linear system.

1.2.6 Time Averaging or Integration

Given a signal $z(t)$, its averaged version $w(t)$ in a time interval τ_i is equal to the integral (1.46) without the limiting operation and with $T = \tau_i$

$$w(t) = \frac{1}{\tau_i}\int_{t-\tau_i}^{t} z(t')dt' \tag{1.105}$$

where τ_i is commonly referred to as "integration time." This expression can be rewritten exactly as (1.95) provided the impulse response is defined as

$$h_i(t) = \frac{1}{\tau_i}\Pi\left(\frac{t-\tau_i/2}{\tau_i}\right) = \begin{cases} \dfrac{1}{\tau_i} & 0 < t < \tau_i \\ 0 & \text{elsewhere} \end{cases} \tag{1.106}$$

where the function $\Pi(t)$ is defined in (A.33). The corresponding frequency response is the Fourier transform of $h_i(t)$ (see Table A.2)

$$H_i(f) = \mathscr{F}\left[h_i(t)\right] = \text{sinc}\,(f\tau_i)e^{-j2\pi f\tau_i/2} \tag{1.107}$$

Figure 1.2 shows a schematic drawing of both the impulse response and the amplitude squared of the frequency response. Note that the filter is casual since the impulse response is null for negative time. In the frequency domain, the bandwidth of the filter, up to the first zero, is equal to $1/\tau_i$. Based on this result, given a

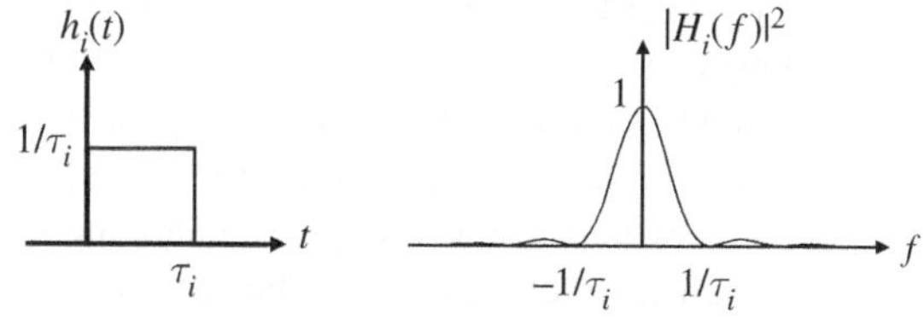

Figure 1.2 Impulse response of an integrator and corresponding amplitude squared of the frequency response.

generic low pass filter, the equivalent integration time is defined as the inverse of the bandwidth, often taken as the half power point (−3 dB). A low pass filter used to average a signal is often referred to as "integrator."

Since $H_i(0) = 1$, equation (1.97) implies that the mean is not affected by the averaging operation.

$$\langle w(t) \rangle = \langle z(t) \rangle \tag{1.108}$$

which is a totally expected result.

The zero-mean signals $w'(t) = w(t) - \langle w(t) \rangle$ and $z'(t) = z(t) - \langle z(t) \rangle$ are obviously related through the same impulse response (1.106). Using (1.101) for these processes, the power spectral density of the filtered signal becomes

$$S_{w'}(f) = S_{z'}(f)\,\mathrm{sinc}^2(f\tau_i) \tag{1.109}$$

and the autocorrelation function is the inverse Fourier transform (see Table A.2)

$$R_{w'}(\tau) = \mathcal{F}^{-1}[S_{z'}(f)\mathrm{sinc}^2(f\tau_i)] \approx \frac{S_{z'}(0)}{\tau_i}\Lambda\left(\frac{\tau}{\tau_i}\right) \tag{1.110}$$

where the approximation holds for signals having a spectrum flat up to frequencies much larger than the inverse of the integration time (first zero of the sinc function). The approximated expression above implies that, not surprisingly, samples of the output signal are partially correlated ($R_{w'}(\tau) \neq 0$) only when the time difference between them is smaller than the integration time. The reader should be aware of the approximation implied for fully understanding this assertion.

According to (1.49), the variance of $w(t)$ is the value of $R_{w'}(\tau)$ at the origin,

$$\sigma_w^2 = R_{w'}(0) = \int_{-\infty}^{\infty} S_{z'}(f)\mathrm{sinc}^2(f\tau_i)\,df \approx \frac{S_{z'}(0)}{\tau_i} \tag{1.111}$$

which states the important result that the variance of the averaged signal is inversely proportional to the integration time. Under the approximation indicated, the proportionality constant is the power spectral density at the origin ($f = 0$) of the input signal *after removing the mean*. Note, furthermore, that the approximated expression is equivalent to

$$\sigma_w^2 \approx \frac{1}{\tau_i}\int_{-\infty}^{\infty} R_{z'}(\tau)\,d\tau = \frac{2}{\tau_i}\int_0^{\infty} \Re e[R_{z'}(\tau)]\,d\tau \tag{1.112}$$

where the symmetry property (1.56) has been used.

Comment: In general, the standard deviation of the averaged signal is lower than that of the original signal. But, unlike averaging uncorrelated discrete samples, there is *not* a proportionality relation between both. In other words, σ_w is *not* equal to a constant (lower than one) multiplied by σ_z, which would imply that changing this one affects always the first. Imagine, for example, that the spectrum of $z(t)$ extends to relatively high frequencies. Including a low pass filter before the integrator does not affect the spectrum at the origin so σ_w is unaffected. However, the filtered signal that enters the integrator has reduced standard deviation due to the filtering.

1.2.6.1 Two signals

The above results can be easily extended to two processes. Assume that $w_1(t)$ and $w_2(t)$ are the time averages of two signals $z_1(t)$ and $z_2(t)$, respectively

$$w_1(t) = \frac{1}{\tau_i}\int_{t-\tau_i}^{t} z_1(t')dt' \quad ; \quad w_2(t) = \frac{1}{\tau_i}\int_{t-\tau_i}^{t} z_2(t')dt' \tag{1.113}$$

in which, for simplicity, the same integration time is assumed for both.[2] Using the same procedure but starting now from (1.104), equation (1.109) becomes

$$S_{w_1'w_2'}(f) = S_{z_1'z_2'}(f)\mathrm{sinc}^2(f\tau_i) \tag{1.114}$$

and the cross-correlation function

$$R_{w_1'w_2'}(\tau) = \mathcal{F}^{-1}[S_{z_1'z_2'}(f)\mathrm{sinc}^2(f\tau_i)] \approx \frac{S_{z_1'z_2'}(0)}{\tau_i}\Lambda\left(\frac{\tau}{\tau_i}\right) \tag{1.115}$$

where, again, the approximation is valid for signals having a cross-power spectrum much wider than the inverse of the integration time. The covariance of the output signals follows now from (1.54)

$$C_{w_1w_2} = R_{w_1'w_2'}(0) = \int_{-\infty}^{\infty} S_{z_1'z_2'}(f)\mathrm{sinc}^2(f\tau_i)\,df \approx \frac{S_{z_1'z_2'}(0)}{\tau_i} \tag{1.116}$$

which can be seen as a generalization of (1.111) for two signals. The corresponding expression of (1.112) is

$$C_{w_1w_2} \approx \frac{1}{\tau_i}\int_{-\infty}^{\infty} R_{z_1'z_2'}(\tau)\,d\tau \tag{1.117}$$

2 Extension to different integration times is straightforward.

1.3 Analytic Signals

The "analytic signal" is a powerful tool aimed at representing any real-valued signal by a related complex function. It can be defined for any kind of signal, either deterministic or random. For a pure sinusoidal tone it is a complex function of time with amplitude and argument equal to those of the sinusoid

$$\mathrm{x}(t) = A\cos(2\pi f_0 t + \phi) \quad \Rightarrow \quad x(t) = Ae^{j\phi}e^{j2\pi f_0 t} \tag{1.118}$$

where $x(t)$ is the analytic signal of $\mathrm{x}(t)$. The convention used throughout this book assigns a positive sign to the time-dependent complex exponential, which is compatible with the definition of Fourier transform in appendix A.7. Some bibliographic sources, as for example [2], use a negative sign so the reader must be careful when comparing expressions in different references. The complex number defined by the amplitude and the phase of the signal is the "phasor":

$$\text{Phasor of } \mathrm{x}(t) = \frac{A}{\sqrt{2}}e^{j\phi} \tag{1.119}$$

where the convention of dividing by $\sqrt{2}$ is useful for simplifying expressions relating mean power. From (1.118),

$$\mathrm{x}(t) = \Re e[x(t)] \; ; \; \langle \mathrm{x}^2(t) \rangle = \frac{1}{2}\langle |x(t)|^2 \rangle = \frac{A^2}{2} \tag{1.120}$$

The original signal is equal to the real part of its analytic representation, and this one has twice average power. Extension of these concepts to other kind of signals, and particularly to random processes, involves the Hilbert transform.

1.3.1 Hilbert Transform and Quadrature Filter

Given a *real* waveform $\mathrm{x}(t)$ its analytic signal representation is a *complex function of time* defined as [13]

$$x(t) = \mathrm{x}(t) + j\hat{\mathrm{x}}(t) \tag{1.121}$$

where the hat denotes Hilbert transform[3]

$$\hat{\mathrm{x}}(t) = \frac{1}{\pi}\int_{-\infty}^{\infty} \frac{\mathrm{x}(t')}{t - t'}\,dt' = \mathrm{x}(t) * \frac{1}{\pi t} \tag{1.122}$$

This transform is the result of passing the signal $\mathrm{x}(t)$ through a "quadrature filter" with the following frequency response:

$$H_q(f) = \mathscr{F}\left[\frac{1}{\pi t}\right] = -j\,\mathrm{sgn}\,f \tag{1.123}$$

3 Cauchy principal value of the integral must be taken to avoid singularities.

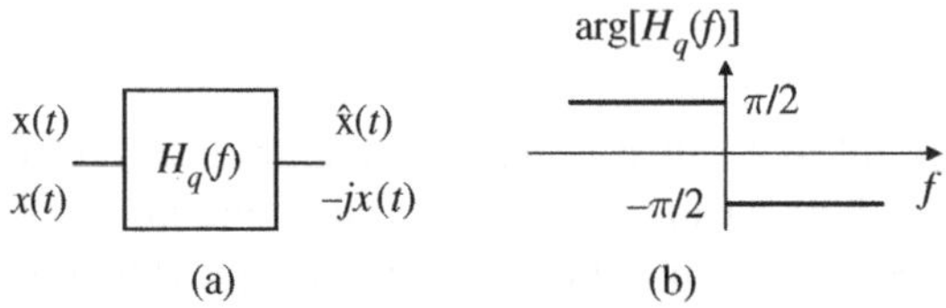

Figure 1.3 Quadrature filter symbol (a) and argument of the frequency response (b). As indicated, the output signal is the Hilbert transform of the input signal. Also, the analytic signal of the output is that of the input multiplied by $-j$.

where the function "signum" is defined in (A.30). The response of this filter is schematized in figure 1.3. It keeps the amplitude unchanged (except at $f = 0$) and shifts the phase by −90° for positive frequencies and 90° for negative frequencies. For example, the Hilbert transform of $\cos\,2\pi f_0 t$ is $\sin\,2\pi f_0 t$ and that of $\sin\,2\pi f_0 t$ is $-\cos\,2\pi f_0 t$.

Conceptually, a quadrature filter is equivalent to a 90° hybrid (see [6] for definition and examples). The real and imaginary parts of the analytic signal (1.121) are the two outputs of an *ideal* 90° hybrid when driven by a single signal. Nevertheless, a perfect quadrature filter cannot be realized by interconnecting circuit elements. The common implementation implies an in-phase and quadrature mixer, properly analyzed in section 5.2.

Introducing (1.122) into (1.121) [see also (A.36)]

$$x(t) = \mathrm{x}(t) * \left[\delta(t) + j\frac{1}{\pi t}\right] \tag{1.124}$$

which means that the analytic signal is the output of a filter driven by the original signal and with frequency response (see Table A.2)

$$\mathscr{F}\left[\delta(t) + j\frac{1}{\pi t}\right] = 2\mathrm{u}(f) \tag{1.125}$$

where the unit step function $\mathrm{u}(f)$ is defined in (A.31). The frequency response that produces the analytic signal is zero for negative frequencies and two for positive ones.

The time domain definition of the analytic signal (1.121) is a bit weird but it takes full meaning in the frequency domain. Its Fourier transform, if it exists, is inferred from (1.124) and (1.125)

$$\mathscr{F}\,[x(t)] = 2\mathscr{F}\,[\mathrm{x}(t)]\,\mathrm{u}(f) \tag{1.126}$$

The analytic signal in the frequency domain is thus twice the Fourier transform of the original signal for positive frequencies and zero for negative frequencies, with the DC component unchanged. This result is schematically depicted in figure 1.4. The reader is invited to demonstrate that (1.126) is consistent with the analytic signal of a pure sinusoid (1.118).

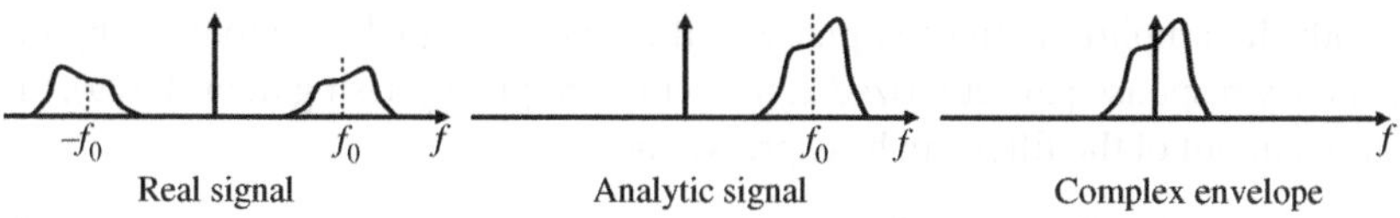

Figure 1.4 Analytic signal and complex envelope in the frequency domain.

An important property of the quadrature filter is now demonstrated. According to the filtering interpretation given above, the analytic signal of the Hilbert transform $\hat{x}(t)$ is the result of passing x(t) through a quadrature filter cascaded by a unit step filter, therefore with an overall frequency response $-j$ sgn f 2u(f) which, except at $f = 0$, is the same as $-j$2u(f). In consequence, *for signals with no DC component*, the following is true

$$\text{Analytic signal of } \hat{x}(t) = -jx(t) \tag{1.127}$$

In other words, *a quadrature filter multiplies the analytic signal of the input by* $-j$. This important property is indicated also in figure 1.3. Consistently, from (1.121) the real part of $-jx(t)$ is equal to the imaginary part of $x(t)$ which is $\hat{x}(t)$. Note that this property implies that $\hat{\hat{x}}(t) = -\mathrm{x}(t)$, which is valid only for signals with no DC component, that is for zero mean processes.

1.3.2 Complex Envelope

The analytic signal is sometimes expressed as

$$x(t) = \tilde{x}(t)e^{j2\pi f_0 t} \tag{1.128}$$

where $\tilde{x}(t)$ is the so-called "complex envelope" with respect to an *arbitrary* reference frequency f_0. According to the frequency shift property of the Fourier transform (table A.1), in the frequency domain the complex envelope is equal to the analytic signal shifted toward the origin by f_0 (see figure 1.4).

Using (1.121) and (1.128), the original waveform and its Hilbert transform can be written as

$$\mathrm{x}(t) = \Re e[\tilde{x}(t)e^{j2\pi f_0 t}] \qquad \hat{x}(t) = \Im m[\tilde{x}(t)e^{j2\pi f_0 t}] \tag{1.129}$$

compatible with the case of sinusoids (1.118) for which the complex envelope with respect to the frequency of the signal is the constant $A \exp j\phi$.

For narrow band-pass signals, if f_0 is a conveniently defined center frequency, the complex envelope becomes a slowly varying function of time and carries the information of signal's amplitude and phase. When passing the signal through a linear system with frequency response $H(f)$, provided it is a smooth function of frequency within the signal bandwidth, the analytic signal at the output becomes

$$x_o(t) \approx H(f_0)x(t) \quad \text{Narrow band-pass signal} \tag{1.130}$$

so both the amplitude and the phase of the analytic signal are modified by the frequency response particularized at the center frequency. Using then (1.129), the real signal out of the filter can be expressed as

$$\mathrm{x}_o(t) \approx \Re e[H(f_0)\tilde{x}(t)e^{j2\pi f_0 t}] \quad \text{Narrow band-pass signal} \tag{1.131}$$

If the frequency response varies within the signal bandwidth, either in amplitude or in phase, the above two expressions are not valid.

1.3.3 Correlation and Spectra

Suppose now that $\mathrm{x}(t)$ is a real stationary random process with power spectral density $S_{\mathrm{x}}(f)$. Since the analytic signal is the output of a filter with frequency response $2\mathrm{u}(f)$, according to (1.101), the power spectral density of the analytic signal is

$$S_x(f) = 4S_{\mathrm{x}}(f)\,\mathrm{u}(f) \tag{1.132}$$

To avoid discontinuities at the origin due to the delta function in (1.82), the analytic signal of random processes is always defined for *zero-mean signals* so $S_{\mathrm{x}}(f)$ is continuous at $f = 0$.

The autocorrelation function of the analytic signal is immediate from (1.83)

$$R_x(\tau) = 4\int_0^{\infty} S_{\mathrm{x}}(f)e^{j2\pi f\tau}df \tag{1.133}$$

Since $S_x(f)$ has only positive frequencies, it is itself an analytic signal so its inverse Fourier transform can be written as

$$R_x(\tau) = 2R_{\mathrm{x}}(\tau) + j2\hat{R}_{\mathrm{x}}(\tau) \tag{1.134}$$

where $R_{\mathrm{x}}(\tau)$ is the autocorrelation function of $\mathrm{x}(t)$ and $\hat{R}_{\mathrm{x}}(\tau)$ its Hilbert transform (1.122). In the frequency domain, this last function is

$$\mathcal{F}\{\hat{R}_{\mathrm{x}}(\tau)\} = S_{\mathrm{x}}(f)(-j\ \mathrm{sgn}\ f) = S_{\hat{\mathrm{x}}\mathrm{x}}(f) \tag{1.135}$$

where the last equality results from applying (1.102) to $\hat{\mathrm{x}}$ and x taking into account that the Hilbert transform is the output of a filter with frequency response (1.123). In time domain this result becomes $\hat{R}_{\mathrm{x}}(\tau) = R_{\hat{\mathrm{x}}\mathrm{x}}(\tau)$ so the auto correlation function of the analytic signal (1.134) becomes

$$R_x(\tau) = 2R_{\mathrm{x}}(\tau) + j2R_{\hat{\mathrm{x}}\mathrm{x}}(\tau) \tag{1.136}$$

The same procedure is applicable to the case of two real zero-mean random processes $\mathrm{x}(t)$ and $\mathrm{y}(t)$ with cross-power spectral density $S_{\mathrm{xy}}(f)$. Using (1.104), the cross-power spectral density of the corresponding analytic signals is

$$S_{xy}(f) = 4S_{\mathrm{xy}}(f)\,\mathrm{u}(f) \tag{1.137}$$

and the cross-correlation function (1.91)

$$R_{xy}(\tau) = 4\int_0^\infty S_{xy}(f)e^{j2\pi f\tau}df \quad (1.138)$$

Again, $S_{xy}(f)$ is the spectrum of an analytic signal and, using the same procedure, the cross-correlation function becomes

$$R_{xy}(\tau) = 2R_{\mathrm{xy}}(\tau) + j2R_{\hat{\mathrm{x}}\mathrm{y}}(\tau) \quad (1.139)$$

which, for $\mathrm{x}(t) = \mathrm{y}(t)$ reduces to (1.136), as expected. Comparing this result with (1.58), the following important properties are deduced:

$$R_{\mathrm{xy}}(\tau) = R_{\hat{\mathrm{x}}\hat{\mathrm{y}}}(\tau)\ ;\ R_{\hat{\mathrm{x}}\mathrm{y}}(\tau) = -R_{\mathrm{x}\hat{\mathrm{y}}}(\tau) \quad (1.140)$$

which can be used to write different versions of (1.139). In case of a single signal $\mathrm{x}(t) = \mathrm{y}(t)$ the above properties become

$$R_{\mathrm{x}}(\tau) = R_{\hat{\mathrm{x}}}(\tau)\ ;\ R_{\hat{\mathrm{x}}\mathrm{x}}(\tau) = -R_{\mathrm{x}\hat{\mathrm{x}}}(\tau) \quad (1.141)$$

The first of the above equations states that the Hilbert transform does not change the autocorrelation function. Using (1.56) in the second one,

$$R_{\hat{\mathrm{x}}\mathrm{x}}(\tau) = -R_{\hat{\mathrm{x}}\mathrm{x}}(-\tau) \Rightarrow R_{\hat{\mathrm{x}}\mathrm{x}}(0) = 0 \quad (1.142)$$

demonstrating that the cross-correlation function of a signal and its Hilbert transform is an odd function of time and, as a consequence, it vanishes at the origin. In consequence, since the mean of $\mathrm{x}(t)$ is null, a signal is always uncorrelated with its Hilbert transform $\langle \mathrm{x}(t)\,\hat{\mathrm{x}}(t)\rangle = 0$.

In some cases, it is convenient to write (1.136) and (1.139) for $\tau = 0$ showing explicitly the averages of the products:

$$\begin{aligned} &\left\langle |x(t)|^2 \right\rangle = 2\left\langle \mathrm{x}^2(t) \right\rangle \\ &\langle x(t)y^*(t)\rangle = 2\langle \mathrm{x}(t)\mathrm{y}(t)\rangle + j2\langle \hat{\mathrm{x}}(t)\mathrm{y}(t)\rangle \end{aligned} \quad (1.143)$$

The first equation makes explicit that the mean power of the analytic signal is twice that of the process it represents. The second equation can also be expressed using different signal pairs based on the properties (1.140) or expanded as (1.58),

$$\langle x(t)y^*(t)\rangle = (\langle \mathrm{x}(t)\mathrm{y}(t)\rangle + \langle \hat{\mathrm{x}}(t)\hat{\mathrm{y}}(t)\rangle) + j(\langle \hat{\mathrm{x}}(t)\mathrm{y}(t)\rangle - \langle \mathrm{x}(t)\hat{\mathrm{y}}(t)\rangle) \quad (1.144)$$

Important Remark: If both $\mathrm{x}(t)$ and $\mathrm{y}(t)$ are zero mean jointly normal processes so they are their Hilbert transforms and their delayed versions. In consequence, by definition, the corresponding analytic signals $x(t)$ and $y(t-\tau)$ are jointly normal for any time delay. Moreover, by comparing (1.140) and (1.141) with (1.71) using the appropriate symbol identifications, it follows that *they satisfy the circularity condition*. Consistently, (1.139) is equivalent to (1.34) and, more important, (1.35) can be used for any two analytic signals and their delayed versions.

1.3.3.1 Complex Envelope of Correlation Functions

Using the complex envelope concept (1.128), the cross-correlation function of two analytic signals becomes

$$R_{xy}(\tau) = \langle x(t)y^*(t-\tau)\rangle = R_{\tilde{x}\tilde{y}}(\tau)e^{j2\pi f_0\tau} \quad (1.145)$$

meaning that, in general, $\tilde{R}_{xy}(\tau) = R_{\tilde{x}\tilde{y}}(\tau)$. Taking the Fourier transform, applying the frequency translation property (Table A.1) and substituting (1.137),

$$S_{\tilde{x}\tilde{y}}(f) = S_{xy}(f+f_0) = 4S_{\mathrm{xy}}(f+f_0)\,\mathrm{u}(f+f_0) \quad (1.146)$$

consistent with the schematic spectra shown in figure 1.4. Furthermore, using (1.139) the cross-correlation function of the two real signals becomes

$$\begin{aligned} R_{\mathrm{xy}}(\tau) &= \frac{1}{2}\Re e[R_{\tilde{x}\tilde{y}}(\tau)e^{j2\pi f_0\tau}] \\ R_{\hat{\mathrm{x}}\mathrm{y}}(\tau) &= \frac{1}{2}\Im m[R_{\tilde{x}\tilde{y}}(\tau)e^{j2\pi f_0\tau}] \end{aligned} \quad (1.147)$$

with $R_{\tilde{x}\tilde{y}}(\tau)$ the inverse Fourier transform of $S_{\tilde{x}\tilde{y}}(f)$ in (1.146).

Particularizing the above expressions for $\mathrm{x}(t) = \mathrm{y}(t)$ is straightforward and given below for completeness

$$\begin{aligned} R_{\mathrm{x}}(\tau) &= \frac{1}{2}\Re e[R_{\tilde{x}}(\tau)e^{j2\pi f_0\tau}] \\ R_{\hat{\mathrm{x}}\mathrm{x}}(\tau) &= \frac{1}{2}\Im m[R_{\tilde{x}}(\tau)e^{j2\pi f_0\tau}] \end{aligned} \quad (1.148)$$

where $R_{\tilde{x}}(\tau)$ is the inverse Fourier transform of

$$S_{\tilde{x}}(f) = S_x(f+f_0) = 4S_{\mathrm{x}}(f+f_0)\,\mathrm{u}(f+f_0) \quad (1.149)$$

1.4 Phasors of Random Signals

In section 13-3 of his classical book [1], Papoulis asserts that "*Stationary processes have, in general, no Fourier transforms.*" As he states in the prologue of the 13th chapter, this is a consequence of the fact that these kind of signals have infinite energy. It is true, but the lack of Fourier transform produces an uncomfortable feeling for an electrical engineer used to solve linear systems in the frequency domain. The power spectral density is not enough, since it fails to describe how the phase is affected by the different components of a system. To use random signals in a way compatible with sinusoidal signals, frequency-dependent *phasors* are here conveniently introduced. Their definition is essentially equivalent to the concept of "spot noise" proposed by H. Bosma in his doctoral thesis [14]. As a matter of fact, phasors are often implicitly assumed in many references, and here they are only made explicit in a formal way.

1.4.1 Concept

Given a real *zero-mean* stationary process $\mathrm{x}(t)$, a new process $\mathrm{x}_{\Delta f}(t)$ is defined by applying an ideal rectangular band-pass filter with bandwidth $\Delta f = f_2 - f_1$. The power spectral density of the filtered process is then equal to $S_{\mathrm{x}}(f)$ for frequencies inside the bandwidth (and their negatives) and zero outside. According to (1.132), the analytic signal of $\mathrm{x}_{\Delta f}(t)$, denoted by $x_{\Delta f}(t)$, has spectral density equal to $4S_{\mathrm{x}}(f)$ for $f_1 < f < f_2$ and zero for all other frequencies including the negatives. Hence, using (1.84) its mean power is

$$\left\langle |x_{\Delta f}(t;\bar{f})|^2 \right\rangle = \int_{f_1}^{f_2} 4S_{\mathrm{x}}(f)df \approx 4S_{\mathrm{x}}(\bar{f})\Delta f \tag{1.150}$$

where $\bar{f} = (f_1 + f_2)/2$ and the approximation holds for narrow band filters. The center frequency has been included as parameter in $x_{\Delta f}(t;\bar{f})$ to make explicit that this signal depends on the choice of this frequency.

Similarly, using (1.137) and (1.92) with an additional process $\mathrm{y}(t)$, the cross-correlation of the two analytic signals of the filtered processes is

$$\left\langle x_{\Delta f}(t;\bar{f}) y^*_{\Delta f}(t;\bar{f}) \right\rangle = \int_{f_1}^{f_2} 4S_{\mathrm{xy}}(f)df \approx 4S_{\mathrm{xy}}(\bar{f})\Delta f \tag{1.151}$$

where $S_{\mathrm{xy}}(f)$ is the cross-power spectral density of the two real processes $\mathrm{x}(t)$ and $\mathrm{y}(t)$.

Definition: The *phasor* of the real zero-mean process $\mathrm{x}(t)$, is here defined as the limit of the analytic signal of the filtered process $x_{\Delta f}(t)$ normalized to the square root of the bandwidth when the bandwidth tends to zero:

$$x(f) \triangleq \lim_{\Delta f \to 0} \frac{x_{\Delta f}(t;\bar{f})}{\sqrt{2\Delta f}} \tag{1.152}$$

where the factor $1/\sqrt{2}$ is included to make power definitions consistent [as in (1.119)]. Alternatively, if the reader is uncomfortable with the limiting operation, the phasor can also be defined as $x_{\Delta f}(t;\bar{f})/\sqrt{2}$ for $\Delta f = 1$ Hz. It is the analytic signal of the filtered process in a 1 Hz bandwidth. Conceptually, both definitions are equivalent.

The phasor just defined is a *complex* random function of *time* having the frequency (*always positive*) as a parameter. It is denoted by $x(f)$, so making time dependence implicit. To simplify notation, the dependency on frequency will be often omitted as well.

1.4.2 Power and Cross-correlation

The mean power of the phasor is obtained from (1.150) and (1.152) interchanging limit and averaging operations

$$\overline{|x(f)|^2} = 2S_x(f)\,\mathrm{u}(f) \tag{1.153}$$

where the overbar instead of angle brackets is here used to denote the averaging operation. This convention will be always adopted when dealing with phasors of random signals. The equation above means that $\overline{|x(f)|^2}$ is a function of frequency equal to twice the power spectral density of $\mathrm{x}(t)$ for positive frequencies. It is essentially the same concept as the power spectral density, so it will be denoted from now on *phasor power* or simply *power* when the context does not lead to confusion. Using (1.132), it can also be expressed as function of the spectral density of the analytic signal

$$\overline{|x(f)|^2} = \frac{1}{2}S_x(f) \tag{1.154}$$

The cross-correlation between two phasors is similarly obtained from (1.151) and (1.152):

$$\overline{x(f)y^*(f)} = 2S_{xy}(f)\,\mathrm{u}(f) \tag{1.155}$$

so it is equal to twice the cross-power spectral density of the signals for positive frequencies. Again, using (1.137), it can be put in terms of the analytic signals' cross-power density

$$\overline{x(f)y^*(f)} = \frac{1}{2}S_{xy}(f) \tag{1.156}$$

Since the phasors are always defined for zero-mean processes, (1.90) implies that, for uncorrelated processes,

$$\overline{x(f)y^*(f)} = 0 \quad \text{(uncorrelated processes)} \tag{1.157}$$

Substituting (1.153) and (1.155) into (1.90) and (1.99), respectively, the auto- and cross- correlation functions of real signals can be written in terms of the phasor power and cross-correlation

$$\begin{aligned} R_x(\tau) &= \int_0^\infty \overline{|x(f)|^2}\cos(2\pi f\tau)df \\ R_{xy}(\tau) &= \int_0^\infty \Re e[\overline{x(f)y^*(f)}e^{j2\pi f\tau}]df \end{aligned} \tag{1.158}$$

which, particularized for $\tau = 0$ become

$$\begin{aligned} \langle \mathrm{x}^2(t)\rangle &= \int_0^\infty \overline{|x(f)|^2}df \\ \langle \mathrm{x}(t)\mathrm{y}(t)\rangle &= \int_0^\infty \Re e[\overline{x(f)y^*(f)}]df \end{aligned} \tag{1.159}$$

The first of the above equations states that the mean power of a signal is equal to the integrated power of the corresponding phasor over all frequencies, considered always positive. This interpretation stresses the idea that the phasor power is a way to describe the power spectral density.

Similarly, the auto- and cross- correlation functions of the *analytic signals* of x(t) and y(t) can be written in terms of the phasor power and cross-correlation. Introducing (1.153) and (1.155) into (1.133) and (1.138) respectively,

$$\begin{aligned} \frac{1}{2}R_x(\tau) &= \int_0^\infty \overline{|x(f)|^2} e^{j2\pi f\tau} df = \mathcal{F}^{-1}\left[\overline{|x(f)|^2}\right] \\ \frac{1}{2}R_{xy}(\tau) &= \int_0^\infty \overline{x(f)y^*(f)} e^{j2\pi f\tau} df = \mathcal{F}^{-1}\left[\overline{x(f)y^*(f)}\right] \end{aligned} \tag{1.160}$$

where the factors 1/2 have been placed at the left-hand side for compatibility with usual notation. Obviously, these equations are equivalent to (1.154) and (1.156). The particularization for $\tau = 0$ is

$$\begin{aligned} \frac{1}{2}\left\langle |x(t)|^2 \right\rangle &= \int_0^\infty \overline{|x(f)|^2} df \\ \frac{1}{2}\langle x(t)y^*(t) \rangle &= \int_0^\infty \overline{x(f)y^*(f)} df \end{aligned} \tag{1.161}$$

As it can be easily checked, these equations are fully compatible with (1.143). Note that the first one is identical to the first of (1.159) as it provides the total power of the signal.

A final remark: The phasor $x(f)$ is NOT the Fourier transform of $x(t)$ or x(t) as these functions of time do not have Fourier transform.

1.4.3 Linear Systems

Since the phasor is the analytic signal of a narrow band-pass process, (1.130) applies. Therefore, if y(t) is the output of a linear system with x(t) as input, the corresponding phasors are related by

$$y(f) = H(f)x(f) \tag{1.162}$$

where $H(f)$ is the frequency response of the system. Phasors of random signal are thus compatible with those of sinusoidal signals (1.119). Moreover, combining (1.153) and (1.101) it is immediately inferred that the mean power of the phasor at the output is related to that at the input by

$$\overline{|y(f)|^2} = |H(f)|^2\, \overline{|x(f)|^2} \tag{1.163}$$

and similarly, given two linear systems of frequency responses $H_1(f)$ and $H_2(f)$ with two different input signals $x_1(t)$ and $x_2(t)$ represented by phasors $x_1(f)$ and

$x_2(f)$, the cross-correlation between the output phasors is obtained by combining (1.155) and (1.104) substituting the corresponding signals

$$\overline{y_1(f)y_2^*(f)} = H_1(f)H_2^*(f)\overline{x_1(f)x_2^*(f)} \tag{1.164}$$

Equations (1.163) and (1.164) can be easily obtained from (1.162) by just applying standard algebra. Other cases, as for example (1.102), are also reproduced with the same operations. In conclusion, *phasors of random signals can be used to analyze linear systems using algebra compatible with the Fourier transform of deterministic signals or phasors of sinusoidal signals.*

1.5 Microwave Networks

A microwave network is an electronic circuit designed to operate at frequencies within the region going nominally from 300 MHz to 300 GHz. Depending on the specific frequency range, they are formed by lumped elements, transmission lines, waveguides, or other propagating structures, in many cases with planar technology [6]. The key concept that marks the difference with respect to low-frequency electronics is the signal propagation delay, which is not only taken into account in the analysis of operation but also used to synthesize circuit elements. This is why the tools used for both analysis and synthesis are based on waves rather than voltage and current levels, typical of low-frequency systems. These waves may represent deterministic or random signals.

1.5.1 Voltage and Current

The two primary variables in circuit theory are the electric potential difference, or *voltage* and the electric *current* flow. They are rigorously defined from the Maxwell equations on the basis of the electromagnetic field, but in circuit theory they are considered as fundamental quantities. In the frame of the present book, both of them are assumed stationary, zero mean, real, and jointly normal random processes denoted by $\mathrm{v}(t)$, in units of Volt $[V]$, and $\mathrm{i}(t)$, in units of Ampere $[A]$, respectively.

An electronic circuit is modeled as a network of interconnected elements with voltages and currents defined in each one. Several rules, particularly the Kirchhoff's and Ohm's laws, are then used to infer the response of the whole system. The voltage and the current defined at the "terminals" of a device, component or system are not independent from each other, as they are related through the electrical properties of the element designated by the terminals. In linear systems, the time-domain voltage $\mathrm{v}(t)$ is the convolution between the current $\mathrm{i}(t)$ and an impulse response $\zeta(t)$

$$\mathrm{v}(t) = \int_{-\infty}^{\infty} \zeta(t')\mathrm{i}(t-t')dt' = \zeta(t) * \mathrm{i}(t) \tag{1.165}$$

Moreover, with the appropriate sign convention, the mean *electric power*, in units of Watt $[W]$, that absorbs[4] the said element is

$$\langle \mathrm{p(t)} \rangle = \langle \mathrm{v}(t)\,\mathrm{i}(t) \rangle = \lim_{T\to\infty} \frac{1}{T} \int_T \mathrm{v}(t)\,\mathrm{i}(t) dt\,[W] \tag{1.166}$$

which is a constant quantity, positive or negative depending on whether the element is "passive" or "active." Using (1.94), the cross-correlation between voltage and current may be written in terms of the cross-power spectral density $S_{\mathrm{vi}}(f)$ as

$$\langle \mathrm{p(t)} \rangle = 2 \int_0^{\infty} \Re e[S_{\mathrm{vi}}(f)] df = \int_0^{\infty} P(f) df \tag{1.167}$$

where $P(f) = 2\Re e[S_{\mathrm{vi}}(f)]$ is the "power density" in $W/Hz = J$. It is a common practice to abbreviate the name to "power," and this is endorsed all along this book. Despite this appellative, though, it is the total *energy* (in Joules) that absorbs the element at a given frequency.

Phasors of voltage and current are now defined according to section 1.4:

$$\begin{aligned} &v(f)\text{: Voltage in } V/\sqrt{Hz} \\ &i(f)\text{: Current in } A/\sqrt{Hz} \end{aligned} \tag{1.168}$$

and taking into account (1.165) and (1.167) together with (1.162) and (1.155)

$$\begin{aligned} &v(f) = Z(f)\,i(f) \quad \text{Ohm's law} \\ &P(f) = \Re e[\overline{v(f) i^*(f)}] \quad [W/Hz \text{ or } J] \end{aligned} \tag{1.169}$$

where $Z(f) = \mathscr{F}\,[\zeta(t)]$ is the "impedance," complex in general, in units of Ohm $[\Omega]$.

Comments: The above expressions are identical[5] to those applicable to phasors of sinusoidal voltage and current (1.119). However, in the sinusoidal case $v(f)$ and $i(f)$ have units of Volts and Amperes respectively, and $P(f)$ is the mean electric power in Watt.

The same equations hold also for the frequency domain voltage and current when the signals have Fourier transform (finite energy signals). In this case, $v(f)$ and $i(f)$ have units of V/Hz and A/Hz, respectively, and $P(f)$ is the energy density in units of J/Hz.

In all cases the impedance $Z(f)$ is the same. As pointed out before, $P(f)$ will always be termed as "power," even though it may not have units of Watt.

4 With different sign conventions this is the power "delivered" by the element.

5 Except for the bar over the correlation product.

1.5.2 Normalized Voltage Waves

In microwave circuit theory, it is advantageous to use the so-called "normalized voltage waves" defined as a linear combination of voltage and current:

$$\begin{aligned} a &= \frac{1}{2}\left(\frac{v}{\sqrt{Z_0}} + i\sqrt{Z_0}\right) \\ b &= \frac{1}{2}\left(\frac{v}{\sqrt{Z_0}} - i\sqrt{Z_0}\right) \end{aligned} \tag{1.170}$$

where Z_0 is an *arbitrary* reference impedance in units of Ohm (Ω) and v and i are the phasors (1.168) with the explicit frequency dependency (f) removed for simplicity. In consequence, both a and b are also frequency-dependent phasors and both have units of $\sqrt{W/Hz}$.

In their original definition [15] Z_0 was allowed to be complex but, in practice, it is always considered real and usually taken as $Z_0 = 50\,\Omega$. The rationale behind the above definition is that a and b are the forward and backward voltage waves (normalized to $\sqrt{Z_0}$) that propagate in a transmission line of characteristic impedance Z_0. But, the definition (1.170) is general and does not need any explicit presence of a transmission line to be valid. The voltage and current phasors are deduced by inverting (1.170):

$$\begin{aligned} v &= \sqrt{Z_0}(a + b) \\ i &= \frac{1}{\sqrt{Z_0}}(a - b) \end{aligned} \tag{1.171}$$

To illustrate the use of the normalized voltage waves, the single circuit loop depicted in figure 1.5 is considered: a source with open-circuited voltage v_s and internal impedance Z_s is connected to a load impedance Z_L. The symbols for a and b are arrows indicating, respectively, *inward* and *outward* waves to/from the load. The sign conventions and arrow directions used in the figure are compatible with (1.170) and (1.171).

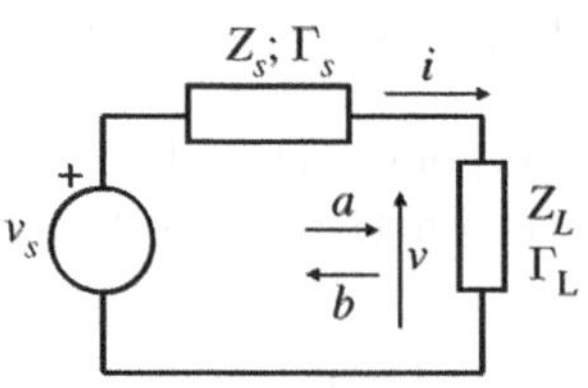

Figure 1.5 A single circuit loop for illustrating the use of voltage waves. Z_s is the internal impedance of the source and Z_L a load.

According to Ohm's law (1.169), the load impedance Z_L is the ratio between voltage and current $Z_L = v/i$. Using (1.171), it follows that

$$Z_L = Z_0 \frac{1+\Gamma_L}{1-\Gamma_L} \quad \Rightarrow \quad \Gamma_L = \frac{Z_L - Z_0}{Z_L + Z_0} \tag{1.172}$$

where $\Gamma_L = b/a$ is the *reflection coefficient* of the load. For a given reference impedance there is a one-to-one correspondence between impedance and reflection coefficient. Moreover, any realizable impedance has $\Re e[Z_L] \geq 0$, which implies that the reflection coefficient is always bounded by the unit circle $|\Gamma_L| \leq 1$. The reflection coefficient of the reference impedance Z_0 is null. A purely reactive load has imaginary impedance, which translates into a reflection coefficient with unit amplitude.

Kirchhoff's voltage law applied to the circuit of figure 1.5 states that $v_s = Z_s i + v$. Substituting v and i by (1.171) the following result is obtained after some algebra:

$$a = \Gamma_s b + b_s \tag{1.173}$$

where $\Gamma_s = (Z_s - Z_0)/(Z_s + Z_0)$ is the reflection coefficient associated to the source impedance and b_s the "source *intrinsic* wave"

$$b_s = \frac{v_s}{2\sqrt{Z_0}}(1 - \Gamma_s) \quad \text{(intrinsic wave)} \tag{1.174}$$

which depends only on the source. The usual convention assigns it to a symbol b instead of a to indicate that it is a wave outward from the source. Equation (1.173) states that the inward wave to the load a (figure 1.5) is the sum of the source intrinsic wave b_s plus the reflection due to the generator reflection coefficient Γ_s of the outward wave b coming from the load.

Taking into account now that $b = \Gamma_L a$ [see definition after (1.172)], the following equations for a and b are found

$$a = \frac{b_s}{1 - \Gamma_s \Gamma_L} \ ; \ b = \frac{b_s \Gamma_L}{1 - \Gamma_s \Gamma_L} \tag{1.175}$$

These general expressions can be used to formulate the following important statement: *The intrinsic wave b_s is the one that any source emanates toward the reference impedance*, as can be easily deduced by making $\Gamma_L = 0$, resulting in $a = b_s$ for any Γ_s. Also, a source with internal impedance equal to the reference (that is, $\Gamma_s = 0$) emanates the intrinsic wave toward any load.

1.5.3 Available Power

According to (1.169), the power delivered to the load is $P_L = \Re e\left[\overline{v\, i^*}\right]$ so substituting (1.171)

$$P_L = \overline{|a|^2} - \overline{|b|^2} = \overline{|a|^2}(1 - |\Gamma_L|^2) \tag{1.176}$$

This power is maximum when the *conjugate matching condition* holds, which is defined as $\Gamma_L = \Gamma_s^*$. In this particular case (1.175) reduces to

$$a = \frac{b_s}{1 - |\Gamma_s|^2} \quad \text{(conjugate matching)} \tag{1.177}$$

and the maximum power that can be delivered by the source, known as "available power," results by substituting this equation into (1.176) taking into account that $|\Gamma_L| = |\Gamma_s|$ in this condition

$$P_{av} = \frac{\overline{|b_s|^2}}{1 - |\Gamma_s|^2} \quad \text{(available power)} \tag{1.178}$$

In the general case, other than conjugate matching, after introducing (1.175) into (1.176) and using the definition above, the power delivered to the load becomes

$$P_L = P_{av} \frac{(1 - |\Gamma_s|^2)(1 - |\Gamma_L|^2)}{|1 - \Gamma_s \Gamma_L|^2} = P_{av}\, C(\Gamma_s, \Gamma_L) \tag{1.179}$$

where the self defined term $C(\Gamma_s, \Gamma_L)$ is the "mismatch coefficient." It is equal to one in conjugate matching condition and otherwise is always lower than one as long as the reflection coefficients' amplitudes are also lower than one, which is true for any passive network in general.

1.5.4 S-parameters and Power Gains

Figure 1.6(a) shows the schematic of a two-port linear network driven at port 1 (considered as "input") by a source and terminated at port 2 (considered as "output") by a load. Both source and load impedances are characterized by their reflection coefficients Γ_s and Γ_L, respectively. Inward and outward waves are defined in each port according to (1.170), with a_i representing always the waves entering into the network and b_i those emerging from it. In general, for a linear

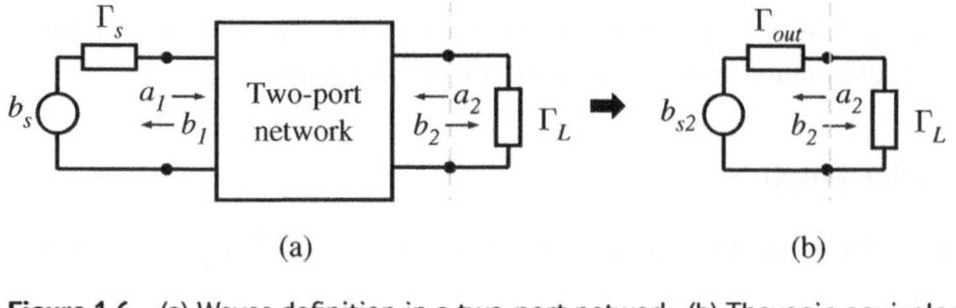

Figure 1.6 (a) Waves definition in a two-port network. (b) Thevenin equivalent circuit at the output port.

network they are mutually related through the so-called *S-parameters*:

$$\begin{aligned} b_1 &= S_{11}a_1 + S_{12}a_2 \\ b_2 &= S_{21}a_1 + S_{22}a_2 \end{aligned} \tag{1.180}$$

which, in general, are frequency dependent.

The *available power gain* is the ratio between the available powers at output and input ports:

$$G_A = \frac{P_{avL}}{P_{avs}} \tag{1.181}$$

The denominator is directly (1.178), and the numerator is the same expression but referred to the Thevenin equivalent circuit at the output port depicted on figure 1.6(b):

$$P_{avL} = \frac{\overline{|b_{s2}|^2}}{1 - |\Gamma_{out}|^2} \tag{1.182}$$

The intrinsic source wave b_{s2} is, by definition, the wave that emanates over the reference impedance, so it can be computed by making $a_2 = 0$ in (1.180). In these conditions $b_1 = S_{11}a_1$ and $b_2 = b_{s2} = S_{21}a_1$ where a_1 is given by (1.175) with Γ_L replaced by S_{11}, which is the input reflection coefficient in this case. The result is

$$b_{s2} = \frac{S_{21}b_s}{1 - S_{11}\Gamma_s} \tag{1.183}$$

Replacing this expression into (1.182) and using (1.178) for P_{avs}, the available power gain (1.181) follows

$$G_A = \frac{P_{avL}}{P_{avs}} = \frac{|S_{21}|^2(1 - |\Gamma_s|^2)}{|1 - S_{11}\Gamma_s|^2\left(1 - |\Gamma_{out}|^2\right)} \quad \text{(available power gain)} \tag{1.184}$$

The output reflection coefficient Γ_{out} is by definition the ratio b_2/a_2 when the voltage generator at the input is removed, that means forcing $a_1 = \Gamma_s b_1$. Directly from (1.180) it is obtained that

$$\Gamma_{out} = S_{22} + \frac{S_{12}S_{21}\Gamma_s}{1 - S_{11}\Gamma_s} \tag{1.185}$$

which when introduced in (1.184) shows that the available power gain is function of the S-parameters and the source reflection coefficient. It does not depend on the load at port 2, Γ_L.

An important property of the available power gain, derived directly from its definition (1.181) and knowing that the available power does not depend on the load, is that it is "cascadable." The available power gain of a chain of two-port networks is equal to the product of the available power gains of the individual networks.

The *transducer power gain* G_T is the ratio between the power delivered to the load and the available power at the input port. Using the concept of mismatch coefficient (1.179), it turns out that $G_T = G_A C(\Gamma_{out}, \Gamma_L)$ resulting in

$$G_T = \frac{P_L}{P_{avs}} = \frac{|S_{21}|^2(1-|\Gamma_s|^2)(1-|\Gamma_L|^2)}{|1-S_{11}\Gamma_s|^2|1-\Gamma_L\Gamma_{out}|^2} \quad \text{(transducer power gain)} \tag{1.186}$$

where again Γ_{out} is (1.185) which, when introduced in (1.186), makes evident that G_T depends on both Γ_s and Γ_L. The transducer power gain is *not* cascadable.

The *power gain* (without adjective) is the ratio between the power delivered to the load and the input power $P_{in} = |a_1|^2 - |b_1|^2$. Although it is seldom used, an expression for it is provided here for completeness

$$G_P = \frac{P_L}{P_{in}} = \frac{|S_{21}|^2(1-|\Gamma_L|^2)}{|1-S_{22}\Gamma_L|^2\left(1-|\Gamma_{in}|^2\right)} \quad \text{(power gain)} \tag{1.187}$$

where Γ_{in} is the input reflection coefficient derived directly from (1.180) with $a_2 = \Gamma_L b_2$,

$$\Gamma_{in} = S_{11} + \frac{S_{12}S_{21}\Gamma_L}{1-S_{22}\Gamma_L} \tag{1.188}$$

The power gain is not cascadable and is independent on the source reflection coefficient Γ_s. It can also be computed as $G_P = G_T/C(\Gamma_{in}, \Gamma_s)$ with $C(\Gamma_{in}, \Gamma_s)$ the mismatch coefficient at the input port.

1.5.5 Noise Waves and Temperature

Thermal noise is the name given to the experimental evidence that any electrical network at temperature above absolute zero generates random signals that propagate through interconnected elements. Noise waves were introduced by Bosma [14] to account for thermal noise in a way compatible with sinusoidal voltage waves propagating in a transmission line.

For a one-port network characterized by a reflection coefficient Γ, thermal noise is modeled by a source emanating an intrinsic noise wave. Using the definitions of figure 1.7 in (1.173)

$$b = \Gamma a + c \tag{1.189}$$

where c is the intrinsic wave generated by the network due to thermal noise. By convention a is always defined pointing toward the network and this is why symbols a and b are swapped with respect to (1.173). The total outward wave b equals to c either if $\Gamma = 0$, corresponding to the reference impedance, or $a = 0$ which is the case of no wave entering from other sources.

The *available power* associated with the thermal noise is then directly (1.178) where c is now the intrinsic wave

$$\mathcal{N}(f) = \frac{\overline{|c(f)|^2}}{1 - |\Gamma(f)|^2} \quad [J] \tag{1.190}$$

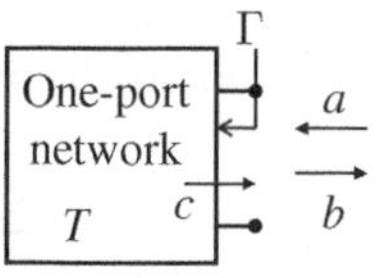

Figure 1.7 One-port linear network to show the thermal noise representation using waves.

where now the explicit dependence on frequency has been included in all terms. Following the definitions (1.169), this "available power" $\mathcal{N}(f)$ has units of Joule, equivalent to W/Hz. To find out its physical significance and deduce an expression for it, some insight into the thermal noise mechanism is needed.

Thermal noise is explained by postulating that heat produces in matter a "latent energy" that eventually can be extracted (all of it or in part) in the form of electromagnetic waves. This latent energy is what has been here named *available power* $\mathcal{N}(f)$, consistent with the units of Joule. As already discovered by the great scientists of the 19th century, this energy can be attributed to the random vibration of electrons, each one with a different energy. Assuming no interaction between them, this model implies that, in equilibrium at an absolute temperature of T, the electron must follow the Boltzmann statistics, predicting that the probability that a given electron has energy $\mathcal{E}$ is proportional to $\exp(-\mathcal{E}/kT)$ where $k = 1.380662 \times 10^{-23} J/K$ is the Boltzmann constant and T the absolute temperature in K. The average energy of the system formed by a very large number of electrons is then

$$\mathcal{N}(f) = \frac{\int_0^{\infty} \mathcal{E} e^{-\frac{\mathcal{E}}{kT}} d\mathcal{E}}{\int_0^{\infty} e^{-\frac{\mathcal{E}}{kT}} d\mathcal{E}} = kT \quad [J] \tag{1.191}$$

which is independent of frequency and often termed as "white noise." This elegant result tells that thermal energy is proportional to the absolute temperature independently of any other parameter, such as the kind of substance matter is made of. Nevertheless, the application of this equation to explain the black body radiation at the beginning of the 20th century led to a paradox coined as "the ultraviolet catastrophe" [16]. It is due to the fact that the mean power of the associated signal, computed by replacing $P(f)$ by $\mathcal{N}(f)$ in (1.167), becomes always infinite, contradicting the original premise that the signal has finite mean power.

Eventually, this contradiction was solved by the Plank's quantum hypothesis, which states that the electrons are harmonic oscillators that can have only discrete energies multiple of the quantum hf where $h = 6.626176 \times 10^{-34} J/Hz$ is the Plank's constant and f the frequency. The Boltzmann statistics still applies, so the average energy can be computed by (1.191) but with discrete sums instead of integrals and replacing $\mathcal{E}$ by ihf with i an integer and f the frequency:

$$\mathcal{N}(f) = \frac{\sum_{i=0}^{\infty} ihf e^{-\frac{ihf}{kT}}}{\sum_{i=0}^{\infty} e^{-\frac{ihf}{kT}}} = \frac{hf}{e^{\frac{hf}{kT}} - 1} \quad [J] \tag{1.192}$$

This equation does predict a dependency of the energy with frequency. At frequencies low enough such that $\exp(hf/kT) \approx 1 + hf/kT$, it reduces to the classical result (1.191) which, in practice, can be used through the whole microwave region. As expected, after inserting (1.192) into (1.167) the mean power remains finite, so solving the paradox. Note, however, that finite mean power signals have infinite total energy when this one is defined as the integral of instantaneous power $\mathrm{p}(t) = \mathrm{v}(t)\,\mathrm{i}(t)$ over time.

Based on the above discussion, the "equivalent noise temperature" T of any one-port *microwave* network is defined on the basis of (1.191) as

$$\text{Available noise power} = k\,T \quad [W/Hz] \tag{1.193}$$

This definition holds for either active or passive networks. In this latter case T is always equal to the physical temperature, as long as it is constant and homogeneous within the network.

The relationship between equivalent noise temperature and available power has been here deduced using hints from fundamental concepts of physics, included mostly without proof. The author is aware of that and expects the interested reader to complement them with more dedicated material on this discipline. Nevertheless they have served as a kind of justification to arrive at equation (1.193), which can now be considered as an educated premise in the development of the theory of noise in electrical networks operating at microwave frequencies.

Substituting (1.191) into (1.190) the power of the intrinsic noise wave of a network with equivalent noise temperature T defined as in (1.193) becomes

$$\overline{|c|^2} = kT(1 - |\Gamma|^2) \tag{1.194}$$

This equation implies that the intrinsic noise wave c vanishes if either the network is at absolute zero ($T = 0$) or if it is lossless ($|\Gamma| = 1$).

1.5.6 Interconnection

Consider two one-port networks with noise equivalent temperatures T_1 and T_2 and reflection coefficients Γ_1 and Γ_2 interconnected as shown in figure 1.8. As long as the reference impedance is the same, wave "a" of one network is "b" of the

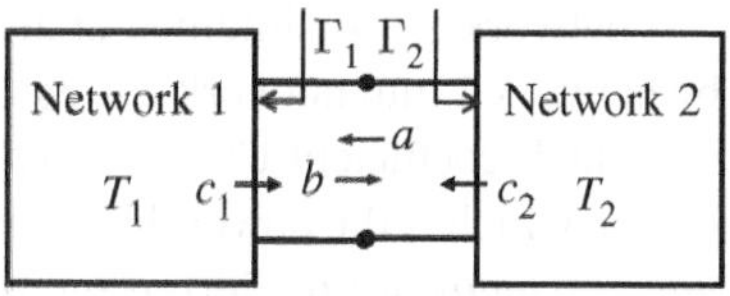

Figure 1.8 Interconnection between two one-port noisy networks at different temperatures.

other and vice-versa. Taken then into account that a is outward from network 2 but inward to network 1 and b the opposite, applying (1.189) to both,

$$a = b\Gamma_2 + c_2\ ; \quad b = a\Gamma_1 + c_1 \tag{1.195}$$

which can be combined to solve for a and b:

$$a = \frac{c_1\Gamma_2 + c_2}{1 - \Gamma_1\Gamma_2}\ ; \quad b = \frac{c_2\Gamma_1 + c_1}{1 - \Gamma_1\Gamma_2} \tag{1.196}$$

Since c_1 and c_2 represent the noise generated by two different networks, they are independent, so they are also uncorrelated and $\overline{c_1 c_2^*} = 0$ [see (1.157)]. Then, the power associated with waves a and b become

$$\overline{|a|^2} = \frac{\overline{|c_1|^2}|\Gamma_2|^2 + \overline{|c_2|^2}}{|1 - \Gamma_1\Gamma_2|^2}\ ; \quad \overline{|b|^2} = \frac{\overline{|c_2|^2}|\Gamma_1|^2 + \overline{|c_1|^2}}{|1 - \Gamma_1\Gamma_2|^2} \tag{1.197}$$

and using (1.194), they are rewritten as

$$\overline{|a|^2} = \frac{kT_1(1 - |\Gamma_1|^2)|\Gamma_2|^2 + kT_2(1 - |\Gamma_2|^2)}{|1 - \Gamma_1\Gamma_2|^2} \tag{1.198}$$

and

$$\overline{|b|^2} = \frac{kT_2(1 - |\Gamma_2|^2)|\Gamma_1|^2 + kT_1(1 - |\Gamma_1|^2)}{|1 - \Gamma_1\Gamma_2|^2} \tag{1.199}$$

According to (1.176), the total power flowing from network 1 to network 2 is the difference between both. After some algebra, it is found that

$$P(T_1, T_2) = \overline{|b|^2} - \overline{|a|^2} = k(T_1 - T_2)\frac{(1 - |\Gamma_1|^2)(1 - |\Gamma_2|^2)}{|1 - \Gamma_1\Gamma_2|^2} \tag{1.200}$$

where the multiplying term is immediately recognized as the mismatch coefficient $C(\Gamma_1, \Gamma_2)$ defined in (1.179).

As expected, power always flows from the hotter network to the colder. If they are in thermal equilibrium ($T_1 = T_2 = T$) then there is no net power flow between them independently of their reflection coefficients. This means that in this condition always $\overline{|a|^2} = \overline{|b|^2}$. A particular case is worth mention at this point: still with both networks at the same temperature, if network 2 is just the reference impedance, then $\overline{|a|^2} = \overline{|c_2|^2} = kT$, so $\overline{|b|^2}$ must also be equal to kT. Obviously the same happens if network 1 is the reference impedance instead, so the following property can be established:

$$\text{If either } \Gamma_1 \text{ or } \Gamma_2 \text{ is zero and both are at } T \text{ then } \overline{|a|^2} = \overline{|b|^2} = kT \tag{1.201}$$

If network 2 is maintained at absolute zero ($T_2 = 0$), then the power delivered by network 1 into network 2 is equal to the available power kT_1 multiplied by the mismatch coefficient $C(\Gamma_1, \Gamma_2)$. This is compatible with the concept of available power, which is the power that would dissipate a conjugate-matched load *at absolute zero temperature*. Furthermore, if keeping $T_2 = 0$ network 2 is made equal to the reference impedance $\Gamma_2 = 0$, then $a = 0$ and

$$P = \overline{|b|^2} = kT_1(1 - |\Gamma_1|^2) \quad (\Gamma_2 = T_2 = 0) \tag{1.202}$$

consistent with (1.194) taking into account that, in this case, $b = c_1$, as deduced from (1.195). This result allows stating the following formal definition:

> **Definition:** The intrinsic noise wave of a one-port network is the outward wave that emanates when it is loaded by the reference impedance Z_0 at absolute zero temperature. This definition is compatible with putting $a = 0$ in (1.189)

1.5.7 Two-port Networks

Noise in two-port networks can also be represented by an equivalent temperature. Consider the diagram depicted on figure 1.9(a). The input port (at left) is connected to a noisy one-port network representing a receiving antenna or a previous stage of the system. The available powers at both input and output ports are represented through (1.193) by equivalent noise temperatures T_s and T_{out}, respectively. Taking into account the definition of available power gain G_A (1.181), and adding the contribution of the noise generated by the network itself,

$$T_{out} = G_A(T_s + T_N) \tag{1.203}$$

where T_N is defined as the equivalent noise temperature of the two-port network *referenced to the input port*. The contribution of the network noise to the available power at the output is kG_AT_N. The above expression holds for any kind of microwave network, either active or passive.

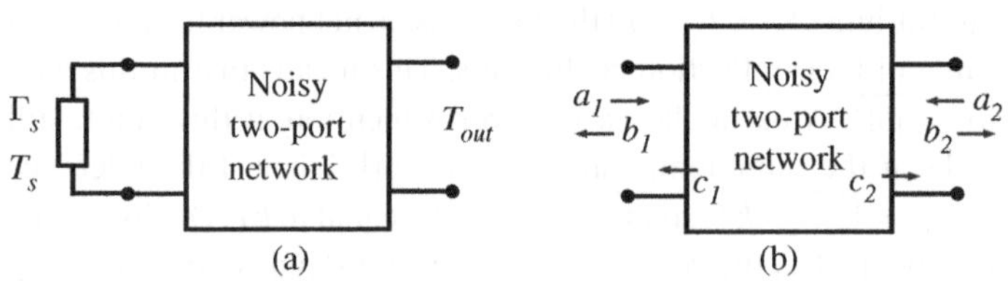

Figure 1.9 Definitions of noise in a two-port network. (a) Equivalent temperature at output. (b) Noise waves.

For a passive network kept at a homogeneous physical temperature, T_N is solely dependent on this temperature and on the available gain. To find an expression for it, assume that both Γ_s and the two-port network are kept at a the same temperature T (thermodynamic equilibrium). In these conditions, the equivalent noise temperature at the output port must also be equal to T by definition. Substituting into (1.203) with $T_s = T$ and solving for T_N,

$$T_N = T\left(\frac{1}{\mathrm{G}_A} - 1\right) \quad \text{(passive network at } T\text{)} \tag{1.204}$$

where it is worth to recall that G_A is the available power gain (1.184) and T the physical temperature of the network in K. Introducing this result in (1.203), the following general relation is found

$$T_{out} - T = \mathrm{G}_A(T_s - T) \quad \text{(passive network at } T\text{)} \tag{1.205}$$

which illustrates that the available power gain is the ratio between noise temperature increments at output and input with respect to the network physical temperature.

Independently of the above interpretation, the general methodology to include noise in two-port networks consists of adding to the S-parameter equations (1.180) intrinsic noise waves at each port

$$\begin{aligned} b_1 &= S_{11}a_1 + S_{12}a_2 + c_1 \\ b_2 &= S_{21}a_1 + S_{22}a_2 + c_2 \end{aligned} \tag{1.206}$$

The conventions for the definition of all waves are given in figure 1.9(b). Consistent with the definition given at the end of section 1.5.6, the intrinsic waves c_1 and c_2 are defined as the outward waves b_1 and b_2 in the particular case of null inward waves ($a_1 = a_2 = 0$), which happens when both ports are connected to the reference impedance Z_0 at absolute zero (i.e. no reflections and no noise). In general c_1 and c_2 are partially correlated and for a passive network at homogeneous temperature their power and cross-correlation are function of the S-parameters and the temperature through the Bosma theorem (1.233).

A general expression for T_N is now derived using (1.206). The procedure starts by determining the equivalent noise temperature at the output port T_{out} and then identifying the result with (1.203). Based on (1.194), T_{out} is directly deduced from the power of the intrinsic noise wave at the output port. According to its definition, this one is in turn obtained by placing at the output port the reference impedance at absolute zero, a situation depicted in figure 1.10. In these conditions b_2 is the intrinsic noise wave so from (1.194) the equivalent noise temperature is

$$kT_{out} = \frac{\overline{|b_2|^2}}{1 - |\Gamma_{out}|^2} \tag{1.207}$$

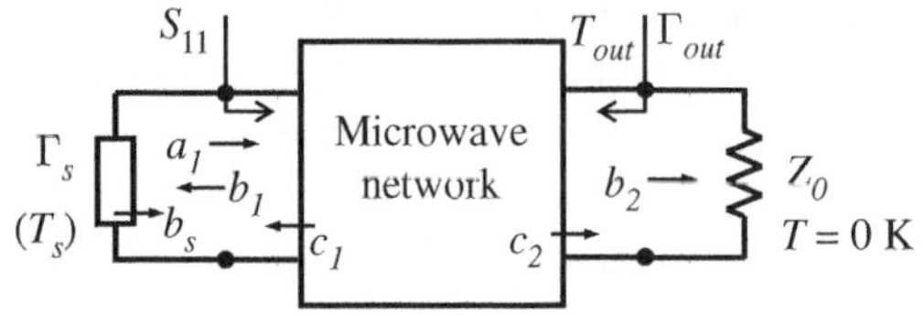

Figure 1.10 Setup for computing the noise equivalent temperature at output T_{out}.

On the other hand, the microwave network is characterized by S-parameters and noise waves according to (1.206). Since the inward wave at the output port a_2 is null, the system of equations reduces to

$$\begin{aligned} b_1 &= S_{11}a_1 + c_1 \\ b_2 &= S_{21}a_1 + c_2 \end{aligned} \tag{1.208}$$

and the inward wave at port 1, a_1, is equal to the expression of b in (1.196) with appropriate symbol identification

$$a_1 = \frac{c_1\Gamma_s + b_s}{1 - S_{11}\Gamma_s} = \Lambda(c_1\Gamma_s + b_s) \tag{1.209}$$

in which the self-defined parameter $\Lambda = 1/(1 - S_{11}\Gamma_s)$ is introduced to simplify the final expressions. Note that the intrinsic noise wave of the source is here denoted by b_s instead of, for example, c_s. This convention will be followed from now on. Substituting this expression into the second one of (1.208)

$$b_2 = S_{21}\Lambda b_s + S_{21}\Lambda\Gamma_s c_1 + c_2 \tag{1.210}$$

and the corresponding power is

$$\begin{aligned} \overline{|b_2|^2} = {} & |S_{21}|^2|\Lambda|^2\overline{|b_s|^2} + |S_{21}|^2|\Lambda|^2|\Gamma_s|^2\overline{|c_1|^2} \\ & + 2\Re e\left[S_{21}\Lambda\Gamma_s\overline{c_1c_2^*}\right] + \overline{|c_2|^2} \end{aligned} \tag{1.211}$$

where it has been taken into account that b_s is uncorrelated with c_1 and c_2 since they are generated by different networks.

This result has to be introduced into (1.207) to get the noise equivalent temperature at the output. To do so, the noise power associated with the source reflection coefficient is written in terms of its noise equivalent temperature using (1.194)

$$\overline{|b_s|^2} = kT_s(1 - |\Gamma_s|^2) \tag{1.212}$$

and the following definitions are introduced:

$$k\mathcal{T}_1 \triangleq \overline{|c_1|^2}; \quad k\mathcal{T}_{12} \triangleq \frac{\overline{c_1c_2{}^*}}{S_{21}^*}; \quad k\mathcal{T}_2 \triangleq \frac{\overline{|c_2|^2}}{|S_{21}|^2} \tag{1.213}$$

Then (1.207) can be written *exactly* as (1.203) with

$$T_N = \frac{|\Lambda|^2|\Gamma_s|^2\mathcal{T}_1 + 2\Re e[\Lambda\Gamma_s\mathcal{T}_{12}] + \mathcal{T}_2}{(1 - |\Gamma_s|^2)|\Lambda|^2} \tag{1.214}$$

The noise equivalent temperature referenced to the input port T_N depends on the source reflection coefficient and on the three equivalent temperatures (1.213). The reader is invited to check that, for a passive network at homogeneous temperature, this equation is equivalent to (1.204) when the Bosma theorem (1.233) is taken into account.[6]

For the special case of a matched input port (in the sense of being equal to the reference impedance $\Gamma_s = 0$), (1.214) reduces to

$$T_N(\Gamma_s = 0) = \mathcal{T}_2 \tag{1.215}$$

which provides a physical significance for the temperature $\mathcal{T}_2$: it is the equivalent noise temperature of the two-port network for matched input. Note that, in this case, $\Lambda = 1$.

1.5.8 Cascade

Microwave radiometers commonly include a number of stages interconnected in cascade, so that the output port of each of them feeds the input of the next one. The first element is always the antenna and the last one drives the power detector or the signal correlator; or the analog to digital converter in digital systems.

Equation (1.203) provides a simple way to take into account cascade interconnections. It indicates that the equivalent noise temperature of one particular network in the chain must be added to the one coming from the previous component and then multiply the sum by the available gain to get the equivalent temperature at the output. This interpretation is easily visualized in the flux diagram of figure 1.11, representing a cascade of three components. The three symbols G_{A1}, G_{A2}, and G_{A3} are the corresponding *available* power gains (1.184) and T_{N1}, T_{N2}, and T_{N3} their noise equivalent temperatures referenced to their input port (1.214). For passive elements (1.204) should be used instead for simplicity. For active components, an expression based on the so called "Noise Figure" F is normally utilized so that, for $i = 1$ to 3,

$$T_{Ni} = \begin{cases} T_i\left(L_i - 1\right) & \text{(Passive component)} \\ T_0(F_i - 1) & \text{(Active component)} \end{cases} \tag{1.216}$$

where $L_i = 1/\mathrm{G}_{Ai}$ is the attenuation of the component, T_i is the corresponding physical temperature, and $T_0 = 290$ K is the reference temperature. An expression for the noise figure can be easily derived by introducing the definition above into (1.214) with $T_N = T_{Ni}$. In practice the noise figure is measured by dedicated instrumentation.

Using the flux diagram of figure 1.11,

$$T_{out} = \{[(T_s + T_{N1})\mathrm{G}_{A1} + T_{N2}]\mathrm{G}_{A2} + T_{N3}\}\mathrm{G}_{A3} \tag{1.217}$$

6 Warning, it is not straightforward. It can be easily checked with a numerical example, though.

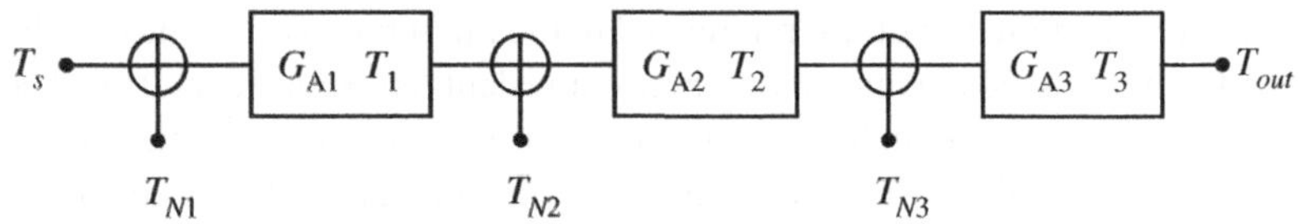

Figure 1.11 Flux diagram to illustrate the cascade connection of networks.

which can be written just as (1.203)

$$T_{out} = G_A(T_s + T_R) \tag{1.218}$$

provided the following identifications are done

$$\begin{aligned} G_A &= G_{A1}G_{A2}G_{A3} \\ T_R &= T_{N1} + \frac{T_{N2}}{G_{A1}} + \frac{T_{N3}}{G_{A1}G_{A2}} \end{aligned} \tag{1.219}$$

where G_A is the total available power gain (1.184) and T_R the noise equivalent temperature of the whole receiver referenced to the input. The above equations can be easily extended by recursion to any number of stages. The second one is known as "Friis formula" and sometimes is expressed in terms of the noise figure using the definition of the second equation (1.216).

It is important to point out that (1.219) is *not* an approximation in the sense that it might neglect the mismatch between units. As long as the correct available gains (1.184) and equivalent temperatures (1.216) are used, these equations are exact. It should be clarified, however, that both parameters depend on the source reflection coefficient of the corresponding component. Therefore, using in (1.219) the values for matched input, usually the outcome of a measuring system, is only an approximation and can lead to an important source of error.

The above results are also valid for systems including frequency conversions, although, in this case, care must be taken to appropriately include the noise contribution of the image bands (see section 5.1). A complete study of receiver noise temperature for heterodyne systems is out of the scope of this book.

1.5.9 High Gain Receiver

The block diagram on figure 1.12(a) represents a whole microwave receiver built from several individual elements in cascade and connected at the input port to a noise source, typically an antenna. The power delivered to the load, with reflection coefficient Γ_L and at physical temperature T_L, is immediate from (1.200)

$$P_L = k(T_{out} - T_L)C(\Gamma_{out}, \Gamma_L) \quad [W/Hz] \tag{1.220}$$

where $C(\Gamma_{out}, \Gamma_L)$ is the mismatch coefficient, Γ_{out} the output reflection coefficient (1.185) and T_{out} the equivalent noise temperature at the output plane (1.218). If

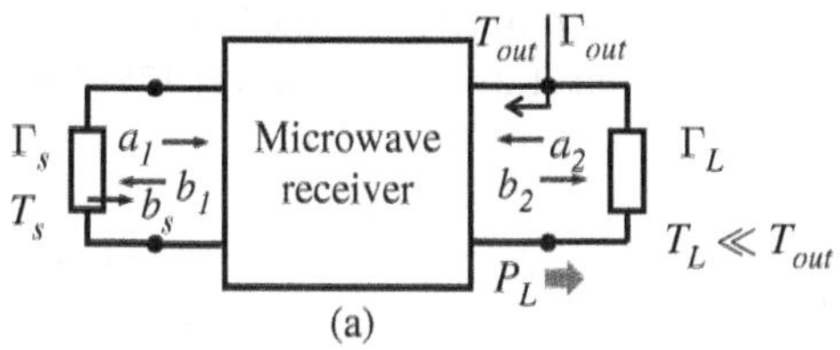

Figure 1.12 Microwave receiver. (a) S-parameter representation. (b) Linear system representation.

the gain of the receiver is high, this one is much larger than the load temperature $T_{out} \gg T_L$ and then the power delivered to the load becomes approximately $P_L = kT_{out}C(\Gamma_{out}, \Gamma_L)$. Substituting (1.218)

$$P_L = k\,G_T(T_s + T_R) \quad \text{(High gain receiver)} \tag{1.221}$$

where $G_T = G_A\,C(\Gamma_{out}, \Gamma_L)$ is the transducer power gain (1.186) for the source and load reflection coefficients Γ_s and Γ_L.

Total power radiometers are based in this simple equation. It is easily interpreted by identifying kT_s as the available noise power coming from the input (antenna) and kT_R the contribution of the receiver to the available noise power referenced to the input plane. The sum of both terms is the total available power at the input, which multiplied by the transducer power gain provides the output power.

It is also interesting to interpret this equation in terms of *signals* that represent each one of the two terms. Using the wave definitions depicted in figure 1.12(a), (1.176) states that the output power is

$$P_L = \overline{|b_2|^2}(1 - |\Gamma_L|^2) \tag{1.222}$$

Using also (1.212), the following signals, interpreted always as phasors (see section 1.4), are conveniently defined

$$\begin{aligned} &\text{Source signal:}\ \ b'_s = \frac{b_s}{\sqrt{1-|\Gamma_s|^2}} \Rightarrow \overline{|b'_s|^2} = kT_s \\ &\text{Output signal:}\ \ b_L = b_2\sqrt{1-|\Gamma_L|^2} \Rightarrow \overline{|b_L|^2} = P_L \end{aligned} \tag{1.223}$$

and then (1.221) is compatible with the following relationship, in which explicit frequency dependence is included,

$$b_L(f) = H(f)\left[b'_s(f) + b_R(f)\right] \tag{1.224}$$

with $|H(f)|^2 = G_T(f)$ and in which $b_R(f)$ represents the noise generated by the receiver referenced to the input port in such a way that

$$\overline{|b_R|^2} = kT_R\ ;\ \overline{b_R b'^*_s} = 0 \tag{1.225}$$

A formal expression of the frequency response $H(f)$ in terms of the S-parameters of the whole receiver can be obtained directly from (1.210) along with (1.175). The result is

$$H(f) = \frac{S_{21}\sqrt{1-|\Gamma_L|^2}\sqrt{1-|\Gamma_s|^2}}{(1-S_{11}\Gamma_s)(1-\Gamma_{out}\Gamma_L)} \tag{1.226}$$

which, as expected, satisfies that $|H(f)|^2 = G_T$. The expression for b_R is

$$b_R = \frac{c_1\Gamma_s}{\sqrt{1-|\Gamma_s|^2}} + \frac{c_2(1-S_{11}\Gamma_s)}{S_{21}\sqrt{1-|\Gamma_s|^2}} \tag{1.227}$$

fully consistent with (1.225) and (1.214) with $T_N = T_R$ in this case. Note that the explicit dependence with frequency has only been included in (1.225) to avoid burdening other expressions. All waves, the transducer power gain and the reflection coefficients are, in general, frequency dependent.

In the particular case of matched input and output ($\Gamma_s = \Gamma_L = 0$) the signals defined in (1.223) become $b'_s = b_s$, $b_L = b_2$ and $b_R = c_2/S_{21}$, while the transducer power gain simplifies to $|S_{21}|^2$. Then (1.224) reduces to $b_2 = S_{21}b_s + c_2$ as expected.

The block diagram of figure 1.12(b) illustrates the representation of the receiver using the signals (1.223) and the frequency response (1.226). Although not explicitly mentioned before, this model holds also when the receiver includes one or several frequency conversions. In this case the convention of section 5.1 for the frequency response must be applied.

1.5.10 The Bosma Theorem

Consider now an N-port microwave network, schematically depicted in figure 1.13(a), where inward and outward noise waves a_i and b_i as well as intrinsic waves c_i are defined at each port as indicated. For any linear network, the system of equations (1.206) is commonly expressed in matrix form

$$\mathbf{b} = \mathbf{S}\mathbf{a} + \mathbf{c} \tag{1.228}$$

where $\mathbf{a}$, $\mathbf{b}$, and $\mathbf{c}$ are column vectors and $\mathbf{S}$ is the S-parameter matrix. For a passive network, the intrinsic waves c_i are function of the S-parameters and the physical temperature through an equation known as Bosma theorem. It is derived by considering first a situation of thermodynamic equilibrium in which the network at temperature T is connected in all ports to respective loads, all equal to the reference impedance Z_0 and also maintained at the same temperature T (see the drawing infigure 1.13(b)). In these conditions, according to (1.201), all waves a_i and b_i have equal power $\overline{|a_i|^2} = \overline{|b_i|^2} = kT\,\forall i$. Furthermore, since waves a_i are solely the intrinsic waves of the corresponding loads, they are all uncorrelated to each other $\overline{a_i a_j^*} = 0$, $\forall i \neq j$. Not so evident is that all waves b_i are also uncorrelated. For a nice

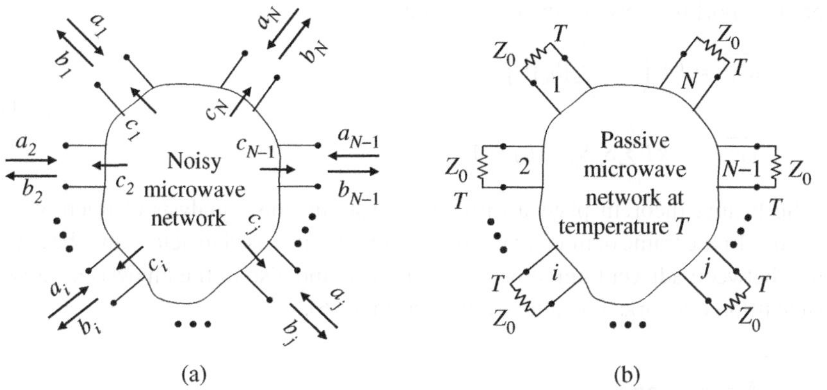

Figure 1.13 (a) Waves definitions in an N-port microwave network. (b) Matched passive network in thermodynamic equilibrium.

demonstration, the reader may consult [17]. In matrix notation, the conditions just stated can be written as

$$\overline{\boldsymbol{a}\boldsymbol{a}^{\dagger}} = \overline{\boldsymbol{b}\boldsymbol{b}^{\dagger}} = kT\mathbf{I} \tag{1.229}$$

where the symbol † indicates Hermitian transpose and **I** is the identity matrix. The remainder of the demonstration follows [17]. From (1.228)

$$\overline{\boldsymbol{b}\boldsymbol{b}^{\dagger}} = \overline{(\boldsymbol{S}\boldsymbol{a}+\boldsymbol{c})(\boldsymbol{a}^{\dagger}\boldsymbol{S}^{\dagger}+\boldsymbol{c}^{\dagger})} = \boldsymbol{S}\overline{\boldsymbol{a}\boldsymbol{a}^{\dagger}}\boldsymbol{S}^{\dagger} + \overline{\boldsymbol{c}\boldsymbol{c}^{\dagger}} \tag{1.230}$$

where $\overline{\boldsymbol{a}\boldsymbol{c}^{\dagger}} = \overline{\boldsymbol{c}\boldsymbol{a}^{\dagger}} = 0$ since the inward waves a_i generated at the loads are uncorrelated with those of the network. Introducing now (1.229), it follows that

$$\overline{\boldsymbol{c}\boldsymbol{c}^{\dagger}} = kT(\mathbf{I} - \boldsymbol{S}\boldsymbol{S}^{\dagger}) \tag{1.231}$$

which is the Bosma theorem, originally deduced in [14]. For a lossless network, the S-parameter matrix is unitary (see appendix A.3), meaning that $\boldsymbol{S}\boldsymbol{S}^{\dagger} = \mathbf{I}$ so all intrinsic noise waves vanish as expected. Independently of this condition, this also happens if the physical temperature is the absolute zero. In general, the intrinsic noise waves emanated by a network in different ports are partially correlated.

For a one-port network (1.231) reduces to (1.194), repeated here for completeness

$$\overline{|c|^2} = kT(1 - |\Gamma|^2) \quad \text{One port network} \tag{1.232}$$

and for a two-port network to

$$\begin{aligned} \overline{|c_1|^2} &= kT(1 - |S_{11}|^2 - |S_{12}|^2) \\ \overline{|c_2|^2} &= kT(1 - |S_{22}|^2 - |S_{21}|^2) \quad \text{Two-port network} \\ \overline{c_1 c_2^*} &= -kT(S_{11}S_{21}^* + S_{12}S_{22}^*) \end{aligned} \tag{1.233}$$

For an N-port network it can be expanded to

$$\begin{aligned}\overline{|c_i|^2} &= kT\left(1 - \sum_{l-1}^{N} |S_{il}|^2\right) \quad (i = 1 \cdots N) \\ \overline{c_i c_j^*} &= -kT \sum_{l=1}^{N} S_{il} S_{jl}^* \quad (i, j = 1 \cdots N)\end{aligned} \tag{1.234}$$

The Bosma theorem plays an important role in noise analysis of microwave systems. In the frame of interferometric radiometry it is used to determine the coupling between adjacent antennas (section 3.2.6) and also in the analysis of noise distribution networks for calibration (section 6.1.6).

1.6 Antennas

Antennas are used in Radiometry to sense the electromagnetic field originated by the thermal emission of distant matter, which leads to determining how an antenna responds to an incoming radiation of this kind. It is also important to understand how it behaves in transmission mode, when it is fed by a thermal noise source. Both transmission and reception parameters are related to each other and they have to be precisely defined for their use in microwave radiometry, either interferometric or not.

1.6.1 Radiated Electric Field and Power Density

Any signal applied to an antenna is radiated in the form of electromagnetic waves in such a way that, at any point with coordinates x, y, z, time-dependent electric and magnetic fields satisfying Maxwell equations are induced. In free space, both fields are related to each other through the wave impedance so only the electric field is needed for a complete description.

The electric field is a three-dimensional real vector $\vec{\mathrm{E}}(x, y, z, t)$ with units of V/m that depends on spatial coordinates and time. Consistent with the signals used in this book, thermal noise is assumed so, for a given point in space, the electric field is characterized by a stationary, zero-mean, normal stochastic process. As such, it will always be represented by the *phasor* for random signals, duly defined in section 1.4. With this convention, the electric field becomes a *complex* vector that depends on space and frequency $\vec{E}(x, y, z, f)$ and has units of $V/\sqrt{Hz}/m$. The explicit frequency dependence will be omitted from now on for simplicity. In section 1.5.1, it is demonstrated that random signal phasors of voltage and current are fully compatible with those defined for pure tones with only a change of units. Here, it happens exactly the same. The phasor defined for a sinusoidal electric field has units of V/m, but otherwise, it is fully compatible with the definition for random signals. Consequently, all antenna equations found in the literature, derived for sinusoidal signals, are directly applicable for random signals as long as phasors are used.

Assuming *far-field distance*[7] the electric field radiated by any antenna propagates in the radial direction as a local spherical wave [9] originated in the antenna position. Assuming the antenna in the center of coordinates and using the convention (A.5) for complex vector notation,

$$\vec{E}(r,\theta,\phi) = E(r,\theta,\phi)\hat{e}_a(\theta,\phi) \quad [V/\sqrt{Hz}/m] \tag{1.235}$$

where r, θ, ϕ are defined in appendix A.4, $E(r,\theta,\phi)$ is the phasor of the field and $\hat{e}_a$ a unit vector, complex in general with only transverse components, known as "antenna polarization vector" and ultimately linked to the particular antenna design. The spatial orientation of this vector is the antenna polarization, coincident with the radiated field polarization.

Consistent with the positive sign in the complex exponential of (1.118), decreasing phase convention is used, so the radiated field phasor depends on the radial coordinate as

$$E(r,\theta,\phi) \propto \frac{e^{-jkr}}{r} \tag{1.236}$$

where $k = 2\pi f/c$ is the wave number, f the frequency, and c the velocity of light.

In free space, the electromagnetic power flow per unit area, also known as *power density* is the amplitude of the "Pointing vector," equal to the norm squared of the electric field divided by the wave impedance $\eta_0 = 120\pi$:

$$\mathscr{P} = \frac{\overline{\vec{E}\cdot\vec{E}}}{\eta_0} = \frac{\overline{|E|^2}}{\eta_0} \propto \frac{1}{r^2} \quad [W/Hz/m^2] \tag{1.237}$$

where the centered dot stands for Hermitian dot product (A.1). Note that, when using random signal phasors, the power flow per unit area has units of $W/Hz/m^2$ instead of W/m^2 applicable to sinusoidal signals. Following the same criterion as in microwave circuits, the name "power density" is kept for compatibility with usual conventions (see comments at the end of section 1.5.1). The reader should not get confused by the qualifier "density" applied to the power in two different contexts: per unit frequency and per unit area.

1.6.2 Antenna Pattern and Directivity

Antennas do not radiate evenly in all directions of space. The antenna *field pattern* is the radiated electric field normalized to its maximum value as a function of angular coordinates at a constant distance

$$\vec{F}(\theta,\phi) = \frac{\vec{E}(r,\theta,\phi)}{E(r,\theta_0,\phi_0)} = F(\theta,\phi)\hat{e}_a(\theta,\phi) \tag{1.238}$$

---

7 For a precise definition of this term, please consult any basic reference on antennas.

where (θ_0, ϕ_0) is the direction of maximum radiation, also known as *boresight*, $F(\theta, \phi)$ a complex number that is often termed field pattern as well, and $\hat{e}_a(\theta, \phi)$ the antenna polarization vector already defined. Depending on the antenna design, and particularly in the case of omnidirectional antennas, the direction of maximum radiation may not be unique but the above definition still holds, although a convention to uniquely define (θ_0, ϕ_0) is needed. By definition, at boresight $\vec{F}(\theta_0, \phi_0) = \hat{e}_a(\theta_0, \phi_0)$, equivalent to $F(\theta_0, \phi_0) = 1$.

Equation (1.238) states that the antenna field pattern is a complex vector with the same polarization as the radiated electric field. For ideal antennas, such as elementary dipoles or small apertures the field has no phase contribution other than the propagation term $\exp -jkr$, so $F(\theta, \phi)$ is real. On the other hand, the radiated field phase of common antennas has additional contributions and therefore the pattern out of boresight is complex in general. More insight on this issue can be found in section 1.6.6.

The antenna *power pattern*, or simply pattern, is the ratio between the power density and its maximum value at constant distance. From (1.237) it turns out to be equal to the amplitude squared of the field pattern:

$$t(\theta, \phi) = \frac{\mathcal{P}(r, \theta, \phi)}{\mathcal{P}(r, \theta_0, \phi_0)} = |F(\theta, \phi)|^2 \tag{1.239}$$

The directional properties of an antenna are quantified by using an ideal isotropic radiator as reference. The "directivity" $D(\theta, \phi)$ is defined such that the power density is

$$\mathcal{P}(r, \theta, \phi) = \frac{P_r}{4\pi r^2} D(\theta, \phi) \tag{1.240}$$

where P_r is the total radiated power in W/Hz and $4\pi r^2$ is the area of a sphere of radius r centered in the antenna position. An isotropic radiator has a directivity equal to one in all directions of space. The above equation is consistent with (1.237) as it also predicts a dependency of the power density with the inverse of the distance squared. Introducing this definition into (1.239),

$$D(\theta, \phi) = D\, t(\theta, \phi) \tag{1.241}$$

where, following the usual convention, D without specifying angular coordinates is the maximum directivity, $D \triangleq D(\theta_0, \phi_0)$.

On the other hand, the total radiated power P_r is the surface integral of the power density $\mathcal{P}(\theta, \phi)$ on a sphere surrounding the antenna:

$$P_r = \iint_{\text{sphere}} \mathcal{P}(\theta, \phi) dS = \iint_{4\pi} \mathcal{P}(\theta, \phi) r^2 d\Omega \tag{1.242}$$

where $d\Omega$ is the differential solid angle [see (A.20) in appendix A.4]. Substituting (1.240) into (1.242) the following consistency condition follows

$$\Omega_a = \iint_{4\pi} t(\theta, \phi) d\Omega = \iint_{4\pi} |F(\theta, \phi)|^2 d\Omega = \frac{4\pi}{D} \tag{1.243}$$

where Ω_a is the so-called "antenna equivalent solid angle," equal to the integral over all space of the antenna power pattern. Justification of this name is evident if D is substituted by $4\pi/\Omega_a$ in (1.240) and it is particularized for (θ_0, ϕ_0). The power density in the direction of maximum radiation is equal to the radiated power divided by the surface subtended by the equivalent solid angle. Of course, this interpretation is visually more intuitive for antennas with a single direction of maximum radiation.

1.6.3 Antenna Polarization

The electric field radiated by an antenna (1.235) has only transverse coordinates, so the field pattern (1.238) is a two-dimensional vector. Using the convention of (A.7) in appendix A.1,

$$\vec{F}(\theta, \phi) = F_\theta(\theta, \phi)\hat{\theta} + F_\phi(\theta, \phi)\hat{\phi} \tag{1.244}$$

where $\hat{\theta}$ and $\hat{\phi}$ are the elevation and azimuth unit vectors defined in figure 1.14(a). Together with the radial vector $\hat{r}$ they define the spherical orthogonal basis

$$\begin{aligned}
\hat{r} &= \sin\theta\cos\phi\,\hat{x} + \sin\theta\sin\phi\,\hat{y} + \cos\theta\,\hat{z} \\
\hat{\theta} &= \cos\theta\cos\phi\,\hat{x} + \cos\theta\sin\phi\,\hat{y} - \sin\theta\,\hat{z} \\
\hat{\phi} &= -\sin\phi\,\hat{x} + \cos\phi\,\hat{y}
\end{aligned} \tag{1.245}$$

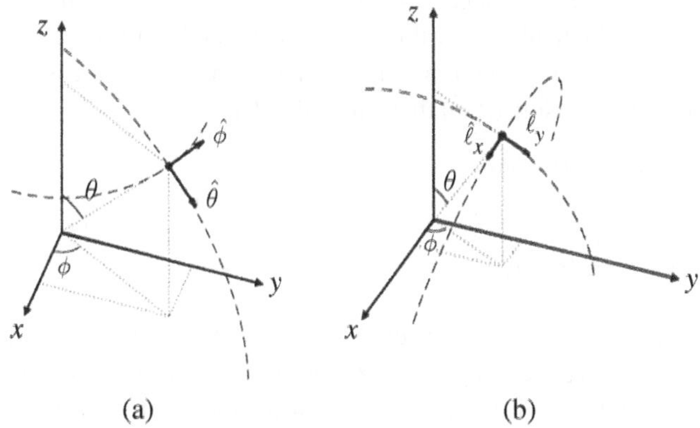

Figure 1.14 (a) Elevation and azimuthal unit vectors. (b) Ludwig third definition of polarization unit vectors.

where $\hat{x}$, $\hat{y}$ and $\hat{z}$ are unit vectors along the three axes. The above definition is consistent with the convention $\hat{\theta} \times \hat{\phi} = \hat{r}$. A linearly polarized antenna has $F_\theta(\theta, \phi)$ and $F_\phi(\theta, \phi)$ always in phase.

1.6.3.1 Ludwig Third Definition of Polarization

Obviously the field pattern can also be represented in any other basis defined by a linear combination of the second and third equations (1.245). The most common one for linearly polarized antennas is a rotation of $(\hat{\theta}, \hat{\phi})$ with respect to the radial direction known as "Ludwig third definition of polarization" [18]

$$\begin{aligned} \hat{\ell}_x &= \cos\phi\,\hat{\theta} - \sin\phi\,\hat{\phi} \\ \hat{\ell}_y &= \sin\phi\,\hat{\theta} + \cos\phi\,\hat{\phi} \end{aligned} \tag{1.246}$$

which satisfy that $\hat{\ell}_x \times \hat{\ell}_y = \hat{r}$. These vectors are drawn in figure 1.14(b). They are both transversal and tangent to slanted circumferences passing through the point [0,0,-r] where r is the radial coordinate. In contrast with $(\hat{\theta}, \hat{\phi})$ they are well defined along the z axis, which must be selected as antenna boresight. In particular, at $\theta = 0$, $\hat{\ell}_x = \hat{x}$, and $\hat{\ell}_y = \hat{y}$. The antenna field pattern (1.244) can then be rewritten as

$$\vec{F}(\theta, \phi) = F_{\ell_x}(\theta, \phi)\hat{\ell}_x + F_{\ell_y}(\theta, \phi)\hat{\ell}_y \tag{1.247}$$

where $F_{\ell_x}(\theta, \phi)$ and $F_{\ell_y}(\theta, \phi)$ are the corresponding components

$$\begin{aligned} F_{\ell_x}(\theta, \phi) &= F_\theta(\theta, \phi)\cos\phi - F_\phi(\theta, \phi)\sin\phi \\ F_{\ell_y}(\theta, \phi) &= F_\theta(\theta, \phi)\sin\phi + F_\phi(\theta, \phi)\cos\phi \end{aligned} \tag{1.248}$$

1.6.3.2 Circular Polarization

A phase difference between the two components of the field pattern (1.246) results in a rotation of the time domain radiated field as it propagates. This feature is characterized by expressing the field in an orthogonal base defined by unit vectors in the so-called "left circular" and "right circular" polarization. For decreasing phase convention [see (1.236)], using $(\hat{\theta}, \hat{\phi})$ they are defined as

$$\begin{aligned} \text{Left circular: } \hat{c}_l &= \frac{1}{\sqrt{2}}(\hat{\theta} + j\hat{\phi}) \\ \text{Right circular: } \hat{c}_r &= \frac{1}{\sqrt{2}}(\hat{\theta} - j\hat{\phi}) \end{aligned} \tag{1.249}$$

from which it is apparent that $\hat{c}_r \times \hat{c}_l = j\hat{r}$. Seen from the center of coordinates, a radiated electric field with *right* circular polarization rotates *clockwise* in time domain as it propagates. A field with left circular polarization rotates counterclockwise. Using this base, the field pattern (1.244) can be written as

$$\vec{F}(\theta, \phi) = F_{c_l}(\theta, \phi)\hat{c}_l + F_{c_r}(\theta, \phi)\hat{c}_r \tag{1.250}$$

where

$$F_{c_l}(\theta,\phi) = \frac{1}{\sqrt{2}}\left(F_\theta(\theta,\phi) - jF_\phi(\theta,\phi)\right)$$
$$F_{c_r}(\theta,\phi) = \frac{1}{\sqrt{2}}\left(F_\theta(\theta,\phi) + jF_\phi(\theta,\phi)\right) \quad (1.251)$$

The circular polarization basis can also be defined using the Ludwig third definition of polarization (1.246)

$$\text{Left circular: } \hat{c}'_l = \frac{1}{\sqrt{2}}(\hat{\ell}_x + j\hat{\ell}_y)$$
$$\text{Right circular: } \hat{c}'_r = \frac{1}{\sqrt{2}}(\hat{\ell}_x - j\hat{\ell}_y) \quad (1.252)$$

which is related to (1.249) by $\hat{c}'_l = \hat{c}_l \exp j\phi$ and $\hat{c}'_r = \hat{c}_r \exp -j\phi$ and consistently $\hat{c}'_r \times \hat{c}'_l = j\hat{r}$.

1.6.3.3 Co-polar and Cross-polar Patterns

In antenna measurement systems it is common to denote the patterns as "co-polar" (or "reference") and "cross-polar." They refer, respectively, to the "wanted" and "unwanted" components of the radiated field so usually the first one is much larger in amplitude than the second one in all (or most) spatial directions. Ideally the cross-polar pattern should be null. The reference and cross-polar unit vectors are here denoted as $\hat{u}_r$ and $\hat{u}_c$, respectively, they are always orthogonal and, by convention $\hat{u}_r \times \hat{u}_c = \hat{r}$. The field pattern can thus be expressed as

$$\vec{F}(\theta,\phi) = R(\theta,\phi)\hat{u}_r + C(\theta,\phi)\hat{u}_c \quad (1.253)$$

where $R(\theta,\phi)$ and $C(\theta,\phi)$ are, respectively, the reference and cross-polar patterns. Since this definition depends on the expected behavior of the antenna, care must be taken to consider appropriate signs for consistency. For example, using Ludwig third definition of polarization,

$$\text{X-polarized antenna} \begin{cases} \text{Co-polar pattern: } R(\theta,\phi) = F_{\ell_x}(\theta,\phi) \\ \text{Cross-polar pattern: } C(\theta,\phi) = F_{\ell_y}(\theta,\phi) \end{cases}$$
$$\text{Y-polarized antenna} \begin{cases} \text{Co-polar pattern: } R(\theta,\phi) = F_{\ell_y}(\theta,\phi) \\ \text{Cross-polar pattern: } C(\theta,\phi) = -F_{\ell_x}(\theta,\phi) \end{cases} \quad (1.254)$$

where the minus sign of the cross-polar pattern of the Y-polarized antenna is needed to match the above-mentioned convention that $\hat{u}_r \times \hat{u}_c = \hat{r}$. Needless to say, a different sign convention is possible but it must be clearly defined in order to correctly interpret the patterns. It is of utmost importance when the antennas are to be used in polarimetric radiometry.

Co-polar and cross-polar patterns can also be defined for circularly polarized antennas depending on the nominal antenna operation. As a matter of fact, antenna measurement systems usually measure $F_\theta(\theta,\phi)$ and $F_\phi(\theta,\phi)$ and then convert to co-polar and cross-polar patterns using (1.248) and (1.254) or others applicable in the particular scenario.

1.6.3.4 The Field Pattern and Its Components

The decomposition of the field pattern into two polarizations can be expressed, in general, as

$$\vec{F}(\theta,\phi) = F_p(\theta,\phi)\hat{p}(\theta,\phi) + F_q(\theta,\phi)\hat{q}(\theta,\phi) \tag{1.255}$$

where $\hat{p}(\theta,\phi)$ and $\hat{q}(\theta,\phi)$ are orthogonal unit vectors, complex in general. Using now the vector component representation (A.10) and the definition of antenna field pattern (1.238),

$$\begin{aligned} F_p(\theta,\phi) &= F(\theta,\phi)e_{a_p}(\theta,\phi) \\ F_q(\theta,\phi) &= F(\theta,\phi)e_{a_q}(\theta,\phi) \end{aligned} \tag{1.256}$$

where $e_{a_p}(\theta,\phi)$ and $e_{a_q}(\theta,\phi)$ are the components of the antenna polarization vector $\hat{e}_a(\theta,\phi)$ in the reference frame defined by $\hat{p}$ and $\hat{q}$. As it is pointed out in section 1.6.2, at boresight $F(\theta_0,\phi_0) = 1$, therefore, the pattern components at boresight are just $e_{a_p}(\theta_0,\phi_0)$ and $e_{a_q}(\theta_0,\phi_0)$, respectively.

For any linearly polarized antenna, an orthogonal basis can always be found at which, for a given direction (θ,ϕ), the polarization vector has a single component, the other vanishing. This basis is usually different at distinct directions.

1.6.4 Thermal Noise Radiation

Consider a thermal noise source connected to a transmitting antenna (figure 1.15). The source is modeled as a generic impedance with reflection coefficient Γ_s and equivalent noise temperature T_s, defined according to (1.193). The antenna has an input reflection coefficient Γ_a and field pattern $\vec{F}(\theta,\phi)$. Assuming lossless antenna

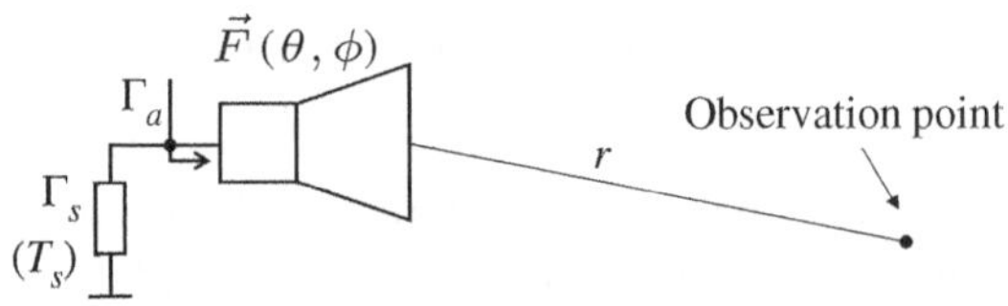

Figure 1.15 An antenna radiating a noise signal.

(see section 1.6.11 for discussion), the radiated power is equal to the power entering the antenna so it is given by (1.179)

$$P_r = CkT_s \tag{1.257}$$

where C is the mismatch coefficient between source and antenna (1.179) and kT_s, with k the Boltzmann constant, is the source available noise power. Substituting (1.237) into the left-hand side of (1.240) and using the above equation, the following relation at boresight is found

$$\frac{\overline{|E(r,\theta_0,\phi_0)|^2}}{\eta_0} = \frac{CkT_s}{4\pi r^2}D \tag{1.258}$$

from which the electric field at a distance r in the direction (θ, ϕ) follows by introducing it into (1.238)

$$\vec{E}(r,\theta,\phi) = \sqrt{\frac{CkT_s\eta_0}{\Omega_a}}\vec{F}(\theta,\phi)\frac{e^{-jkr}}{r} \tag{1.259}$$

To write this result, the propagation phase of (1.236) has been added and also (1.243) has been used to write it in terms of the antenna solid angle instead of the directivity. This expression may include also an arbitrary constant phase that has been set to zero without loss of generality. This phase, along with all other terms in (1.259) except the Boltzmann constant, the wave impedance and the geometric parameters are frequency dependent.

1.6.5 Received Signal

An antenna intercepting an electromagnetic wave responds by producing a signal proportional to the strength of the received radiation. As intuitively expected and experimentally corroborated, the received signal peaks when the electric field polarization is aligned with the antenna polarization vector $\hat{e}_a$ (1.238). Polarization mismatch will be taken into account later (section 1.6.7) so perfect alignment is considered now.

The electromagnetic field an antenna is immersed in is necessarily perturbed by its presence, but this difficulty is bypassed by expressing the antenna parameters in terms of *the field that would be present in the absence of the antenna*. With this model, the *available power* collected by an antenna is proportional to the power density of the unperturbed wave it intercepts. Assuming the antenna at the center of coordinates (see appendix A.4), if some radiation arrives from a direction with angular coordinates (θ, ϕ), the *available* power at the antenna terminals is

$$P_{av} = A_{eff}(\theta,\phi)\mathscr{P}(\theta,\phi) = A_{eff}(\theta,\phi)\frac{\overline{|E(\theta,\phi)|^2}}{\eta_0} \quad [W/Hz] \tag{1.260}$$

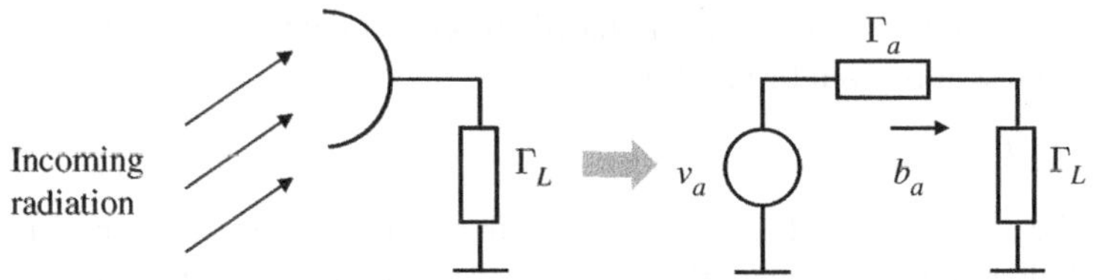

Figure 1.16 Antenna model in reception using a voltage generator with internal impedance. The model is fully described by the intrinsic wave b_a and the internal reflection coefficient Γ_a.

where $A_{eff}(\theta, \phi)$ is the *antenna effective area* and $E(\theta, \phi)$ is the incoming electric field neglecting the effect of the antenna itself. Expression (1.237) for the power density $\mathscr{P}$ has been used. The noise equivalent temperature that results from applying the definition (1.193) is here known as "antenna temperature" $P_{av} = kT_a$. This concept is further elaborated in section 2.3.

A receiving antenna is a source and can thus be modeled as shown in figure 1.16. According to (1.179) the power delivered to the load is

$$P_L = CP_{av} \tag{1.261}$$

where C is the mismatch coefficient and the available power P_{av} is given by (1.260).

Using the voltage waves defined in section 1.5.2 and the model of figure 1.16, the antenna can be fully described by an intrinsic wave (1.174) and an internal reflection coefficient, here both denoted by b_a and Γ_a, respectively. According to Thevenin's theorem, the impedance, and hence the reflection coefficient, of any antenna is the same whether it is used in transmission or in reception. Using (1.178) for the available power, it is now convenient to introduce the following definition

$$b'_a = \frac{b_a}{\sqrt{1 - |\Gamma_a|^2}} \tag{1.262}$$

in such a way that $\overline{|b'_a|^2} = P_{av} = kT_a$. This is consistent with (1.223), and it is useful to simplify some expressions. Note, however, that b'_a does *not* correspond to any wave defined by (1.170). Using now (1.260),

$$b'_a(\theta, \phi) = \sqrt{A_{eff}(\theta, \phi)}\frac{E(\theta, \phi)}{\sqrt{\eta_0}}e^{j\varphi'(\theta,\phi)} \tag{1.263}$$

where $\varphi'(\theta, \phi)$ is an arbitrary phase. Later in section 1.6.9, the general relation (1.279) between effective area and power pattern is demonstrated. Introducing it into the above equation and using (1.239) to put it in terms of the field antenna pattern,

$$b'_a(\theta, \phi) = \frac{1}{\sqrt{\Omega_a}}F(\theta, \phi)\frac{\lambda}{\sqrt{\eta_0}}E(\theta, \phi)e^{j\varphi} \tag{1.264}$$

where Ω_a is the antenna equivalent solid angle (1.243) and φ is now a *constant phase* independent of the angular coordinates. Any dependency on them is included in either the antenna pattern or the electric field. Without loss of generality φ can be considered null since it only sets an arbitrary phase reference, and it can be associated to the receiver the antenna is connected to. The above equation states that, for a given arrival direction, the phase of the output signal is equal to the sum of those of the pattern and the electric field.

1.6.6 Phase Center

Conceptually, the antenna field pattern can be measured in reception by means of a distant probe emitting a constant field and moving around the antenna under test at constant distance. The pattern is then equal to the received signal (1.264) at each probe position normalized to its value at boresight.

$$F(\theta, \phi) = \frac{b'_a(\theta, \phi)}{b'_a(\theta_0, \phi_0)} \tag{1.265}$$

where θ and ϕ are defined with respect to a fixed system of coordinates, nominally centered in the antenna under test. However, its position in the mechanical structure is not well defined a priori since it depends on the particular shape. A real antenna has always finite size.

Assume that the electric field emitted by the probe is a spherical wave originated at its own position. Then, the field that impinges the receiving antenna is

$$E(\theta, \phi) \propto e^{-jkr_a} = e^{-jkr}e^{jk\Delta r} \tag{1.266}$$

where r_a is the distance between the antenna under test and the probe antenna, r its radial coordinate in the predefined system of coordinates and $\Delta r = r - r_a$. The probe antenna rotates about the center so r is constant. Therefore, using (1.264) and (1.265) the pattern measured with this coordinate system is

$$F'(\theta, \phi) = F(\theta, \phi)e^{jk\Delta r} \tag{1.267}$$

where $F(\theta, \phi)$ is the pattern that would be measured with $\Delta r = 0$ and where the probe position is assumed far enough to approximate the elevation and azimuth angles of the probe as seen from the antenna under test equal to the angular coordinates. Otherwise, a correction should be applied.

The above equation shows that the phase of the measured antenna pattern depends on the reference point (center of coordinates) with respect to which it is measured. Moreover, using (1.267) any pattern can be referenced to a center point different from the one used in the actual measurement setup.

The *phase center* of an antenna is the location for which the measured phase of the pattern is most insensitive to the angular coordinates. In other words, if the

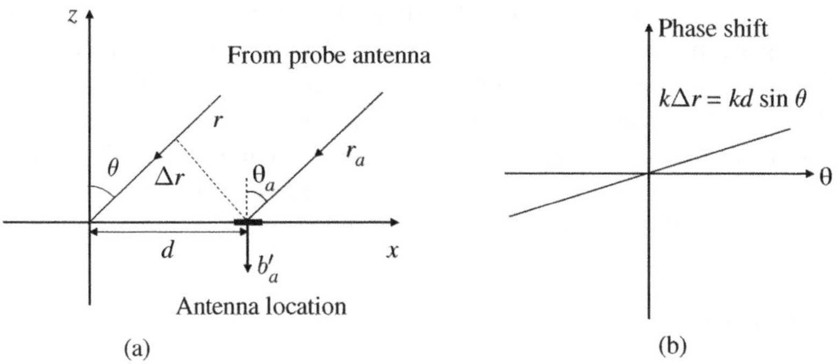

Figure 1.17 (a) Antenna located in the x-axis off the center of coordinates of the measurement system. (b) Phase of measured antenna pattern as a function of polar angle θ for small angles.

center of the system of coordinates is the phase center, then the antenna pattern phase has the minimum variation with respect to the spatial directions and therefore it is approximately real. This concept is useful to provide a precise meaning to the sentence "*an antenna is located at the center of coordinates*" for an actual antenna with finite dimensions: it means defining the coordinates with the origin at the antenna phase center. In consequence, unless specified differently, the location of an antenna is commonly understood as that of its phase center. It represents the point from where the field is emitted when the antenna is used as a transmitter.

An example would maybe clarify this concept. Consider the situation of figure 1.17(a), in which the antenna under test is located in the x-axis close to the center and the probe antenna rotates about the center an angle θ. Then

$$\Delta r \approx d \sin\,\theta \;;\; \theta_a \approx \theta \tag{1.268}$$

where d is the antenna separation from the center. Assuming an antenna with real pattern and using (1.267), the measured pattern has a direction-dependent phase equal to $kd \sin\,\theta$, which is approximately linear for small angles, as schematically shown in figure 1.17(b).

1.6.7 Polarization Misalignment

In general the incoming electric field and the antenna polarization vector are not aligned. If both vectors are decomposed into two orthogonal polarizations, the linearity property can be applied to express the total response as a linear combination of the responses to each one. For each polarization separately, field and polarization vector components are aligned so (1.264) applies. Disregarding the constant phase φ (which on the other hand can be different in each polarization), the total output signal can thus be written as

$$b'_a(\theta,\phi) = \frac{1}{\sqrt{\Omega_a}} \vec{F}(\theta,\phi) \odot \vec{e}(\theta,\phi) \tag{1.269}$$

where the operator $\odot$ stands for the non-Hermitian dot product (A.2), $\vec{F}(\theta,\phi)$ is the field antenna pattern (1.238) and $\vec{e}$ is a *normalized* electric field, in units of $\sqrt{W/Hz}$, defined as

$$\vec{e} = \frac{\lambda}{\sqrt{\eta_0}} \vec{E} \quad [\sqrt{W/Hz}] \tag{1.270}$$

It is a complex vector with the same polarization as the electric field $\vec{E}$ and with power $\overline{|e|^2} = \lambda^2 \mathscr{P}$ where $\mathscr{P}$ is the power density of the electromagnetic radiation in $W/Hz/m^2$. This normalized electric field plays an important role in the equations for microwave radiometry presented in this book. As shown in section 2.2, it is directly related to the "brightness temperature" of a natural radiation.

> **Comment:** It is important to clarify that $\vec{e}(\theta,\phi)$ in (1.269) is the normalized electric field that would be present at the center of coordinates, in the absence of the receiving antenna, due to a spherical wave originated at (r,θ,ϕ)

1.6.8 Transmission Link

If the incoming field is radiated by a distant antenna, substitution of (1.259) into (1.269) results in

$$b'_a = \sqrt{\frac{CkT_s}{\Omega_{at}\Omega_a}} \frac{\lambda}{r} \vec{F}(\theta,\phi) \odot \vec{F}_t(\theta_t,\phi_t) e^{-jkr} \tag{1.271}$$

where r is the distance between antennas, the subscript t is used to indicate the transmitting antenna parameters and (θ_t,ϕ_t) are the angular coordinates of the receiving antenna in a reference frame centered in the transmitting antenna. The non-Hermitian dot product of the patterns can be conveniently written in terms of the corresponding antenna polarization vectors (1.238)

$$\vec{F}(\theta,\phi) \odot \vec{F}_t(\theta_t,\phi_t) = F(\theta,\phi) F_t(\theta_t,\phi_t) \hat{e}_a(\theta,\phi) \odot \hat{e}_{at}(\theta_t,\phi_t) \tag{1.272}$$

Furthermore, using (1.262), the antenna temperature $T_a = \overline{|b'_a|^2}/k$ becomes

$$T_a = \frac{CT_s}{\Omega_{at}\Omega_a} \frac{\lambda^2}{r^2} t(\theta,\phi) t_t(\theta_t,\phi_t) C_p \tag{1.273}$$

where the constant $C_p = |\hat{e}_a \odot \hat{e}_{at}|^2$ is the "polarization mismatch coefficient."

1.6.9 Reciprocity

The Lorentz reciprocity theorem is generally formulated in terms of the electric and magnetic fields in an isotropic medium and is direct consequence of the Maxwell equations (see for example [8]). Particularized for circuit theory it states that, for any passive linear system not including anisotropic materials such as magnetized ferrites or plasmas, the scattering[8] matrix is always symmetrical, $S_{ij} = S_{ji}\, \forall i \neq j$. A system formed by two antennas and the space around them can be modeled as a two-port linear network with ports defined at the corresponding antenna terminals. Therefore, for this system the reciprocity condition must hold.

Consider two different antennas pointing to each other, perfectly aligned and separated a large distance so that they can both be considered in the far field of the other. A generator is applied to one antenna while the other is loaded by a generic impedance (see figure 1.18).

The power delivered to the load (1.261) is equal to the available power at antenna terminals multiplied by the mismatch coefficient. In turn the available power collected by the receiving antenna is equal to its effective area multiplied by the power density of the incoming field (1.260), which in turn is function of the radiated power through (1.240). Finally, the radiated power is equal to the available power of the source multiplied by the mismatch coefficient. Putting all this together results in the following expression for the transducer power gain (1.186) of the link:

$$G_T = \frac{P_L}{P_{av_s}} = C_1 C_2 \frac{D_1 A_{eff_2}}{4\pi r^2} \tag{1.274}$$

where C_1 and C_2 are the mismatch coefficients at ports 1 (transmitting antenna) and 2 (receiving antenna), respectively.

If both source and load impedances are equal to the reference impedance ($\Gamma_s = \Gamma_L = 0$), according to the definitions of mismatch coefficient (1.179), S-parameters

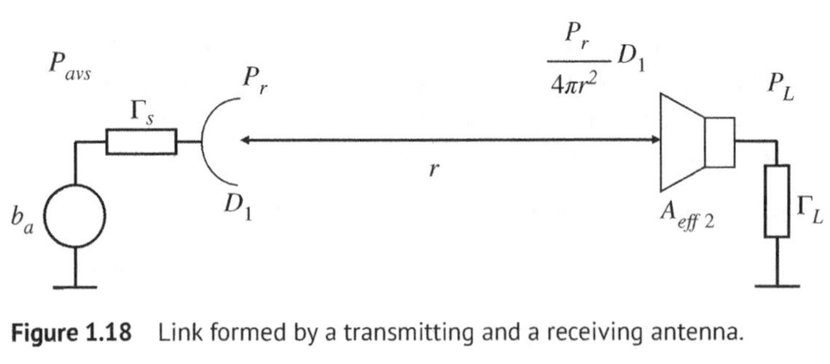

Figure 1.18 Link formed by a transmitting and a receiving antenna.

8 And also the Z and Y parameters, not used here.

(1.180) and transducer power gain (1.186), the above equation changes to

$$|S_{21}|^2 = (1 - |S_{11}|^2)(1 - |S_{22}|^2)\frac{D_1 A_{eff_2}}{4\pi r^2} \tag{1.275}$$

If now the generator and the load are interchanged, the same equation holds but with subscripts 1 and 2 swapped. The reciprocity condition $S_{12} = S_{21}$ forces then the following identity:

$$D_1 A_{eff_2} = D_2 A_{eff_1} \Rightarrow \frac{A_{eff_1}}{D_1} = \frac{A_{eff_2}}{D_2} \tag{1.276}$$

In other words, for any antenna, no matter how it is designed, the ratio of effective area to directivity is a constant. Obviously, this result is not restricted to having both antennas pointing to each other in the direction of maximum radiation. The same result is obtained at any angle, meaning that the effective area is proportional to the directivity *at any spatial direction* and thus varying as the power pattern. It can then be expressed as

$$A_{eff}(\theta, \phi) = A_{eff}\, t(\theta, \phi) \tag{1.277}$$

where $t(\theta, \phi)$ is the antenna power pattern (1.239) defined for the antenna in transmission mode. Consistent with standard convention, A_{eff} without explicit angle dependence is the effective area in the direction of maximum radiation. The above expression states that the antenna power pattern measured in reception is the same as the one measured in transmission. Moreover, since the reciprocity is not limited to the amplitude of the S-parameters but also to the phase, using a similar methodology, the same property is demonstrated for the antenna field pattern (1.238). It is the same in transmission or reception in both amplitude and phase.

Since all antennas have the same ratio (1.276), it suffices to derive it for just one particular case. When done analytically for a small wire antenna driven by a constant current [19], the following result is obtained:

$$\frac{A_{eff}}{D} = \frac{\lambda^2}{4\pi} \tag{1.278}$$

where λ is the wavelength. This is a fundamental relation between the transmitting and receiving parameters and *holds for any antenna*, no matter how it is designed. Combining this result with (1.243), equation (1.277) becomes

$$A_{eff}(\theta, \phi) = \frac{\lambda^2}{\Omega_a} t(\theta, \phi) \tag{1.279}$$

where Ω_a is the antenna equivalent solid angle. This is the expression used to derive (1.264).

1.6.10 Other Definitions

The standard definition of directivity (1.240) is related to the total radiated power, which is equal to the antenna input power. Similarly, the effective area (1.260) is defined from the available collected power. But in some cases, particularly related to a multiple antenna scenario, it is convenient to redefine these two parameters in relation to voltage wave power.

Consider first the transmitting antenna of figure 1.15. Using directly (1.176), the radiated power (1.257) can be expressed as $P_r = \overline{|a|^2}(1 - |\Gamma_a|^2)$ where a is the inward wave at antenna terminals. Equation (1.259) becomes then

$$\frac{\vec{E}(r,\theta,\phi)}{\sqrt{\eta_0}} = a\sqrt{\frac{D'}{4\pi}}\vec{F}(\theta,\phi)\frac{e^{-jkr}}{r} \tag{1.280}$$

with

$$D' = D(1 - |\Gamma_a|^2) \tag{1.281}$$

This new definition of directivity is compatible with (1.240) as long as the radiated power is substituted by the inward wave power $\overline{|a|^2}$. Note that (1.280) can be interpreted as an S-parameter relating a properly defined normalized voltage wave at a given location $(E\,r/\sqrt{\eta_0})$ to the inward wave at antenna terminals.

For the receiving antenna of figure 1.16, substituting (1.262) into the left hand side of (1.263)

$$b_a(\theta,\phi) = \sqrt{A'_{eff}(\theta,\phi)}\frac{E(\theta,\phi)}{\sqrt{\eta_0}}e^{j\varphi'(\theta,\phi)} \tag{1.282}$$

where the newly defined effective area is

$$A'_{eff} = A_{eff}(1 - |\Gamma_a|^2) \tag{1.283}$$

compatible with (1.260) if the available power in the left hand side is replaced by the intrinsic wave power deduced from (1.178).

As it is apparent, the general relation (1.278) still holds also for these definitions of directivity and effective area. For consistency, the antenna solid angle (1.243) is also redefined as

$$\Omega'_a = \frac{\Omega_a}{1 - |\Gamma_a|^2} \tag{1.284}$$

so as to keep its nominal relations with the new definitions of both directivity and effective area. Then, the intrinsic wave out of the receiving antenna can be written as (1.264) with only replacing Ω_a by Ω'_a. An S-parameter interpretation is also valid by defining the normalized voltage wave of the impinging field as $E\,\lambda/\sqrt{\eta_0}$. Also (1.280) can be expressed using Ω'_a instead of D' so the S-parameters are consistent in both cases.

The parameters defined in this section were implicitly used in both [20] and [21].

1.6.11 Antenna Loss

Up to this point, the antenna has been implicitly assumed lossless. In consequence, the radiated power has been made equal to the input power and, in reception, the power delivered to the load has been computed as the available power multiplied by the mismatch coefficient. However, in actual antennas this is not true and there is always some power lost in the antenna itself. The standard parameter to account for antenna loss is the "antenna ohmic efficiency," here denoted by the symbol η_a and always lower than or equal to one. Depending on whether the antenna is transmitting or receiving, this parameter has a different interpretation:

$$\begin{aligned} \text{Transmission: } \eta_a &= \frac{\text{Radiated power}}{\text{Input power}} \\ \text{Reception: } \eta_a &= \frac{\text{Available output power}}{\text{Available collected power}} \end{aligned} \tag{1.285}$$

but the concept is the same, so the value of η_a for a given antenna is identical in both cases.

Loss does not affect the directional characteristics of the antenna. Therefore it can be taken into account by modeling the antenna as shown in figure 1.19, a cascade of a lossless antenna with the same pattern and a lossy passive network. From (1.285), the antenna ohmic efficiency in transmission is the *power gain* of this network when loaded by the lossless antenna impedance and in reception is the *available power gain* when the source reflection coefficient is that of the lossless antenna. Using (1.184) and (1.187), it is easily demonstrated that in both cases the antenna efficiency is exactly the same, as expected.

When S-parameter formalism is used, the most convenient choice for the network representing the antenna loss is one having the following S-parameters

$$[S] = \begin{bmatrix} 0 & \alpha \\ \alpha & 0 \end{bmatrix} \tag{1.286}$$

with $\alpha \leq 1$ a positive real constant. In this case the antenna efficiency is

$$\eta_a = \frac{\alpha^2(1 - |\Gamma_a|^2)}{1 - \alpha^4|\Gamma_a|^2} \tag{1.287}$$

where Γ_a is the reflection coefficient of the lossless antenna. The actual reflection coefficient of the antenna including loss is obtained from (1.185) or (1.188)

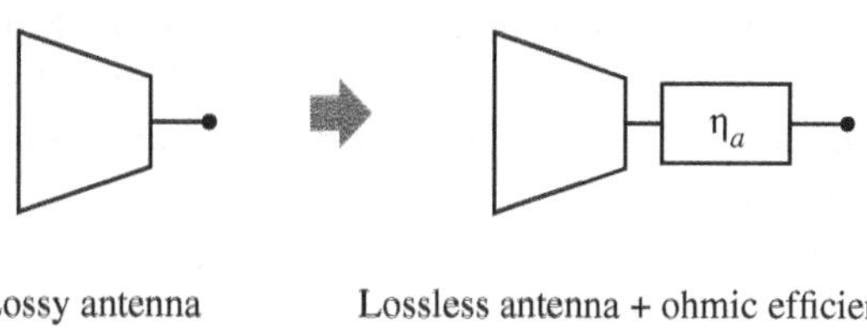

Figure 1.19 Modeling a lossy antenna with a lossless antenna and a passive network.

resulting in $\alpha^2\Gamma_a$. Note that, with this model, a matched antenna ($\Gamma_a = 0$) maintains this condition independently of whether losses are considered or not. In antenna textbooks the lossy network is usually modeled by a series resistance named "loss resistance" and the radiated power is associated with the "radiation resistance." This model does not keep matching condition when loss is included so, even though it is very common, is not further used in this book.

Including loss in the analysis of a receiving antenna of section 1.6.5 means multiplying the output signal (1.269) by $\sqrt{\eta_a}$. In addition, the thermal noise generated by the two-port network must be added to the output signal using (1.204) with $G_A = \eta_a$. Using the flux diagram method of section 1.5.8, the equivalent noise temperature at the antenna terminals is readily deduced

$$T'_a = T_a\eta_a + T_p(1 - \eta_a) \tag{1.288}$$

where T_a is the equivalent temperature at the output of the lossless antenna (antenna temperature) and T_p the physical temperature of the antenna, assumed constant.

Another option is to associate the lossy two-port network with the receiver, effectively increasing its noise equivalent temperature and reducing its gain. In this case, the formalism developed for lossless antennas can be used with no change. But, this option only solves the theory and provides confidence in its validity. In practice, antenna loss is a strong challenge in microwave radiometry when calibration is done internally, as in many practical systems. Antenna loss is very difficult to measure accurately, and its effect has to be removed from the measurements. The reader could refer to section 6.1.2 for more details.

References

1 Athanasios Papoulis. *Probability, Random Variables and Stochastic Processes*. McGraw-Hill, 1965.

2 Joseph W. Goodman. *Statistical Optics*. John Wiley & Sons, Inc., New York, 1st edition, 1985. ISBN 0-471-01502-4.

3 A. Bruce Carlson, Paul B. Crilly, and Janet C. Rutledge. *Communication Systems: An Introduction to Signals and Noise in Electrical Communication*. McGraw Hill, New York, 4th edition, 2002. ISBN 0-07-011127-8.

4 L. E. Franks. *Signal Theory*. Prentice Hall, Inc., 1969.

5 David Middleton. *Introduction to Statistical Communication Theory*. Peninsula Publishing, 1987.

6 David M. Pozar. *Microwave Engineering*. Addison Wesley, 1st edition, 1990. ISBN 0-201-50418-9.

7 Max W. Medley. *Microwave and RF Circuits: Analysis, Synthesis and Design.* Artech House, Boston, MA, 1993. ISBN 0-89006-546-2.

8 Simon Ramo, John R. Whinnery, and Theodore Van Duzer. *Fields and Waves in Communication Electronics.* John Wiley & Sons, Inc., 3rd edition, 1994. ISBN 978-0471-58551-3.

9 Constantine A. Balanis. *Antena Theory: Analysis and Design.* John Wiley & Sons, New York, 1997.

10 E. C. Jordan and K. G. Balmain. *Electromagnetic Waves and Radiating Systems.* Prentice-Hall, Englewood Cliffs, NJ, 1968.

11 I. S. Reed. On a moment theorem for complex Gaussian processes. *IEEE Transactions on Information Theory*, 8(3):194–195, April 1962.

12 Adriano Camps, Francesc Torres, Ignasi Corbella, Javier Bará, and J. A. Lluch. Threshold and timing errors of 1 bit/2 level digital correlators in Earth observation synthetic aperture radiometry. *Electronics Letters*, 33(9):812–814, April 1997.

13 J. Dugundji. Envelopes and pre-envelopes of real waveforms. *IRE Transactions on Information Theory*, 4(1):53–57, March 1958.

14 Hendrik Bosma. *On the theory of linear noisy systems.* PhD thesis, Technische Hogeschool Te Eindhoven, The Netherlands, 1967.

15 Kaneyuki Kurokawa. Power waves and the scattering matrix. *IEEE Transactions on Microwave Theory and Techniques*, MTT-13(2):194–202, March 1965.

16 Joseph T. Verdeyen. *Laser Electronics.* Prentice-Hall, 2nd edition, 1989.

17 Scott W. Wedge and David B. Rutledge. Noise waves and passive linear multiports. *IEEE Microwave and Guided Wave Letters*, 1(5):117–119, May 1991.

18 Arthur C. Ludwig. The definition of cross polarization. *IEEE Transactions on Antennas and Propagation*, AP-21(1):116–119, January 1973.

19 Ángel Cardama Aznar, Jordi Romeu Robert, Juan Manuel Rius Casals, Lluís Jofre Roca, Sebastián Blanch Boris, and Miguel Ferrando Bataller. *Antenas.* Edicions UPC, Barcelona, 2nd edition, 2002. ISBN 84-8301-625-7.

20 Sebastián Blanch, Jordi Romeu, and Ignasi Corbella. An exact representation of an antenna system diversity performance from its input parameter description. *IEE Electronics Letters*, 39(9):705–706, May 2003.

21 Ignasi Corbella, Nuria Duffo, Mercè Vall-llossera, Adriano Camps, and Francesc Torres. The visibility function in interferometric aperture synthesis radiometry. *IEEE Transactions on Geoscience and Remote Sensing*, 42(8):1677–1682, August 2004.

2

Microwave Radiometry

According to chapter 4 of [1], *the term "radiometry" means the measurement of incoherent radiant electromagnetic energy*. It relies on the experimental evidence that matter at temperature above absolute zero radiates electromagnetic waves. Applications of radiometry are very broad and comprise the remote sensing of the Earth, including atmosphere, and celestial bodies. Depending on the particular application and the parameter to observe, the frequency of interest may range from the low microwave to the optical region. In all cases the fundamental concepts are essentially the same, with small nuances and differences mainly related to the technology used. This chapter presents the radiometry theoretical basis in the microwave region of the spectrum. To facilitate reading, some key concepts are defined in chapter 1, which the reader should consult for a better understanding of the main topics of this chapter.

2.1 Thermal Emission

Around the end of the 19th century, several attempts were made to explain thermal emission observations. The problem was challenging and it was solved in 1901 by the Planck's quantum hypothesis leading to the well-known and elegant black body radiation law. It was the gate to quantum mechanics, which was fully developed during the first half of the 20th century.

The Planck's black body radiation law is classically demonstrated by considering a very large electromagnetic cavity (see for example section 7.2 of [2]). For a parallelepiped, the number of resonant modes per unit volume in a frequency range between f and $f + df$ can be quite straightforwardly computed using standard electromagnetic theory. The resultant expression is then divided by df and multiplied by the average energy per mode, computed according to Planck's quantum hypothesis (1.192). The result is the average energy of the system per unit frequency and volume in units of $J/m^3/Hz$. The radiation intensity passing

Principles of Interferometric and Polarimetric Radiometry, First Edition. Ignasi Corbella.

through a small hole open in one of the walls of the cavity and radiated at a particular direction per unit solid angle ($J/m^2/sr$) is equal to this average energy multiplied by the velocity of light and divided by 4π steradians. The result is the Planck's black body radiation law.

Certainly, this approach is correct, but it is somewhat artificial when trying to explain the spontaneous emission of, for example, the Earth surface. In this chapter, thermal radiation is reformulated using concepts and terminology commonly encountered in electrical and remote sensing engineering, particularly those related to random signals, electromagnetic waves, and antenna theory. Themes such as Kirchhoff's law (relation between absorption and emission), emissivity, wave polarization, and others are incorporated naturally in the formulation from the beginning. In contrast with other treatises on radiometry, here, Planck's law is *not* stated as an initial postulate, but it is derived as a side result. It is found as a direct consequence of the reciprocity of the Maxwell equations combined with the quantum hypothesis and the Boltzmann statistics.

Moreover, for historical reasons, magnitudes describing electromagnetic radiation are given different names depending on the context (radiance, intensity, brightness, etc.), making it sometimes difficult to comprehend the basic concepts of radiometry. This chapter uses a consistent naming convention based only on power and energy, while other specific names are avoided. Punctually, references to other nomenclatures are given for illustration purposes.

2.1.1 Emissivity and Brightness Temperature

Consider an extended source of thermal radiation and a distant observation point, as schematically depicted in figure 2.1(a). Two spherical systems of coordinates following the convention of appendix A.4 are defined: (r', θ', ϕ') centered in a selected source location and displayed in figure 2.1(b), and (r, θ, ϕ) centered in the observation point (not explicit in the figure) such that the radial unit vectors (1.245) satisfy $\hat{r} = -\hat{r}'$.

The first assumption is that, for far-field distance, a small region of the source, highlighted in the figure, induces in the observation point a local spherical electromagnetic wave. Consistent with this postulate and using the concepts of section 1.6.1, the electric field at this point is represented by a frequency-dependent complex vector written as a linear combination of two orthogonal polarizations denoted p and q

$$\vec{E}(r, \theta', \phi', f) = E_p(r, \theta', \phi', f)\hat{p} + E_q(r, \theta', \phi', f)\hat{q} \quad [V/\sqrt{Hz}/m] \tag{2.1}$$

where $E_p(r, \theta', \phi', f)$ and $E_q(r, \theta', \phi', f)$ are the corresponding components expressed as complex *phasors* duly defined in section 1.4. From now on, the explicit dependence on coordinates and frequency will be omitted for simplicity

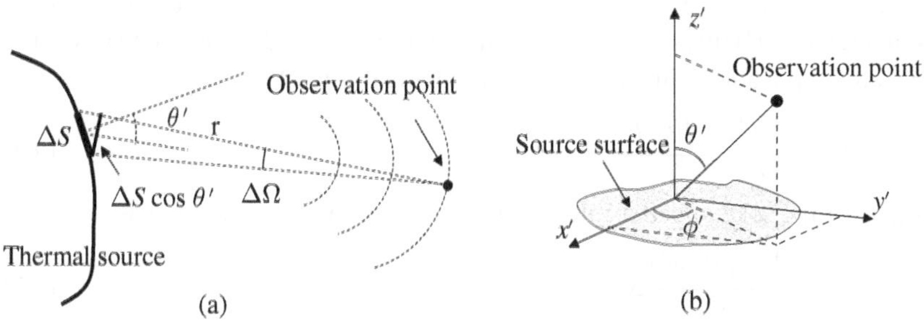

Figure 2.1 (a) Extended source of thermal radiation observed at a distant point and spherical coordinates referenced to the emission point. (b) The z' axis is defined as the local normal direction.

so the above expression reduces to $\vec{E} = E_p\hat{p} + E_q\hat{q}$. The unit vectors $\hat{p}$ and $\hat{q}$ define an orthogonal basis in the reference frame of the *observation point*. They are arbitrary and, for natural sources, no coordinate rotation or change of basis can be found that leads to the vanishing of one field component. Their specific choice depends ultimately on the polarization of the observation system, either linearly or circularly based (see section 1.6.3). More details about polarization are given later in section 2.2.

The power density (1.237) associated with the wave propagation can then be split into the contributions of both polarizations [see (A.4)]

$$\mathscr{P} = \frac{\overline{\vec{E}\cdot\vec{E}}}{\eta_0} = \frac{\overline{|E_p|^2}}{\eta_0} + \frac{\overline{|E_q|^2}}{\eta_0} \quad [W/Hz/m^2] \tag{2.2}$$

Based on this power separation, the second important premise consists of modeling the thermal emission of a small region in the source as *two antennas* with polarization vectors $\hat{p}$ and $\hat{q}$, respectively. Using then (1.240) and the system of coordinates outlined in figure 2.1, the power density at, for example, polarization p arriving at the observation point is:

$$\mathscr{P}_p = \frac{\overline{|E_p|^2}}{\eta_0} = \frac{\mathcal{N}C_p}{4\pi r^2}D_p(\theta',\phi') \tag{2.3}$$

where $D_p(\theta',\phi')$ is the *directivity* of the equivalent antenna and the product $\mathcal{N}C_p$ is the total radiated power at this polarization. It is written as the product of the *available* power associated with thermal emission $\mathcal{N}$, introduced in section 1.5.5 and given by (1.191) or (1.192) depending on whether quantum effects are neglected or not, and a mismatch coefficient C_p [see (1.179)], basically related to the different wave impedances of the medium and the free space. The total radiated power is always lower than (or equal to) $\mathcal{N}$. Note that the explicit dependence on frequency is not included for clarity.

The directivity $D_p(\theta', \phi')$ is in turn related to the *effective area* of the radiating element $A_p(\theta', \phi')$ through the reciprocity relation (1.278) demonstrated in section 1.6.9 and applicable to any radiator. Introducing it into (2.3)

$$\mathscr{P}_p = \frac{\mathcal{N} C_p}{\lambda^2} \frac{A_p(\theta', \phi')}{r^2} \tag{2.4}$$

which, for the sake of better interpretation, is conveniently rewritten as

$$\mathscr{P}_p = \frac{\mathcal{N}\, \varepsilon_p(\theta', \phi')}{\lambda^2} \Delta\Omega \tag{2.5}$$

where $\Delta\Omega = \Delta S \cos\,\theta'/r^2$ is the solid angle subtended by the source element as seen from the observation point (see figure 2.1) and $\varepsilon_p(\theta', \phi')$, a dimensionless parameter known as "emissivity" equal to

$$\varepsilon_p(\theta', \phi') = \frac{C_p A_p(\theta', \phi')}{\Delta S \cos\,\theta'} \quad \text{(emissivity)} \tag{2.6}$$

To have more insight into the significance of this important parameter, consider now an *external* thermal radiation impinging on the surface element from the same direction. The element is now a receiving antenna so, according to (1.260) and (1.261), the total absorbed power P_{abs} is equal to the incident power density $\mathscr{P}_i(\theta', \phi')$ multiplied by the effective area A_p and by the mismatch coefficient C_p. Using (2.6) it can thus be written as

$$P_{abs} = \varepsilon_p(\theta', \phi') \Delta S \cos\,\theta' \mathscr{P}_i(\theta', \phi') = \varepsilon_p(\theta', \phi') P_i \tag{2.7}$$

where P_i, the product of power density times projected area, is easily interpreted as total incident power crossing the normal surface. The above equation provides the power absorbed by the element. Since the power not absorbed must be reflected,

$$P_{refl} = P_i - P_{abs} = P_i[1 - \varepsilon_p(\theta', \phi')] = P_i \Gamma_p(\theta', \phi') \tag{2.8}$$

where $\Gamma_p(\theta', \phi')$[1] is the power reflection coefficient relating the total reflected power to the one incident from a given direction at polarization p. This relationship between emissivity and reflection coefficient provides a practical means to compute the former in specific scenarios, either analytically or numerically.

A *Lambertian surface* has emissivity independent of the direction. The effective area at any polarization is equal to the physical area multiplied by $\cos\,\theta$ and $\varepsilon_{p,q}(\theta', \phi') = C_{p,q}\ \forall(\theta', \phi')$. This is the case of a rough surface and a typical example is the *black body*, with $C_p = C_q = 1$ so unit emissivity and null reflection coefficient at both polarizations.

1 Not to be confused with the reflection coefficient defined in section 1.5.2.

A perfect reflector, for example a metal plate, has unit reflection coefficient so null emissivity. Therefore, the effective area is also null in all directions.

2.1.1.1 Receiving System

If a receiving antenna with polarization p and effective area A_a is placed at the observation point, the collected available power is found by substituting (2.5) into the first part of (1.260),

$$P_a = \mathscr{P}_p A_a = \frac{\mathcal{N}\,\varepsilon_p(\theta', \phi')}{\lambda^2} \Delta\Omega A_a \quad [W/Hz] \quad \text{(receiver view)} \tag{2.9}$$

which, at microwave frequencies, substituting $\mathcal{N} = kT$ (1.191) with T the source physical temperature, becomes

$$P_a = \frac{kT_{B_p}(\theta', \phi')}{\lambda^2} \Delta\Omega A_a \tag{2.10}$$

where T_{B_p} is the *brightness temperature* of the source at polarization p:

$$T_{B_p}(\theta', \phi') = T\,\varepsilon_p(\theta', \phi') \quad \text{Brightness temperature} \tag{2.11}$$

from which it is apparent that the emissivity at a given polarization is the ratio between brightness temperature and physical temperature. A similar definition holds for the q polarization and, in case there is no need to be specified, the brightness temperature is just denoted by $T_B(\theta', \phi')$. A microwave radiometer is a device intended to measure brightness temperature.

2.1.2 Planck and Rayleigh–Jeans Laws

Just for completeness, although not strictly needed, the Planck's law will now be derived from the concepts just defined. As indicated by the label "receiver view," equation (2.9) is written in terms of the receiving antenna effective area A_a and the solid angle subtended by the source as seen from the receiver $\Delta\Omega$. It is the most practical form for computing what a receiver would detect, and particularly

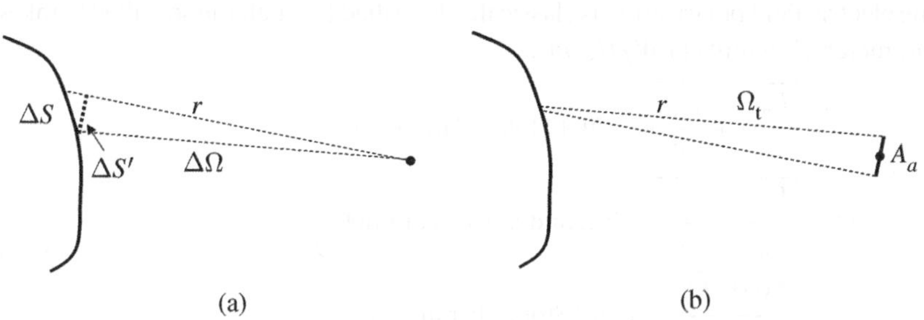

Figure 2.2 Solid angles and areas used to define the spectral brightness. (a) Receiver view. (b) Source view.

to derive the expression for antenna noise temperature in section 2.3. Nevertheless, it can also be written in terms of the emitting area and the solid angle subtended by the receiver as seen from the source. From figure 2.2, it is apparent that:

$$\Delta\Omega = \frac{\Delta S'}{r^2} \ ; \ \Omega_t = \frac{A_a}{r^2} \ \Rightarrow \ \Delta\Omega A_a = \Omega_t \Delta S' \tag{2.12}$$

so the available power (2.9) can also be written as

$$P_a = \frac{\mathcal{N}\, \varepsilon_p(\theta', \phi')}{\lambda^2} \Omega_t \Delta S' \quad [W/Hz] \quad \text{source view} \tag{2.13}$$

which is the form normally used to describe the radiation of the (black) body.

The *spectral brightness* (or spectral radiance or radiation intensity or specific intensity or just intensity) is the available power per unit area per unit solid angle, either considering the source or the receiver points of view

$$B_p(\theta', \phi') = \frac{P_a}{\Delta\Omega A_a} = \frac{P_a}{\Omega_t \Delta S'} = \frac{\mathcal{N}\, \varepsilon_p(\theta', \phi')}{\lambda^2} \quad [W/Hz/m^2/sr] \tag{2.14}$$

The spectral brightness B_p just defined is *exactly* the same parameter as the "specific intensity" I_ν defined in [3] to derive the radiative transfer equation and the theory behind it. Substituting $\mathcal{N}$ by (1.192), the spectral brightness becomes

$$B_p(\theta', \phi') = \frac{\varepsilon_p(\theta', \phi')}{c^2} \frac{hf^3}{e^{\frac{hf}{kT}} - 1} \quad \text{Planck's law} \tag{2.15}$$

which for $\varepsilon_p(\theta', \phi') = 1$ is the Planck's law of black body radiation for a single polarization. It should be multiplied by two to consider both. The Rayleigh–Jeans law is obtained from this equation by approximating $\exp\,(hf/kT) \approx 1 + hf/kT$ or just by making $\mathcal{N} = kT$ and $\varepsilon_p = 1$ in (2.14).

2.2 Polarization

2.2.1 Stokes Parameters and Polarimetric Brightness Temperature

The electric field polarization is classically described through the so-called "Stokes parameters," in units of $W/Hz/m^2$,

$$\begin{aligned} I &= \frac{\overline{|E_p|^2}}{\eta_0} + \frac{\overline{|E_q|^2}}{\eta_0} \quad \text{First Stokes Parameter} \\ Q &= \frac{\overline{|E_p|^2}}{\eta_0} - \frac{\overline{|E_q|^2}}{\eta_0} \quad \text{Second Stokes Parameter} \\ U &= \frac{2\Re e[\overline{E_p E_q^*}]}{\eta_0} \quad \text{Third Stokes Parameter} \\ V &= \frac{2\Im m[\overline{E_p E_q^*}]}{\eta_0} \quad \text{Fourth Stokes Parameter} \end{aligned} \tag{2.16}$$

where E_p and E_q are the two components of the electric field (2.1). From (2.2), it is apparent that the first Stokes parameter is the total power density $\mathscr{P}$.

In microwave radiometry, the Stokes parameters are preferably expressed in terms of brightness temperatures. From (2.5), taking into account (2.2), substituting $\mathcal{N} = kT$ and using the definition of brightness temperature (2.11), the power densities associated with each polarization p or q can be written as

$$\frac{\overline{|E_p|^2}}{\eta_0} = \frac{kT_{B_p}}{\lambda^2}\Delta\Omega \qquad \frac{\overline{|E_q|^2}}{\eta_0} = \frac{kT_{B_q}}{\lambda^2}\Delta\Omega \tag{2.17}$$

Note that the dependency of the electric field on $1/r$ is implicitly contained in the solid angle, which is inversely proportional to r^2 when the same source area is considered. The complex *polarimetric brightness temperature* $T_{B_{pq}}$ is similarly defined such that

$$\frac{\overline{E_p E_q^*}}{\eta_0} = \frac{kT_{B_{pq}}}{\lambda^2}\Delta\Omega \tag{2.18}$$

The above equations are more conveniently written in terms of the normalized electric field (1.270)

$$\begin{aligned} \overline{|e_p|^2} &= kT_{B_p}\Delta\Omega \\ \overline{|e_q|^2} &= kT_{B_q}\Delta\Omega \\ \overline{e_p e_q^*} &= kT_{B_{pq}}\Delta\Omega \end{aligned} \tag{2.19}$$

An *unpolarized field*, such as for example a black body, has equal power in both polarizations while the corresponding field components are uncorrelated

$$T_{B_p} = T_{B_q} = T_B \ ; \ T_{B_{pq}} = 0 \quad \text{(unpolarized field)} \tag{2.20}$$

The *modified* Stokes parameters, in units of K, are those that result of combining brightness temperatures as in (2.16)

$$\begin{aligned} T_1 &= T_{B_p} + T_{B_q} && \text{First Stokes Parameter} \\ T_2 &= T_{B_p} - T_{B_q} && \text{Second Stokes Parameter} \\ T_3 &= 2\Re e[T_{B_{pq}}] && \text{Third Stokes Parameter} \\ T_4 &= 2\Im m[T_{B_{pq}}] && \text{Fourth Stokes Parameter} \end{aligned} \tag{2.21}$$

from which the following properties are derived:

$$T_3 = T_{B_{pq}} + T_{B_{qp}} \ ; \ T_4 = -j(T_{B_{pq}} - T_{B_{qp}}) \tag{2.22}$$

Note that, for an unpolarized field, all Stokes parameters other than the first one are null.

Even though they are quite commonly found in literature, these parameters are seldom used in this book. The polarimetric description of a radiation will normally be based directly on brightness temperatures.

2.2.2 Change of Polarization Frame

Consider now a different polarization frame described by unit vectors $(\hat{p}', \hat{q}')$ related to the old ones by

$$\begin{aligned} \hat{p}' &= A\hat{p} + B\hat{q} \\ \hat{q}' &= C\hat{p} + D\hat{q} \end{aligned} \tag{2.23}$$

where A, B, C, and D are complex in general. The electric field is

$$\vec{E} = E_p\hat{p} + E_q\hat{q} = E_{p'}\hat{p}' + E_{q'}\hat{q}' \tag{2.24}$$

from which, substituting (2.23) the following linear relationship is easily found

$$\begin{bmatrix} E_p \\ E_q \end{bmatrix} = \begin{bmatrix} A & C \\ B & D \end{bmatrix} \begin{bmatrix} E_{p'} \\ E_{q'} \end{bmatrix} \tag{2.25}$$

The square matrix relating the field components is sometimes referred to as "change of basis matrix." Since the power density (2.2) must be the same regardless of the reference frame, applying the method of appendix A.3, it follows that this matrix must be unitary

$$\begin{bmatrix} A & C \\ B & D \end{bmatrix} \begin{bmatrix} A^* & B^* \\ C^* & D^* \end{bmatrix} = \begin{bmatrix} A^* & B^* \\ C^* & D^* \end{bmatrix} \begin{bmatrix} A & C \\ B & D \end{bmatrix} = \begin{bmatrix} 1 & 0 \\ 0 & 1 \end{bmatrix} \tag{2.26}$$

from which the following properties are easily deduced,

$$\begin{aligned} &|A| = |D| \; ; \; |B| = |C| \; ; \; |A|^2 + |B|^2 = 1 \\ &AB^* = -CD^* \; ; \; AC^* = -BD^* \end{aligned} \tag{2.27}$$

The inverse of (2.25) becomes then

$$\begin{bmatrix} E_{p'} \\ E_{q'} \end{bmatrix} = \begin{bmatrix} A^* & B^* \\ C^* & D^* \end{bmatrix} \begin{bmatrix} E_p \\ E_q \end{bmatrix} \tag{2.28}$$

The reader may want to check that this result is consistent with (A.8), which states that the field components in the new frame are equal to the Hermitian scalar product of the field vector times the corresponding unit vectors $E_{p'} = \vec{E} \cdot \hat{p}'$ and $E_{q'} = \vec{E} \cdot \hat{q}'$ with $\hat{p}'$ and $\hat{q}'$ given in (2.23).

The matrix relating the *brightness temperatures* in both reference frames is obtained by introducing the above conversion into (2.17) and (2.18). Taking into account also (2.27),

$$\begin{bmatrix} T_{B_{p'}} \\ T_{B_{q'}} \\ T_{B_{p'q'}} \\ T_{B_{q'p'}} \end{bmatrix} = \begin{bmatrix} |A|^2 & |B|^2 & A^*B & AB^* \\ |B|^2 & |A|^2 & -A^*B & -AB^* \\ A^*C & -A^*C & A^*D & B^*C \\ AC^* & -AC^* & BC^* & AD^* \end{bmatrix} \begin{bmatrix} T_{B_p} \\ T_{B_q} \\ T_{B_{pq}} \\ T_{B_{qp}} \end{bmatrix} \tag{2.29}$$

where, compatible with the definition of polarimetric brightness temperature (2.18), the third and fourth rows are conjugate of each other. According to (2.22), the sum of these two rows is the third Stokes parameter while the difference divided by j is the fourth. Similarly, the sum of the first and second rows is the first Stokes parameter and their difference is the second. Taking into account (2.27), the first Stokes parameter turns out to be invariant to any change of polarization frame: $T_{B_{p'}} + T_{B_{q'}} = T_{B_p} + T_{B_q}$, which is expected since the total radiated power must be independent of the chosen coordinate frame.

The conversion matrix of the modified Stokes parameters (2.21) can be deduced from the above system by adding and subtracting rows and columns. The result is given here below for completeness

$$\begin{bmatrix} T_1' \\ T_2' \\ T_3' \\ T_4' \end{bmatrix} = \begin{bmatrix} 1 & 0 & 0 & 0 \\ 0 & |A|^2 - |B|^2 & 2\Re e[A^*B] & -2\Im m[A^*B] \\ 0 & 2\Re e[A^*C] & \Re e[\Sigma] & -\Im m[\Delta] \\ 0 & 2\Im m[A^*C] & \Im m[\Sigma] & \Re e[\Delta] \end{bmatrix} \begin{bmatrix} T_1 \\ T_2 \\ T_3 \\ T_4 \end{bmatrix} \tag{2.30}$$

where $\Sigma = A^*D + B^*C$ and $\Delta = A^*D - B^*C$.

2.2.3 Linear Axis Rotation

Figure 2.3 depicts a counterclockwise rotation[2] from a linear polarization frame (x, y) to another (x', y') by an angle α. In this figure, α is positive and changing the sign just reverses the rotation sense.

The unit vectors that define the rotated frame are

$$\begin{aligned} \hat{x}' &= \cos\,\alpha\,\hat{x} + \sin\,\alpha\,\hat{y} \\ \hat{y}' &= -\sin\,\alpha\,\hat{x} + \cos\,\alpha\,\hat{y} \end{aligned} \tag{2.31}$$

so, identifying with (2.23),

$$A = D = \cos\,\alpha\ ;\ B = -C = \sin\,\alpha \tag{2.32}$$

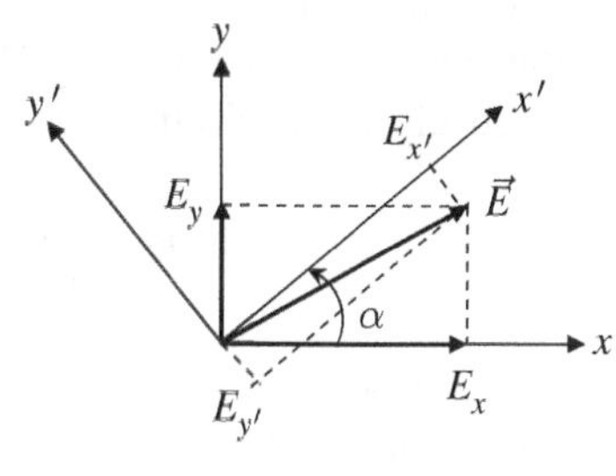

Figure 2.3 Linear polarization frame rotation.

2 Equivalent to a clockwise rotation of the electric field vector.

which introduced into (2.29) result in

$$\begin{bmatrix} T_{B_{x'}} \\ T_{B_{y'}} \\ T_{B_{x'y'}} \\ T_{B_{y'x'}} \end{bmatrix} = \begin{bmatrix} \cos^2\alpha & \sin^2\alpha & \frac{1}{2}\sin 2\alpha & \frac{1}{2}\sin 2\alpha \\ \sin^2\alpha & \cos^2\alpha & \frac{-1}{2}\sin 2\alpha & \frac{-1}{2}\sin 2\alpha \\ \frac{-1}{2}\sin 2\alpha & \frac{1}{2}\sin 2\alpha & \cos^2\alpha & -\sin^2\alpha \\ \frac{-1}{2}\sin 2\alpha & \frac{1}{2}\sin 2\alpha & -\sin^2\alpha & \cos^2\alpha \end{bmatrix} \begin{bmatrix} T_{B_x} \\ T_{B_y} \\ T_{B_{xy}} \\ T_{B_{yx}} \end{bmatrix} \tag{2.33}$$

where a trigonometric identity has been applied to simplify some expressions. Subtracting the fourth row from the third shows that $T_{B_{x'y'}} - T_{B_{y'x'}} = T_{B_{xy}} - T_{B_{yx}}$ which, using (2.22), means that the *fourth Stokes parameter is independent of rotation*. This is also deduced from (2.30), which, since now all coefficients A to D are real, has all elements in the last row and column equal to zero except the one in the diagonal that is equal to 1.

A particularly interesting case is a rotation of an angle $\alpha = 45°$ for which $\cos\ \alpha = \sin\ \alpha = 1/\sqrt{2}$. The Stokes parameter conversion matrix (2.30) simplifies then to

$$\begin{bmatrix} T'_1 \\ T'_2 \\ T'_3 \\ T'_4 \end{bmatrix} = \begin{bmatrix} 1 & 0 & 0 & 0 \\ 0 & 0 & 1 & 0 \\ 0 & -1 & 0 & 0 \\ 0 & 0 & 0 & 1 \end{bmatrix} \begin{bmatrix} T_1 \\ T_2 \\ T_3 \\ T_4 \end{bmatrix} \tag{2.34}$$

so the second and third Stokes parameters convert to each other when rotating the axes 45°, while the rest remain unchanged. This property is sometimes used to measure the third Stokes parameter without the need of any correlation measurement. It suffices to physically rotate the antenna 45° and get the difference between the power measurements at both orthogonal polarizations.

2.2.4 Horizontal and Vertical Polarization

The most common electric field components in Earth observation are the so called "horizontal" and "vertical" polarizations, schematically defined in figure 2.4 with the convention $\hat{h} \times \hat{v} = \hat{r}$ where $\vec{r}$ is the radial vector *from* the observation point *to* the impinging point in Earth. Horizontal and vertical polarizations correspond, respectively, to the electric field being perpendicular or parallel to the incidence

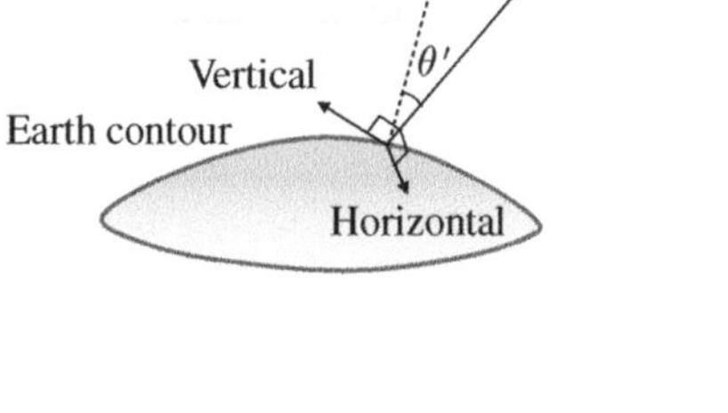

Figure 2.4 Definition of the ground reference plane. The incidence angle is θ'.

plane. For a specular (or flat) surface, the power reflection coefficients at these polarizations depend only on the incident angle θ' and can be analytically derived from the Maxwell equations, giving rise to the known Fresnel coefficients [1]

$$\Gamma_h = \left| \frac{\cos\,\theta' - \sqrt{\epsilon_{eff} - \sin^2\,\theta'}}{\cos\,\theta' + \sqrt{\epsilon_{eff} - \sin^2\,\theta'}} \right|^2$$
$$\Gamma_v = \left| \frac{\epsilon_{eff}\cos\,\theta' - \sqrt{\epsilon_{eff} - \sin^2\,\theta'}}{\epsilon_{eff}\cos\,\theta' + \sqrt{\epsilon_{eff} - \sin^2\,\theta'}} \right|^2 \qquad (2.35)$$

where ε_{eff} is the complex permittivity of the surface. From these equations, the corresponding emissivities are immediate, $\varepsilon_{h,v} = 1 - \Gamma_{h,v}$. Microwave radiometry for Earth observation is ultimately based on the dependence of complex permittivity with geophysical parameters: soil composition, water content, salinity, amount of ice, snow, and others. Most Earth surfaces can be described by a flat surface with added corrections to account for roughness.

Assume a linearly polarized antenna in the observation point. The polarization of its field pattern in the direction of the point on ground is defined by vectors (1.246) based on Ludwig's third definition of polarization. In general, they are rotated a certain angle α with respect to the horizontal and vertical directions just defined. Looking from Earth to antenna, the two polarization frames become those of figure 2.3 with the following unit vector identifications

$$\hat{x} \to \hat{h} \quad \hat{y} \to \hat{v} \quad \hat{x}' \to \hat{\ell}_x \quad \hat{y}' \to \hat{\ell}_y \qquad (2.36)$$

and equation (2.33) is directly applicable. The expression is not rewritten here to avoid excessive repetition but a special case is highlighted.

In most natural sources the horizontal and vertical components of the emitted electric field are known to be uncorrelated, so $T_{B_{hv}} = 0$. Equation (2.33) is then simplified to

$$\begin{bmatrix} T_{B_x} \\ T_{B_y} \\ 2T_{B_{xy}} \end{bmatrix} = \begin{bmatrix} \cos^2\alpha & \sin^2\alpha \\ \sin^2\alpha & \cos^2\alpha \\ -\sin\,2\alpha & \sin\,2\alpha \end{bmatrix} \begin{bmatrix} T_{B_h} \\ T_{B_v} \end{bmatrix} \qquad (2.37)$$

where the subscripts x and y refer to the corresponding antenna polarization vectors $\hat{\ell}_x$ and $\hat{\ell}_y$, respectively. The polarimetric brightness temperature in the antenna reference frame is now real and generated by the polarization axes rotation but not from ground emission. The above equation is useful, for example, to estimate the polarimetric measurement of an antenna out of the simulated brightness temperatures at horizontal and vertical polarizations obtained, for example, by an emission model.

Depending on the frequency, particularly at L-band, the angle α must include also the Faraday rotation affecting the signal in the propagation from the Earth surface to the antenna. Conversely, the above system can be used for Faraday rotation angle retrieval from radiometric measurements [4]. Solving for α from the third row of (2.37) and using also the other two to get the brightness temperature difference,

$$\alpha_F = -\alpha_G - \frac{1}{2} \arctan \frac{2T_{B_{xy}}}{T_{B_x} - T_{B_y}} \tag{2.38}$$

where α_G is the geometric rotation angle and α_F the contribution of the Faraday effect. Vertical total electron content of the ionosphere has been successfully estimated from the above retrieval using data from the Soil Moisture and Ocean Salinity (SMOS) mission [5].

The submatrix formed by the first and second rows of (2.37) relates the ground emission at horizontal and vertical polarizations to the antenna measurements at x and y polarizations. This matrix is singular at $\alpha = 45°$ but for *small rotation angles* it can be inverted to compute the ground-based brightness temperatures from the dual polarization measurements:

$$\begin{bmatrix} T_{B_h} \\ T_{B_v} \end{bmatrix} = \frac{1}{\cos 2\alpha} \begin{bmatrix} \cos^2\alpha & -\sin^2\alpha \\ -\sin^2\alpha & \cos^2\alpha \end{bmatrix} \begin{bmatrix} T_{B_x} \\ T_{B_y} \end{bmatrix} \tag{2.39}$$

2.2.5 Circular Polarization

Another interesting case is the circular polarization. Conversion from linear $(\hat{\ell}_x, \hat{\ell}_y)$ to circular polarization is defined by the left and right circular polarization unit vectors (1.252). The corresponding values of A and B in (2.28) are then

$$A = C = 1/\sqrt{2}\,;\; B = j/\sqrt{2}\,;\; D = -j/\sqrt{2} \tag{2.40}$$

which, when inserted in (2.29) result in

$$\begin{bmatrix} T_{B_l} \\ T_{B_r} \\ T_{B_{lr}} \\ T_{B_{rl}} \end{bmatrix} = \frac{1}{2} \begin{bmatrix} 1 & 1 & j & -j \\ 1 & 1 & -j & j \\ 1 & -1 & -j & -j \\ 1 & -1 & j & j \end{bmatrix} \begin{bmatrix} T_{B_x} \\ T_{B_y} \\ T_{B_{xy}} \\ T_{B_{yx}} \end{bmatrix} \tag{2.41}$$

where the subscripts l and r stand for left and right circular polarization.

Subtracting the first and second rows, it follows that

$$T_{B_l} - T_{B_r} = 2\Im m[T_{B_{xy}}] \tag{2.42}$$

which is useful to measure the fourth Stokes parameter defined for linear polarization (2.21) just from the difference of the measurements at right and left circular polarizations. Other interesting properties can be easily derived from the above matrix equation or from the one relating directly to the Stokes parameters (2.30), which becomes,

$$\begin{bmatrix} T'_1 \\ T'_2 \\ T'_3 \\ T'_4 \end{bmatrix} = \begin{bmatrix} 1 & 0 & 0 & 0 \\ 0 & 0 & 0 & -1 \\ 0 & 1 & 0 & 0 \\ 0 & 0 & -1 & 0 \end{bmatrix} \begin{bmatrix} T_1 \\ T_2 \\ T_3 \\ T_4 \end{bmatrix} \tag{2.43}$$

where the primed parameters refer to the circular polarization and the unprimed to linear.

2.3 Antenna Temperature

2.3.1 Concept

The response of an antenna to an incoming electromagnetic wave arriving from a *single* spatial direction is derived in section 1.6.5. In general, a receiving antenna collects contributions from all directions in the whole space around it, so the overall response is the sum of all of them. Figure 2.5 shows an antenna and a generic source of radiation. The antenna is located in the center of a spherical system of coordinates defined in accordance to appendix A.4. The individual contribution of a single elementary source at (θ, ϕ) is directly (1.269) with the normalized electric field $d\vec{e}(\theta, \phi)$ originated in it and arriving to the antenna from the corresponding direction. The total collected signal is then

$$b'_a = \frac{1}{\sqrt{\Omega_a}} \iint \vec{F}(\theta, \phi) \odot d\vec{e}(\theta, \phi) \tag{2.44}$$

where the integration region includes all sources. In this equation, $\vec{F}(\theta, \phi)$ is the antenna field pattern (1.238), Ω_a the antenna equivalent solid angle (1.243), the normalized electric field is defined in (1.270) and the output signal b'_a in (1.262), such that $\overline{|b'_a|^2}$ is the available power. The operator $\odot$ is the non-Hermitian dot product (A.2).

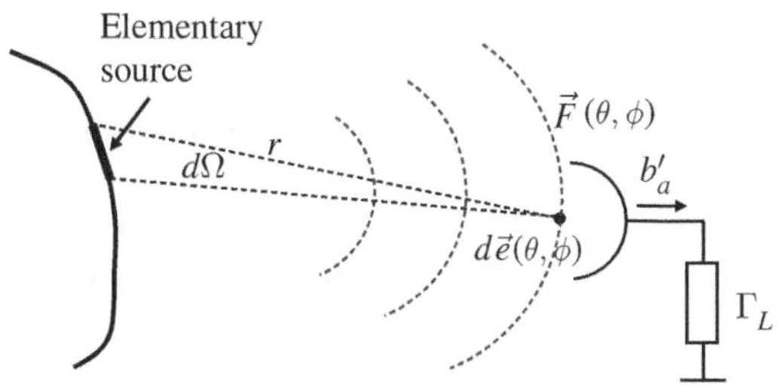

Figure 2.5 An antenna collecting energy from an extended source of radiation.

The equivalent noise temperature is defined in (1.193) on the basis of the available power. Consistently, the antenna noise equivalent temperature, or simply *antenna temperature*,[3] is defined in such a way that

$$kT_a = \overline{|b'_a|^2} = \frac{\overline{|b_a|^2}}{1 - |\Gamma_a|^2} \quad \text{Antenna temperature} \tag{2.45}$$

where b_a is the intrinsic noise wave and Γ_a the reflection coefficient of the equivalent source that models the antenna in reception (see figure 1.16). Therefore, assuming uncorrelated elementary radiators in (2.44),

$$kT_a = \frac{1}{\Omega_a} \iint \overline{|\vec{F}(\theta, \phi) \odot d\vec{e}(\theta, \phi)|^2} \tag{2.46}$$

which is the general expression for the antenna temperature. It can be expanded if the antenna pattern is decomposed into *the same* two components of the incoming electric field (2.1) based on polarization vectors $\hat{p}$ and $\hat{q}$

$$\begin{aligned} \vec{F}(\theta, \phi) &= F_p(\theta, \phi)\hat{p} + F_q(\theta, \phi)\hat{q} \\ d\vec{e}(\theta, \phi) &= de_p(\theta, \phi)\hat{p} + de_q(\theta, \phi)\hat{q} \end{aligned} \tag{2.47}$$

In practice, the polarization basis is selected according to that of the antenna. For example, $\hat{p}$ and $\hat{q}$ can be the unit vectors of the Ludwig third definition of polarization (1.246) for linearly polarized antennas. Or they can represent a circular polarization system based on (1.249). General symbols $\hat{p}$ and $\hat{q}$ are kept for generality.

Introducing the relationship between normalized electric fields and brightness temperatures (2.19), equation (2.46) becomes

$$T_a = \frac{1}{\Omega_a} \iint_{4\pi} \left[|F_p(\theta, \phi)|^2 T_{B_p} + |F_q(\theta, \phi)|^2 T_{B_q} + 2\Re e[F_p(\theta, \phi) F_q^*(\theta, \phi) T_{B_{pq}}] \right] d\Omega \tag{2.48}$$

where the integration region is extended to all the sphere of 4π steradians to include sources coming from all directions in space. For example, for a linearly polarized antenna, using (1.238) the antenna pattern components can be written as $F_p(\theta, \phi) = F(\theta, \phi) \cos\, \alpha$ and $F_q(\theta, \phi) = F(\theta, \phi) \sin\, \alpha$ where α is the angle between the antenna polarization vector $\hat{e}_a$ and the unit vector $\hat{p}$. Introducing these expressions into (2.48)

$$T_a = \frac{1}{\Omega_a} \iint_{4\pi} t(\theta, \phi) \left[T_{B_p} \cos^2 \alpha + T_{B_q} \sin^2 \alpha + \Re e[T_{B_{pq}}] \sin\, 2\alpha \right] d\Omega \tag{2.49}$$

where $t(\theta, \phi) = |F(\theta, \phi)|^2$ is the antenna power pattern (1.239) and a trigonometric identity has been used to simplify the expression. The reader can check

---

3 Not to be confused with the antenna physical temperature.

using (2.33) that, not surprisingly, the term inside the brackets is exactly the brightness temperature at a linear polarization direction rotated an angle α with respect to $\hat{p}$.

This same result could have been obtained by expressing the antenna field pattern in (2.46) as $\vec{F}(\theta,\phi) = F(\theta,\phi)\hat{e}_a(\theta,\phi)$ with $\hat{e}_a$ the antenna polarization vector and assuming the electric field polarization defined with $\hat{p} = \hat{e}_a$ and $\hat{q}$ the corresponding orthogonal direction. In any case, the antenna temperature is

$$T_a = \frac{1}{\Omega_a}\iint_{4\pi} T_B(\theta,\phi)t(\theta,\phi)d\Omega \tag{2.50}$$

where $t(\theta,\phi)$ is the antenna power pattern and $T_B(\theta,\phi)$ the brightness temperature *in the antenna polarization basis.* No subscript to denote polarization has been included to simplify notation. It is important to point out that this expression holds for any polarization schema. For example, for a circularly polarized antenna T_B would be T_{B_l} or T_{B_r}, depending on the sense of rotation of the antenna polarization vector.

Substituting the antenna equivalent solid angle Ω_a by (1.243), the antenna temperature (2.50) is interestingly interpreted as the average brightness temperature weighted at each direction by the power antenna pattern.

Three special cases are considered now for evaluating (2.50): a black body enclosing the antenna, a point source, and a narrow beam antenna pointing to an extended source.

2.3.2 Flat Target

A "flat target" is a completely unpolarized thermal source having equal brightness temperature observed from any direction in space. The most typical example is an "anechoic chamber," a box having all walls lined by a black body at constant temperature (figure 2.6(a)). The microwave cosmic background is also (almost) a flat target, although celestial bodies make overall radiation depart from it.

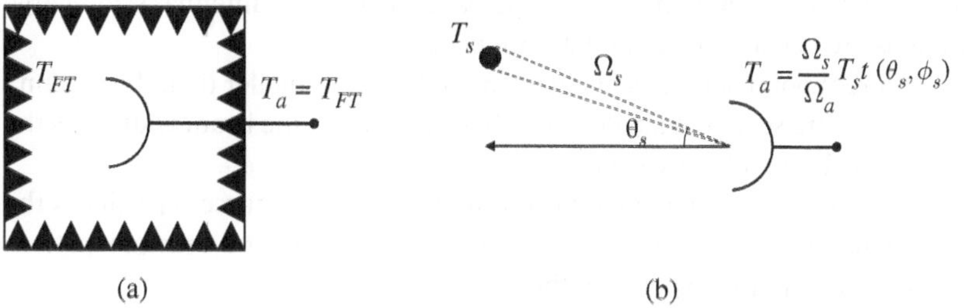

Figure 2.6 (a) Antenna surrounded by a constant source (flat target). (b) Antenna pointing to a point source.

The polarimetric brightness temperatures of a flat target at temperature T_{FT} are then, by definition

$$T_{B_p} = T_{B_q} = T_{FT} \quad T_{B_{pq}} = 0 \tag{2.51}$$

Therefore, from (2.48), taking into account (1.243) and the definition of power pattern (1.239)

$$T_a = T_{FT} \qquad \text{Flat target} \tag{2.52}$$

independently of the antenna pattern and its position within the chamber. This result is consistent with the fact that the whole system viewed from the antenna connector is a one-port microwave network at a constant temperature, so its noise equivalent temperature must be equal to the physical temperature.

2.3.3 Point Source

Assume now the second case of figure 2.6 in which the source is a distant body subtending a solid angle Ω_s from the receiving position much smaller than the antenna solid angle ($\Omega_s \ll \Omega_a$). In this case the antenna temperature (2.50) becomes

$$T_a = \frac{\Omega_s}{\Omega_a} T_s t(\theta_s, \phi_s) \qquad \text{Distant point source} \tag{2.53}$$

where T_s is the source brightness temperature at coordinates (θ_s, ϕ_s). If the antenna boresight is correctly oriented toward the source, then $t(\theta_s, \phi_s) = 1$, so the antenna temperature reduces to just $T_a = \Omega_s/\Omega_a T_s$. In other words, it is equal to the source brightness temperature weighted by the ratio of source- to antenna-solid angles. The following properties are derived from (2.53):

- Since $\Omega_s \propto 1/r^2$ where r is the distance to the source, the antenna temperature depends inversely on the square of the distance. Therefore, for the same antenna, more distant bodies are seen fainter.
- Since $\Omega_a \propto 1/A_a$ where A_a is the effective area of the antenna, for a given source, the antenna temperature increases with the antenna size. Large antennas are needed to measure distant bodies.
- Two different point sources separated an angle much smaller than the antenna beam width are seen as a single point. In other words, the beam width sets the angular resolution of the system.
- The output of a scanning antenna passing through the source reproduces the antenna pattern shape. In terms of signal processing, the antenna pattern becomes the impulse response of the system.

Clearly this is the typical situation of Radioastronomy applications where large dish antennas are used to perform measurements of distant stars.

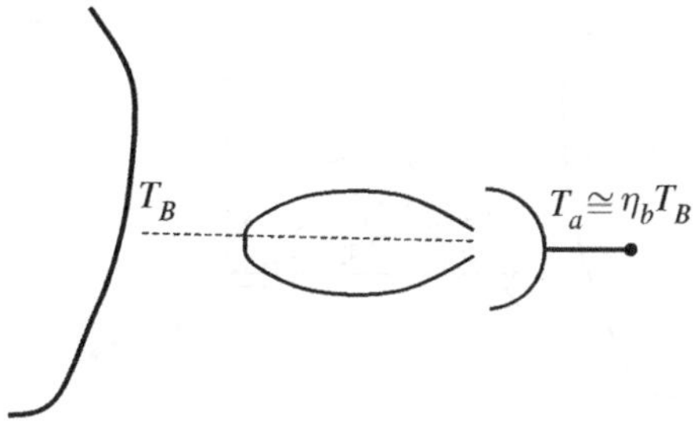

Figure 2.7 Narrow beam antenna pointing to and extended source.

2.3.4 Extended Source

Consider an antenna pattern with a clearly defined main lobe around boresight pointing to an extended source of radiation (figure 2.7). In this case, the integral of (2.50) can be split into

$$T_a = \frac{1}{\Omega_a} \iint_{\mathrm{MB}} T_B(\theta, \phi) t(\theta, \phi) d\Omega + \frac{1}{\Omega_a} \iint_{other} T_B(\theta, \phi) t(\theta, \phi) d\Omega \tag{2.54}$$

where MB stands for "Main Beam." If it is narrow enough so that the brightness temperature within it can be assumed constant, then

$$T_a = T_B(\theta_0, \phi_0)\eta_b + T_{SL} \tag{2.55}$$

where T_{SL} is the contribution of the side (including back) lobes and η_b the so-called "antenna main beam efficiency"

$$\eta_b = \frac{1}{\Omega_a} \iint_{\mathrm{MB}} t(\theta, \phi) d\Omega \tag{2.56}$$

which should be close to unity for a well-designed pencil-beam antenna. Depending on the scenario, particularly regarding other sources, often the contribution from side lobes can be neglected. Otherwise, it must be either estimated theoretically or characterized experimentally to correctly interpret the antenna temperature measurements. This kind of processing is out of the scope of this book.

Equation (2.55) implies that for extended sources and narrow beam antennas the antenna temperature is independent on the distance to the source and it is essentially equal to the brightness temperature of the source zone to which the antenna boresight points to. The typical application of this equation is in Earth observation, in which directive antennas, either mounted on airplanes or satellites, scan the surface for measuring the brightness temperature over different terrains.

2.3.5 Angular Resolution

Brightness temperature images can be formed by a mechanical scan of a highly directive antenna. This is achieved, for example, by rotating the antenna with

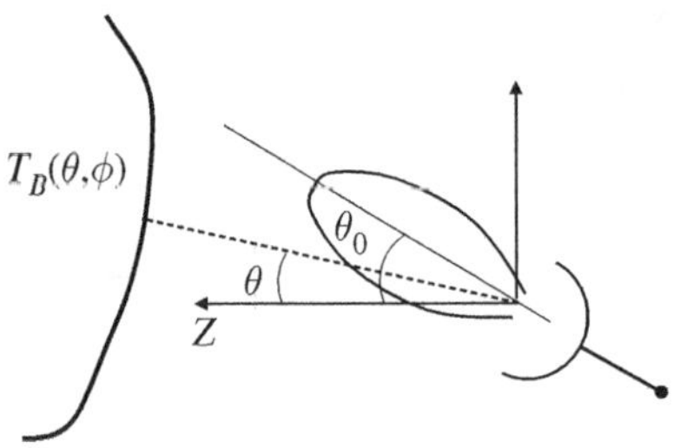

Figure 2.8 A scanning directive antenna.

respect to a fixed coordinate system, as shown schematically in figure 2.8. The antenna temperature (2.50) can then be rewritten as

$$T_a(\theta_0, \phi_0) = \frac{1}{\Omega_a} \int_0^{2\pi} \int_0^{\pi} T_B(\theta, \phi) t(\theta_0 - \theta, \phi_0 - \phi) \sin\ \theta d\theta d\phi \tag{2.57}$$

where (A.20) has been used for differential solid angle. This equation is recognized as a two-dimensional convolution integral [see (A.40)] between the brightness temperature image and the antenna power pattern. This result is consistent with the last bullet of section 2.3.3 referring to a point source and stating that the antenna pattern is the system impulse response. This interpretation holds also for an extended source of radiation.

The beam width (defined for example at half power) of an antenna is inversely proportional to its size and, for an aperture antenna, it depends on the field illumination pattern. The following examples are taken from [6]:

- Rectangular aperture of size (a,b) with uniform illumination: $\Delta\theta = 50.4^\circ \lambda/a$.
- Circular aperture of radius r with uniform illumination: $\Delta\theta = 58.36^\circ \lambda/r$.
- Piramidal horn of size (a,b) $\Delta\theta = 78^\circ \lambda/a$ and $\Delta\theta = 56^\circ \lambda/b$.

Other antenna illumination strategies provide different antenna widths but in general the angular resolution of an antenna is always proportional to the ratio of wavelength and maximum size. For example, a circular antenna of 4 meters diameter at a frequency of 1.4 GHz has an angular resolution of 3°. Larger values are obtained when the field is tapered within the aperture.

More information on imaging radiometers and angular resolution can be found, for example in [1, 7].

2.4 Total Power Radiometers

A microwave radiometer is a passive remote sensing instrument designed to measure the natural radiation emitted by a body in the microwave region. The term "total power" refers to measuring the *antenna temperature* just defined in section 2.3. The *brightness temperature* associated with the emission is then

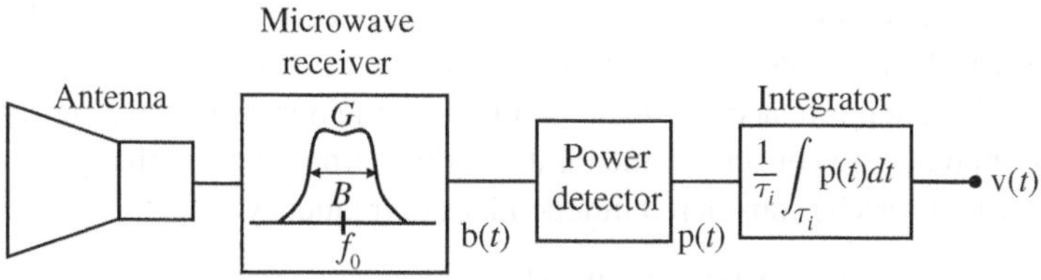

Figure 2.9 Conceptual block diagram of a total power radiometer. The receiver output signal b(t) is zero-mean Gaussian noise. The power detector output p(t) is the *instantaneous power* and v(t) is the output *voltage*.

retrieved from (2.55) provided the antenna main beam efficiency and side lobes contribution are known, either trough modeling or measurement.

Antenna temperature is measured by a microwave receiver, a power detector and an integrator arranged as depicted in figure 2.9. The receiver has a limited bandwidth B around a center frequency f_0, both parameters in principle fixed by system specifications based on scientific requirements and/or regulations. Different applications need measurements at different frequency bands ranging from the low microwave region to millimeter or submillimeter waves. Nevertheless, often, the receiver includes one or more frequency conversions (see section 5.1) and in this case f_0 is interpreted as the last intermediate frequency. Double sideband conversion to DC is not covered here so it is always assumed that $f_0 > B/2$.

2.4.1 Received Signal

The microwave receiver of figure 2.9 is modeled, as described in section 1.5.9, particularly using the block diagram on figure 1.12(b). Directly from (1.223) and (1.224), replacing T_s by the antenna temperature T_a and taking into account that the frequency response (1.226) is defined in terms of transducer power gain, the output power is given by

$$P_L(f) = \overline{|b_L(f)|^2} = k\,\mathrm{G}\,|\overline{H}(f)|^2\,[T_a(f) + T_R(f)] \quad [W/Hz] \tag{2.58}$$

where $\mathrm{G} = \mathrm{G}_T(f_0)$ is the transducer power gain at the center frequency f_0 (subscript "T" dropped to simplify notation) and $\overline{H}(f) = H(f)/|H(f_0)|$ the *normalized* frequency response. The precise selection of the center frequency f_0 does not alter (2.58) since it is consistently used in the definitions of both G and $\overline{H}(f)$. Note that the normalization of the frequency response is defined with respect to the absolute value, so the phase of $\overline{H}(f_0)$ is not necessarily zero.

The sum of T_R plus T_a, that is, the noise equivalent temperature of the whole receiver referenced to the antenna plane, is known as "System Temperature." Microwave radiometers are usually designed to measure the incoming radiation in a relatively narrow bandwidth around the center frequency. It is then reasonable to assume that the system temperature is frequency independent within the receiver bandwidth. If needed, an eventual evolution of this parameter with frequency may be taken into account by assigning it to the normalized frequency response and then using a constant value for it. Anyhow, from now on, the system temperature will always be considered independent of frequency,

$$T_{sys} = T_a + T_R \quad \text{System Temperature}\,[K] \tag{2.59}$$

The *time-domain* receiver output signal $\mathrm{b}(t)$ indicated in figure 2.9 is a real, stationary (see section 1.2.1), zero mean, and normal random process. For consistency with (2.58), its phasor is $b_L(f)$, as defined in (1.223). Using now the first equation (1.159), the mean power of $\mathrm{b}(t)$, denoted from now on by the symbol $\mathcal{P}$, is

$$\mathcal{P} = \langle \mathrm{b}^2(t) \rangle = \int_0^{\infty} \overline{|b_L(f)|^2} df \quad [W] \tag{2.60}$$

This is the total integrated power, in Watts, that the receiver delivers to the power detector. It is fully consistent with (1.167) in which the mean output power is expressed as the integration of the power spectrum [$P_L(f)$ in this case] over all positive frequencies. Substituting (2.58) and assuming frequency independent system temperature,

$$\mathcal{P} = kT_{sys}\mathrm{G}B \tag{2.61}$$

where B is the *noise equivalent bandwidth*, in Hz

$$B = \int_0^{\infty} |\overline{H}(f)|^2 df = \frac{1}{\mathrm{G}_T(f_0)} \int_0^{\infty} \mathrm{G}_T(f) df \tag{2.62}$$

The power spectral density of $\mathrm{b}(t)$ is easily deduced from (2.58) and (1.153), taking into account that it must be an even function of frequency,

$$S_{\mathrm{b}}(f) = \frac{1}{2}kT_{sys}\mathrm{G}|\overline{H}(f)|^2 = \frac{\mathcal{P}}{2B}|\overline{H}(f)|^2 \tag{2.63}$$

where (2.61) has been used in the second expression. As expected, integrating this spectral density over the whole frequency range, positive and negative, results in the total power $\mathcal{P}$.

The autocorrelation function of $\mathrm{b}(t)$ is the inverse Fourier transform of (2.63). Since it is a band-pass signal, it is advantageous to use the concept of *complex envelope*. Directly from the first equation (1.148), identifying $\mathrm{x}(t) = \mathrm{b}(t)$ and using (2.63) in (1.149),

$$R_{\mathrm{b}}(\tau) = \mathcal{P}\,\Re e[\tilde{r}(\tau)e^{j2\pi f_0 \tau}] \tag{2.64}$$

where $\tilde{r}(t)$ is a function of time that depends only on the normalized frequency response,

$$\tilde{r}(t) = \frac{1}{B}\mathcal{F}^{-1}[|\overline{H}(f+f_0)|^2 \mathrm{u}(f+f_0)] \tag{2.65}$$

in which $u(f)$ is the unit step function (A.31). The above equation defines a special case of the so-called "fringe washing function," duly generalized for two receivers in section 3.3.2. From (2.64), $\tilde{r}(\tau)$ is recognized as the normalized complex envelope of the autocorrelation function of $b(t)$, $\tilde{R}_b(\tau)/\tilde{R}_b(0)$ and this is why the symbol is a lowercase r with a tilde. Nevertheless, $\tilde{r}(t)$ is related to the receiver, not to the signal. In general, it is a complex function, although it turns into real if the frequency response is symmetrical around the center frequency, which is quite a common situation. Note that always $\tilde{r}(0) = 1$ in virtue of the definition of noise equivalent bandwidth (2.62). Therefore, as expected, the autocorrelation function (2.64) particularized at the origin is the total integrated power

$$R_b(0) = \langle b^2(t) \rangle = \sigma_b^2 = \mathcal{P} \tag{2.66}$$

2.4.1.1 Analytic Signal

The analytic signal of $b(t)$ (see section 1.3) is here denoted by $b(t)$. Its power spectral density is directly obtained from (1.132) after substituting (2.63),

$$S_b(f) = 4S_b(f)u(f) = \frac{2\mathcal{P}}{B}|\overline{H}(f)|^2 u(f) \tag{2.67}$$

and, directly from Fourier inversion, the corresponding autocorrelation function follows

$$R_b(\tau) = 2\mathcal{P}\,\tilde{r}(\tau)\,e^{j2\pi f_0 \tau} \tag{2.68}$$

where $\tilde{r}(\tau)$ is given by (2.65). The value at the origin is

$$R_b(0) = \langle |b(t)|^2 \rangle = \sigma_b^2 = 2\mathcal{P} \tag{2.69}$$

which is twice the total integrated power in agreement with the properties of the analytic signal. The reader is invited to check that the autocorrelation functions (2.64) and (2.68) are consistent with the first of equations (1.148).

2.4.1.2 Frequency Response

In all examples given in this book in relation to radiometric signals, two frequency responses are used: an ideal rectangular filter and a more realistic shape based on the measured response of an actual filter installed in a receiver of the Microwave Imaging Radiometer with Aperture Synthesis (MIRAS) instrument on board satellite SMOS [8]. For the first case, ideal rectangular filter,

$$\overline{H}(f)u(f) = \Pi\left(\frac{f - f_0}{B}\right) \Rightarrow \tilde{r}(t) = \operatorname{sinc} Bt \tag{2.70}$$

where the rectangle function $\Pi(f)$ is defined in (A.33). For the second case $\tilde{r}(\tau)$ is numerically computed from (2.65) using the *measured* normalized response $\overline{H}(f)$.

Figure 2.10 shows the absolute value of both normalized frequency responses as well as the corresponding amplitude of the fringe washing functions. Note that,

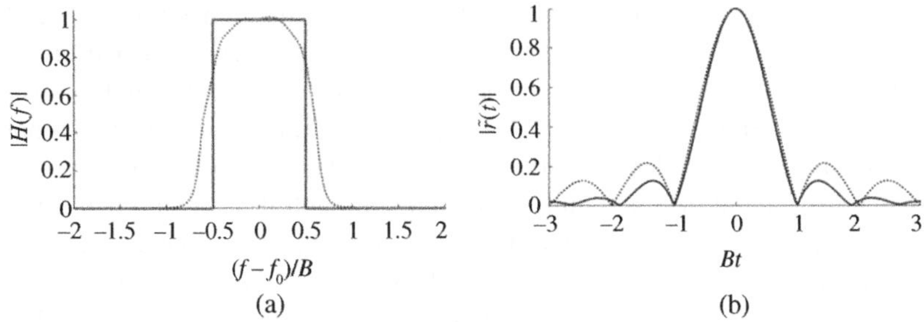

Figure 2.10 (a) Normalized frequency response (amplitude). (b) Amplitude of the fringe washing function (2.65). In both plots, the solid line corresponds to an ideal rectangular filter while the dotted line to a measured frequency response.

even though the measured response is far from rectangular, the main lobe of $\tilde{r}(t)$ approximates well by a sinc function and the first zeroes lay in similar coordinates.

2.4.2 Power Measurement and Sensitivity

The power detector followed by the integrator in figure 2.9 are used to provide at the output port, a random signal (voltage) with a mean proportional to the total integrated power $\mathcal{P}$

$$\langle \mathrm{v}(t) \rangle = C_d \mathcal{P} + \mathrm{v}_o \quad [V] \tag{2.71}$$

where C_d is the detector sensitivity in units of V/W and v_o is the instrumental offset in V. Both parameters depend on the specific hardware used. Substituting (2.61)

$$\langle \mathrm{v}(t) \rangle = \mathcal{G} T_{sys} + \mathrm{v}_o = \mathcal{G}(T_a + T_R) + \mathrm{v}_o \tag{2.72}$$

where the *total radiometric gain* $\mathcal{G}$, in units of V/K, includes all multiplicative contributions

$$\mathcal{G} = k G B C_d \quad \text{Radiometric gain } [V/K] \tag{2.73}$$

Note that, in practice, C_d does not play any role in itself and can always be set equal to one by englobing it in the receiver gain. It will be kept in the subsequent analyses for historical purposes, but in some other parts of the book it will be omitted.

The three receiver parameters ($\mathcal{G}$, T_R, and v_o) are estimated through a process known as "radiometer calibration" covered in section 6.2.6. Once retrieved, they become known *constants*, and the antenna temperature follows by solving (2.72) out of the actual instantaneous measurement

$$\mathrm{T}_a(t) = \frac{\mathrm{v}(t) - \mathrm{v}_o}{\mathcal{G}} - T_R \tag{2.74}$$

where $\mathrm{T}_a(t)$ is a random signal with mean T_a. The uncertainty in the antenna temperature retrieval is the standard deviation of $\mathrm{T}_a(t)$, usually denoted by ΔT,

$$\Delta T = \frac{\sigma_\mathrm{v}}{G} \quad [K] \tag{2.75}$$

where σ_v is the standard deviation of v(t), deduced later [see (2.81)]. This formula, in units of Kelvin, defines the "radiometric sensitivity" or "radiometric resolution," also known as "Noise-Equivalent Delta T" (NEDT or NEΔT). It is interpreted as the minimum change in antenna temperature necessary to produce a detectable change in the output. The impact on ΔT of receiver instability is assessed at the end of this section.

If a directive antenna is used to measure the brightness temperature of an extended source, this one is retrieved by inverting (2.55). Therefore, neglecting the side lobe contribution, the uncertainty in brightness temperature measurement becomes

$$\Delta T_B = \frac{\Delta T}{\eta_b} \tag{2.76}$$

where ΔT is (2.75). Poor main beam efficiency degrades; thus, the radiometric sensitivity. In what follows, this effect will be omitted for simplicity, but it has to be taken into account in practice for precise radiometer characterization.

As indicated in figure 2.9, the output voltage v(t) is the result of averaging the instantaneous power p(t) during a finite integration time τ_i. Time averaging in general is covered in section 1.2.6, where it is shown that the variance of the averaged signal v(t) depends on the power spectral density of the unfiltered signal with mean removed, which in this case is

$$\mathrm{p}'(t) = \mathrm{p}(t) - \langle \mathrm{p}(t) \rangle \tag{2.77}$$

Specifically, assuming an integration time sufficiently large so that $B\tau_i \gg 1$, (1.111) states that the variance of the filtered signal is

$$\sigma_\mathrm{v}^2 \approx \frac{S_{\mathrm{p}'}(0)}{\tau_i} \tag{2.78}$$

The power spectral density of the instantaneous power $S_{\mathrm{p}'}(f)$ depends on the kind of power detector used, either a square law device (section 2.4.3) or a quadratic detector (section 2.4.4). However, as it is demonstrated in these two sections, its value at the origin is the same in both cases and equal to

$$S_{\mathrm{p}'}(0) = \frac{C_d^2 P^2}{B} \alpha_F^2 \tag{2.79}$$

where α_F^2 depends on the filter shape,

$$\alpha_F^2 = \frac{1}{B} \int_0^\infty |\overline{H}(f)|^4 \, df \tag{2.80}$$

For an ideal rectangular filter, $|\overline{H}(f)|$ is equal to one inside the bandwidth and zero otherwise so the integral above is equal to B and then $\alpha_F = 1$. For the real filter with response shown as dotted line in figure 2.10, $\alpha_F^2 = 0.8459$.

Substituting (2.79) into (2.78) and applying the square root results in the following expression for the standard deviation of the output voltage

$$\sigma_{\mathrm{V}} = \frac{C_d P}{\sqrt{B\tau_i}}\alpha_F \tag{2.81}$$

Introducing this result into (2.75), knowing that $C_d P = GT_{sys}$, the following expression for the radiometric sensitivity is obtained:

$$\Delta T = \frac{T_{sys}}{\sqrt{B\tau_i}}\alpha_F = \frac{T_a + T_R}{\sqrt{B\tau_i}}\alpha_F \tag{2.82}$$

This is a fundamental result of microwave radiometers, normally approximated with $\alpha_F = 1$ (rectangular filter). It is used to infer general design guidelines: to reduce ΔT the receiver noise temperature must be kept low and the product of the bandwidth times the integration time large. Nevertheless, untracked long-term evolution of either gain, noise temperature, or offset; or simply the change in brightness temperature due to antenna movement, can put an important limitation in setting the integration time arbitrarily large. In turn, the bandwidth is usually specified by other requirements at system level. An interesting discussion on the interrelation between these parameters and others in imaging radiometers can be found in [7].

2.4.2.1 Dependence on $B\tau_i$

The impact of the approximation $B\tau_i \gg 1$ applied to (2.78) can be assessed by a numerical computation based on the exact version of (1.111), which in this case is

$$\sigma_{\mathrm{V}}^2 = \int_{-\infty}^{\infty} S_{\mathrm{p'}}(f)\,\mathrm{sinc}^2\,(f\tau_i)\,df \tag{2.83}$$

with $S_{\mathrm{p'}}(f)$ derived below in section 2.4.3 or 2.4.4 depending on the type of power detector. As it is seen there, both the square law device and the quadratic detector have the same low-frequency spectrum and only the first one has an additional band at twice the center frequency. Assuming that the integrator frequency response $\mathrm{sinc}^2(f\tau_i)$ washes out the high-frequency band, in both cases the result of (2.83) is exactly the same and it is presented here.

The graphic of figure 2.11(a) shows the low-frequency band of the spectrum of $\mathrm{p}'(t)$. The plot is normalized to have a value equal to α_F^2 at the origin $f = 0$ (1 for the ideal filter and 0.8459 for the real filter).

The spectrum $S_{\mathrm{p'}}(f)$ is now substituted into (2.83), and the integral computed numerically. The result is given in figure 2.11(b), which shows the ratio between

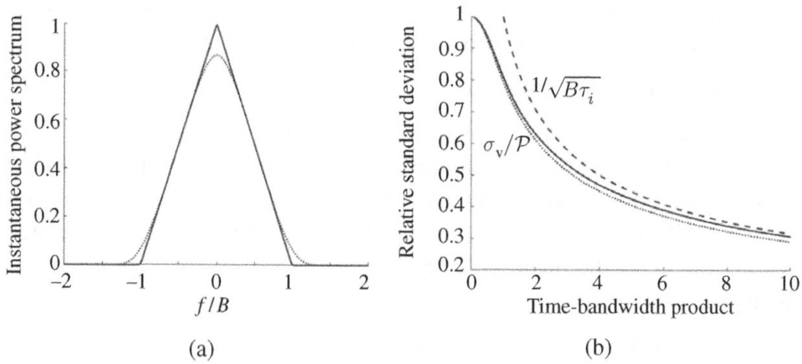

Figure 2.11 (a) Spectrum of instantaneous power. (b) Relative standard deviation as a function of time-bandwidth product. The dashed line is the approximation for large values of time-bandwidth product. In both plots the solid line corresponds to an ideal rectangular filter and the dotted line to a real, measured filter.

the standard deviation of v and the total power $\mathcal{P}$ as a function of $B\tau_i$ for both filters. Also shown in the figure, as a dashed line, is the approximation $1/\sqrt{B\tau_i}$ ($C_d = 1$ has been used without loss of generality). As seen, for a time-bandwidth product of 10 the approximation is already reasonably good in this example. As expected, the behavior is similar in both filters with slightly lower values for the nonideal filter due to a lower value of the constant α_F.

2.4.2.2 On Standard Deviation Reduction

The function of the integrator is to reduce the standard deviation of the detected signal. To quantify this reduction, the standard deviation of the output voltage (2.81) is put in terms of that of the instantaneous power, which depends on the kind of power detector used. The corresponding expressions are (2.92) and (2.103) for a square law device or a quadratic detector, respectively. Then,

$$\sigma_v = \begin{cases} \dfrac{\sigma_p}{\sqrt{2B\tau_i}}\alpha_F & \text{Square law device} \\ \dfrac{\sigma_p}{\sqrt{B\tau_i}}\alpha_F & \text{Quadratic Detector} \end{cases} \tag{2.84}$$

which means that the reduction is larger when a quadratic detector is used. But, in this case, the detected signal is noisier, so the final result is the same: the radiometric resolution is *always* given by the general expression (2.82).

2.4.2.3 Receiver Instability

Random fluctuations of gain, receiver noise temperature and instrumental offset have an impact on measurement uncertainty since they are responsible for output voltage variations, which can be wrongly interpreted as changes in antenna temperature. They can be taken into account by modeling the output voltage as

$$v(t) = g(t)[T_a(t) + T_R(t)] + v_o(t) \tag{2.85}$$

where $T_a(t)$ is the antenna temperature retrieval for a stable receiver (2.74) and $g(t)$, $T_R(t)$, and $v_o(t)$ are real random processes characterizing the receiver parameters' instability. Their respective means are $\langle g(t)\rangle = G$, $\langle T_R(t)\rangle = T_R$ and $\langle v_o(t)\rangle = v_o$. All four processes are assumed independent, hence uncorrelated, so according to (1.5), the mean of (2.85) reproduces (2.72). The variance of $v(t)$ follows by combining (1.14) and (1.24),

$$\sigma_v^2 \approx G^2[(\Delta T)^2 + \sigma_{T_R}^2] + T_{sys}^2\sigma_g^2 + \sigma_{v_o}^2 \tag{2.86}$$

where, as already indicated, ΔT is the standard deviation of $T_a(t)$, the radiometric sensitivity in the absence of receiver fluctuations. The measurement uncertainty (2.75) now becomes

$$\frac{\sigma_v}{G} = \sqrt{(\Delta T)^2 + \sigma_{T_R}^2 + T_{sys}^2\frac{\sigma_g^2}{G^2} + \frac{\sigma_{v_o}^2}{G^2}} \tag{2.87}$$

which, obviously, reverts to (2.75) when the variances of all receiver parameters are set to zero. In general, they are all decreased if receivers are kept on a temperature-controlled environment, and care is taken to reduce bias voltage fluctuations, which are major sources of receiver unstableness. In any case, unless the instrumental offset is very noisy, the last term can be neglected for high-gain receivers. Of the other two, the most important contribution is that of the gain. Keeping only this one and substituting (2.82), equation (2.87) becomes

$$\frac{\sigma_v}{G} = T_{sys}\sqrt{\frac{\alpha_F^2}{B\tau_i} + \left(\frac{\sigma_g}{G}\right)^2} \tag{2.88}$$

which is interpreted as the radiometric sensitivity when random gain fluctuations are included.

Methods to reduce the impact of gain unstability at the expense of worsening the radiometric sensitivity were proposed in the 1940s. They are all based in the so-called "Dicke radiometer" and include fast switching of the receiver input port between the antenna and a reference load at a known temperature. The interested reader may consult for example [1] for a good analysis of the different flavors of this technique, which is completely out of the scope of this book.

2.4.3 Square Law Device

If the power detector of figure 2.9 is a square law device, the instantaneous power $p(t)$ is

$$p(t) = C_d b^2(t) \tag{2.89}$$

with mean proportional to the total integrated power (2.60):

$$\langle \mathrm{p}(t)\rangle = C_d \left\langle \mathrm{b}^2(t)\right\rangle = C_d\mathcal{P} = GT_{sys} \tag{2.90}$$

Analog square law devices do not exist in practice. However this option is applicable to digital radiometers, in which the signal $\mathrm{b}(t)$ is digitized and its square computed numerically [see equation (5.55)].

Since $\mathrm{b}(t)$ is a zero mean normal process, the autocorrelation function of $\mathrm{p}(t)$ can be computed with the help of (1.66) identifying $\mathrm{z}(t) = \mathrm{w}(t) = \mathrm{b}(t)$ and including the constant C_d. Defining $\mathrm{p}'(t) = \mathrm{p}(t) - \langle \mathrm{p}(t)\rangle$ and using (1.49)

$$R_{\mathrm{p}'}(\tau) = 2C_d^2 R_{\mathrm{b}}^2(\tau) \tag{2.91}$$

from which the standard deviation of $\mathrm{p}(t)$ is immediate given that $\sigma_p^2 = R_{\mathrm{p}'}(0)$ and $R_{\mathrm{b}}(0) = \mathcal{P}$,

$$\sigma_{\mathrm{p}} = \sqrt{2}C_d\mathcal{P} \quad \text{(square law device)} \tag{2.92}$$

The power spectral density of $\mathrm{p}'(t)$ is the Fourier transform of (2.91). Using the multiplication property of the Fourier transform (Table A.1) and the definition of the convolution integral applied to functions of frequency (A.42),

$$S_{\mathrm{p}'}(f) = 2C_d^2 S_{\mathrm{b}}(f) * S_{\mathrm{b}}(f) = 2C_d^2 \int_{-\infty}^{\infty} S_{\mathrm{b}}(f')S_{\mathrm{b}}(f - f')\, df' \tag{2.93}$$

Particularizing the above expression for $f = 0$ and taking into account that $S_{\mathrm{b}}(f)$ is an even function of frequency

$$S_{\mathrm{p}'}(0) = 4C_d^2 \int_0^{\infty} S_{\mathrm{b}}^2(f)\, df \tag{2.94}$$

where the dummy variable f' has been substituted by f. Introducing (2.63) in the above equation, (2.79) follows immediately.

A convenient expression for the autocorrelation function of $\mathrm{p}'(t)$ is obtained by introducing (2.64) into (2.91) and expanding the square of the real part [see (A.11)],

$$R_{\mathrm{p}'}(\tau) = C_d^2\mathcal{P}^2|\tilde{r}(\tau)|^2 + C_d^2\mathcal{P}^2\Re e[\tilde{r}^2(\tau)e^{j4\pi f_0\tau}] \tag{2.95}$$

which is consistent with (2.92) when particularized at $\tau = 0$. From this expression it is apparent that the spectrum of $\mathrm{p}'(t)$ has two terms corresponding to the two summands above. The first one is proportional to the Fourier transform of $|\tilde{r}(\tau)|^2$ and therefore is limited to a relatively low-frequency band. The second term, due to the complex exponential multiplying the fringe washing function, corresponds to a band centered at twice the center frequency. This conclusion is confirmed here below using an ideal rectangular filter.

2.4.3.1 Ideal Rectangular Filter

As an academic exercise, useful to understand the concepts just learned, the power spectral densities of the signals involved are here analytically derived for

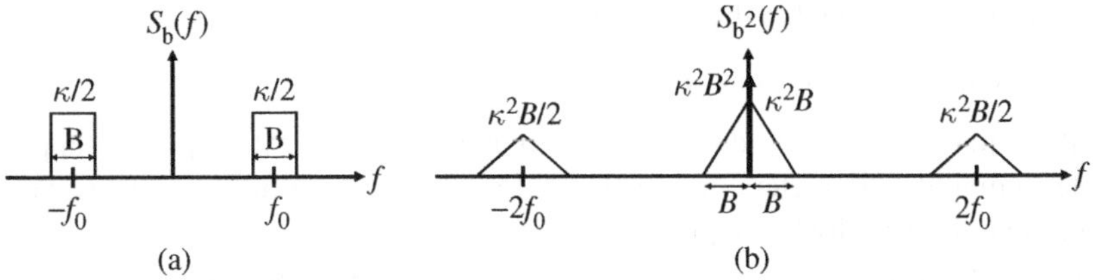

Figure 2.12 Power spectral density of the receiver output signal (a) and of its square (b), assuming an ideal rectangular filter.

a radiometer implemented with an ideal rectangular filter with a response given by (2.70).

The plot of figure 2.12(a) shows the power spectral density of the receiver output signal $S_b(f)$ (2.63). To simplify the labels, the symbol κ is used to denote the spectrum amplitude. In accordance with (2.61),

$$\kappa \triangleq kT_{sys}G = \frac{\mathcal{P}}{B} \tag{2.96}$$

The autocorrelation function of b(t) is (2.64) with $\tilde{r}(\tau) = \text{sinc}\ B\tau$ (2.70)

$$R_b(\tau) = \kappa B\ \text{sinc}\ B\tau\ \cos\ 2\pi f_0\tau \quad \text{(rectangular filter)} \tag{2.97}$$

The autocorrelation function of the instantaneous power (2.89) is obtained by substituting $\tilde{r}(\tau)$ into (2.95) and adding the square of the mean $\langle \text{p}(t)\rangle^2$. Taking for simplicity $C_d = 1$ without loss of generality

$$R_p(\tau) = \kappa^2B^2 + \kappa^2B^2\ \text{sinc}^2\ B\tau + \kappa^2B^2\ \text{sinc}^2\ B\tau \cos\ 4\pi f_0\tau \tag{2.98}$$

where (2.90) and (2.96) have been used. The power spectral density $S_p(f)$ is obtained by Fourier transform of the above expression with the aid of Table A.2 and using the modulation property of Table A.1

$$S_p(f) = \kappa^2B^2\delta(f) + \kappa^2B\Lambda\left(\frac{f}{B}\right) + \frac{\kappa^2B}{2}\Lambda\left(\frac{f-2f_0}{B}\right) + \frac{\kappa^2B}{2}\Lambda\left(\frac{f+2f_0}{B}\right) \tag{2.99}$$

where $\Lambda(f)$ is the triangle function defined in (A.34).

The spectral density (2.99) is drawn in figure 2.12(b). It includes a Dirac delta function at the origin, representing the mean value, a low frequency spectrum and high-frequency bands at twice the center frequency. All bands have a triangular shape of width $2B$. The low-frequency spectrum coincides with the one depicted as a solid line in the plot of figure 2.11(a).

With the aid of this figure and using simple geometry, it is easy to infer that the mean power of b(t) is $\mathcal{P} = \kappa B$ and the mean power of p(t) = b^2(t) is $3\mathcal{P}^2$, equally distributed into DC, low-frequency and high-frequency components ($\mathcal{P}^2$ in each one). The variance of p(t) is the "AC" power, so it is equal to $2\mathcal{P}^2$, fully consistent with (2.92) with $C_d = 1$. Moreover, the spectrum at the origin discarding the "DC" component is equal to $\kappa^2 B = \mathcal{P}^2/B$, consistent with (2.79) with $\alpha_F = 1$ and $C_d = 1$.

2.4.4 Quadratic Detector

In analog systems, the component labeled "power detector" of figure 2.9 is often implemented by a *quadratic detector*, a device that provides a signal proportional to the amplitude squared of the slowly varying envelope of the narrow band input signal. Figure 2.13 shows the block that would substitute the one in figure 2.9 for this component and a practical implementation using a Schottky diode as a half-wave rectifier. For a given range of input signal amplitudes, the output voltage of this device turns out to be proportional to the square of the amplitude, so this circuit implements the block of figure 2.13(a). More sophisticated components can be found in commerce but as long as they implement a quadratic detector, the analysis that follows applies. With this implementation, the detector output (instantaneous power) is

$$\mathrm{p}(t) = C_d \frac{1}{2} |b(t)|^2 \tag{2.100}$$

where $b(t)$ is the analytic signal of $\mathrm{b}(t)$ and the factor $1/2$ is included to take into account that the analytic signal has twice as much power as the original waveform it represents. According to (2.69), the mean value of the detected signal is

$$\langle \mathrm{p}(t) \rangle = C_d \frac{1}{2} \left\langle |b(t)|^2 \right\rangle = C_d \mathcal{P} \tag{2.101}$$

which is the same as (2.90).

In section 1.3.3 it is pointed out that the analytic signals of jointly normal processes and their delayed versions satisfy the circularity condition [(1.33) or (1.71)]. Therefore, the autocorrelation function of $\mathrm{p}(t)$ can be computed with the aid of (1.63) identifying $z(t) = w(t) = b(t)$ and including the factor $C_d/2$. Defining also $\mathrm{p}'(t) = \mathrm{p}(t) - \langle \mathrm{p}(t) \rangle$,

$$R_{\mathrm{p}'}(\tau) = \frac{C_d^2}{4} |R_b(\tau)|^2 \tag{2.102}$$

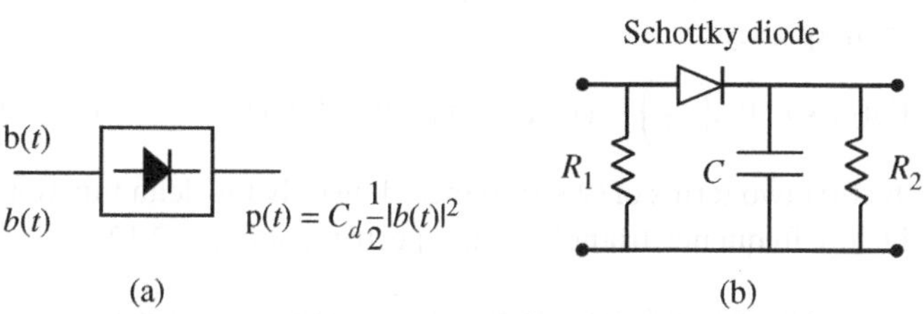

Figure 2.13 A quadratic detector. (a) Block diagram. (b) Example of circuit implementation using a half wave rectifier.

with $R_b(\tau)$ given in (2.68). Using (2.69), the standard deviation of p(t) is equal to the square root of $R_{p'}(0)$, turns out to be

$$\sigma_p = C_d P \quad \text{(quadratic detector)} \tag{2.103}$$

which compared to (2.92) is $\sqrt{2}$ times lower.

The spectrum of p′(t) is the Fourier transform of (2.102)

$$S_{p'}(f) = \frac{C_d^2}{4} S_b(f) * S_b^*(-f) = \frac{C_d^2}{4} \int_{-\infty}^{\infty} S_b(f') S_b^*(-f+f')\, df' \tag{2.104}$$

where the multiplication property of the Fourier transform (table A.1) has been applied. Its value at the origin is

$$S_{p'}(0) = \frac{C_d^2}{4} \int_0^{\infty} |S_b(f)|^2\, df \tag{2.105}$$

where the dummy variable f' has been replaced by f, and it has been taken into account that the spectrum of analytic signals is always zero for negative frequencies. Substituting (2.67) into the above expression, (2.79) follows immediately.

The autocorrelation function of p′(t) is easily obtained by introducing (2.68) into (2.102)

$$R_{p'}(\tau) = C_d^2 P^2 |\tilde{r}(\tau)|^2 \tag{2.106}$$

which is *exactly the same* as the first term of (2.95), corresponding to the low-frequency band of the spectrum when a square law device is used. In other words, in regard to output signal, *a quadratic detector is equivalent to a square law device followed by a low-pass filter* that rejects the high-frequency bands.[4] The standard deviation of the output voltage depends on the spectrum of p′(t) at the origin $f = 0$, which is unchanged whether or not the high-frequency bands are present.

Equation (2.106) also justifies why the standard deviation of the instantaneous power is lower when a quadratic detector is used. It is due to the absence of the high-frequency bands that are responsible for increased "noise" in a square law detector. For example, if the filter has an ideal rectangular response (2.70), the spectrum of the instantaneous correlation is equal to the Fourier transform of (2.106) plus the zero-frequency term

$$S_p(f) = \kappa^2 B^2 \delta(f) + \kappa^2 B \Lambda\left(\frac{f}{B}\right) \quad \text{(Ideal rectangular filter)} \tag{2.107}$$

corresponding to the first two terms of (2.99) so including only the delta function at the origin and the low-frequency triangle of the spectrum of figure 2.12.

4 For this interpretation, a sufficiently high center frequency is assumed to avoid bands overlap.

2.4.4.1 Using a Quadrature Filter

Another way to implement a quadratic detector is by separating the analytic signal into the real and imaginary parts and adding the corresponding powers. Directly from the definition of the analytic signal (1.121), equation (2.100) can be expressed as

$$\mathrm{p}(t) = C_d \frac{1}{2} \mathrm{b}^2(t) + C_d \frac{1}{2} \hat{\mathrm{b}}^2(t) = \mathrm{p}_1(t) + \mathrm{p}_2(t) \tag{2.108}$$

where the hat means Hilbert transform and $\mathrm{p}_1(t)$ and $\mathrm{p}_2(t)$ are self-defined. Since the Hilbert transform does not change the mean power [see (1.141) at $\tau = 0$], it follows that $\langle \mathrm{p}_1(t) \rangle = \langle \mathrm{p}_2(t) \rangle$. Then, directly from (2.60), $\langle \mathrm{p}(t) \rangle = C_d \mathcal{P}$ fully consistent with (2.101).

According to (1.142), a signal and its Hilbert transform are uncorrelated. Knowing that $\mathrm{b}(t)$ and $\hat{\mathrm{b}}(t)$ are jointly normal processes, they are also independent. Hence, their squares are also independent, and the variance of the sum is equal to the sum of variances

$$\sigma_{\mathrm{p}}^2 = \sigma_{\mathrm{p}_1}^2 + \sigma_{\mathrm{p}_2}^2 \tag{2.109}$$

The first term comes directly from (2.92) taking into account that the definition of $\mathrm{p}_1(t)$ in (2.108) is the same as that of $\mathrm{p}(t)$ in (2.89) with an additional factor of $1/2$. Therefore $\sigma_{\mathrm{p}_1}^2 = C_d^2 \mathcal{P}^2/2$. This result is obtained using as starting point the autocorrelation function of $\mathrm{b}(t)$. Since, according to (1.141), the autocorrelation function of the Hilbert transform is equal to that of the signal itself, the same result is obtained for $\sigma_{\mathrm{p}_2}^2$, so (2.109) becomes

$$\sigma_{\mathrm{p}}^2 = \frac{1}{2} C_d^2 \mathcal{P}^2 + \frac{1}{2} C_d^2 \mathcal{P}^2 = C_d^2 \mathcal{P}^2 \tag{2.110}$$

consistent with (2.103), as expected since $\mathrm{p}(t)$ is the same in both (2.100) and (2.108). For the same reason, equation (2.106) also holds in this case.

It is interesting to point out that even though the spectrum of each term in (2.108) includes a high frequency band – in agreement to (2.95) – the spectrum of their sum is limited to low frequency, as deduced from (2.106). An explanation for this apparent contradiction is found by first using (1.136) to write

$$|R_b(\tau)|^2 = 4R_{\mathrm{b}}^2(\tau) + 4R_{\hat{\mathrm{b}}\mathrm{b}}^2(\tau) \tag{2.111}$$

and then expressing each summand in terms of the autocorrelation function of the complex envelope as in (1.148)

$$\begin{aligned} R_{\mathrm{b}}(\tau) &= \frac{1}{2} \mathfrak{Re}[R_{\tilde{b}}(\tau) e^{j2\pi f_0 \tau}] \\ R_{\hat{\mathrm{b}}\mathrm{b}}(\tau) &= \frac{1}{2} \mathfrak{Im}[R_{\tilde{b}}(\tau) e^{j2\pi f_0 \tau}] \end{aligned} \tag{2.112}$$

Substituting these expressions into (2.111) and using identities (A.11) and (A.12) it is found that the components centered at twice the center frequency in both summands cancel each other. The interested reader may want to check it in more detail.

References

1 Fawwaz T. Ulaby, Richard K. Moore, and Adrian K. Fung. *Microwave Remote Sensing. Active and Passive: Fundamentals and Radiometry*, volume I. Artech House, 1981.

2 Joseph T. Verdeyen. *Laser Electronics*. Prentice-Hall, 2nd edition, 1989.

3 Subrahmanyan Chandrasekhar. *Radiative Transfer*. Dover Publications, Inc., 1960.

4 Ignasi Corbella, Lin Wu, Francesc Torres, Nuria Duffo, and Manuel MartNeira. Faraday rotation retrieval using SMOS radiometric data. *IEEE Geoscience and Remote Sensing Letters*, 12(3):458–461, March 2015.

5 Roselena Rubino, Nuria Duffo, Vera Gonzz-Gambau, Ignasi Corbella, Francesc Torres, Israel Dur and Manuel Martin-Neira. Deriving VTEC maps from SMOS radiometric data. *Remote Sensing*, 12(10):1604, May 2020. doi: 10.3390/rs12101604.

6 Ángel Cardama Aznar, Jordi Romeu Robert, Juan Manuel Rius Casals, Lluís Jofre Roca, Sebastián Blanch Boris, and Miguel Ferrando Bataller. *Antenas*. Edicions UPC, Barcelona, 2nd edition, 2002. ISBN 84-8301-625-7.

7 Niels Skou and David M. Le Vine. *Microwave Radiometer Systems: Design and Analysis*. Artech House Publishers, 2nd edition, 2006. ISBN 1-58053-974-2.

8 Kevin McMullan, Michael Brown, Manuel MartNeira, W. Rits, S. Ekholm, J. Marti, and Jerzky Lemanzyk. SMOS: The payload. *IEEE Transactions on Geoscience and Remote Sensing*, 46(3):594–605, March 2008.

3

Interferometry and Polarimetry

Interferometric radiometry is a technique aimed at enhancing the spatial resolution of a radiometer without increasing the size of its antenna. Instruments based on this technique consist of several small, and hence low directive, antennas grouped in pairs, each one collecting simultaneously the incoming radiation and connected to corresponding receivers to measure the cross-correlation between their output signals. Antenna pairs can be formed either by moving or static elements depending on the specific design and application. Correlation measurements are transformed to brightness temperature images through mathematical algorithms.

Polarimetric radiometers are instruments intended to measure the polarimetric brightness temperature of a source (2.18). They are usually formed by a single dual polarization antenna having thus two output signals, and a system to measure the cross-correlation between them. Polarimetric radiometers can also use the interferometric technique to replace the single (large) antenna by a number of smaller ones.

The fundamental measurement in both cases is the cross-correlation between the output signals of two receivers, and this is the main topic of this chapter. Cross-correlation measurement is historically linked to the research on signal interference patterns. Therefore, before dealing with the most technical aspects, a section devoted to the history of interferometry is included.

3.1 Historical Perspective

Interferometry refers to the measurement of interference patterns created by the superposition of two harmonic waves and it is ultimately rooted on the early work of A. Michelson at the end of 19th century who used interferometric measurements to perform astronomical observations at optical wavelengths. When applied to passive remote sensing, both signals are acquired by two sensors

Principles of Interferometric and Polarimetric Radiometry, First Edition. Ignasi Corbella.

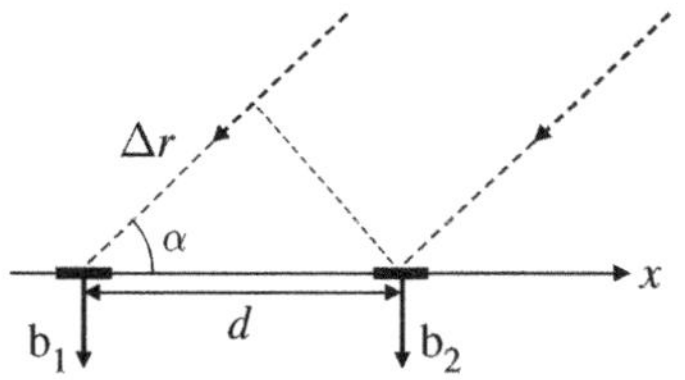

Figure 3.1 Interferometer geometry.

separated at certain length – this set known as a *baseline* – both pointing to the same distant source. In reference to figure 3.1, assuming a quasi-monochromatic wave of frequency f_0, the time-domain signals measured at each observation point are $\mathrm{b}_1(t) = \mathrm{A}\cos 2\pi f_0(t - \Delta r/c)$ and $\mathrm{b}_2(t) = \mathrm{A}\cos 2\pi f_0 t$ where c is the velocity of light and A the peak amplitude, which is allowed to have small random fluctuations in general. The cross-correlation between these two signals (average of their product) is immediate using a trigonometric identity

$$\langle \mathrm{b}_1(t)\mathrm{b}_2(t)\rangle = \frac{\langle \mathrm{A}^2\rangle}{2}\cos 2\pi f_0 \frac{\Delta r}{c} \tag{3.1}$$

Since $\Delta r = d\cos\alpha$, the correlation shows an oscillatory behavior as a function of the angle of arrival α with alternating maxima and minima known as *fringes*. For a given source location, fringes are observed by rotating the baseline. They can also be produced by inserting a variable phase shifter in one of the branches or by other means such as, for example in optical systems, using two holes as sensors and an observation screen behind to implement different spatial phase shifts.

If the source has nonnegligible size, it is modeled by multiple elementary point sources and the correlation becomes the sum of the contributions of all of them, assumed independent from each other. Consider, for example a small (but not point) source located at $\alpha = 90°$. Assuming independent elementary radiators, all with the same mean amplitude, and defining $\xi = \cos\alpha$ (director cosine) and $u = d/\lambda_0$ with $\lambda_0 = c/f_0$ the wavelength, the cross-correlation between output signals becomes

$$\langle \mathrm{b}_1(t)\mathrm{b}_2(t)\rangle = I\int_{-\Delta\xi/2}^{\Delta\xi/2} \cos 2\pi u\xi \, d\xi = I\Delta\xi \,\mathrm{sinc}\, u\Delta\xi \tag{3.2}$$

where $\Delta\xi$ is the total extent of the source, centered at $\xi = 0$, and I is the "intensity," defined such that the signal amplitude of an elementary radiator satisfies $\langle A^2\rangle/2 = I\,d\xi$. Comparing this result with the one for a point source (3.1) at the same location [$\alpha = 90°$, so $\langle \mathrm{b}_1(t)\mathrm{b}_2(t)\rangle = I\,d\xi$], now the correlation becomes multiplied by a factor $\mathrm{sinc}\, u\Delta\xi$, always lower than unity. Therefore, the fringes observed by rotating the baseline (or by other means) have lower amplitude. Eventually, if $u\Delta\xi = 1$, they disappear. In consequence, changing the antenna separation until observing the fringes vanish provides a measurement of the source angular extent, as $\Delta\xi = 1/u$ in this case. This idea was successfully used by Michelson and Pease in 1921 to measure the size of Betelgeuse [1] (second-brightest in the constellation of Orion) and other distant celestial bodies. To this end, Michelson defined the *fringe visibility* as the normalized fringes' amplitude for a given sensor separation. This name –"*visibility*"– is still used today to denote the cross-correlation between the two incoming signals as a function of the antenna relative distance.

A similar effect is observed when the signal is nonmonochromatic. If it has finite bandwidth B, the measured correlation can be expressed as the sum of contributions from all frequency components, assumed independent from each other. If the intensity of one such component is $I = I_f\, df$ with I_f the "spectral intensity," then for a given static geometry,

$$\langle \mathrm{b}_1(t)\mathrm{b}_2(t)\rangle = I_f \int_{f_0-B/2}^{f_0+B/2} \cos 2\pi f \tau \, df = I_f\, B \,\mathrm{sinc}\, B\tau \cos\, 2\pi f_0 \tau \tag{3.3}$$

where $\tau = \Delta r/c$ is the delay in arrival time and f_0 the center frequency. For a nonmonochromatic signal the fringes have no longer a constant amplitude but they are modulated with a function that decreases with the delay between signals, which depends on the angle. In relatively long baselines (large distance between sensors) the correlation measurement may drop rapidly when rotating the system to visualize fringes if the signal bandwidth is large. This is why the modulating term $\mathrm{sinc}\, B\tau$ was originally termed as "fringe washing function," and this name is still used today to include the effect of signal bandwidth in correlation measurements (see section 3.3.2).

The observation of fringes is interesting for understanding the basics behind interferometry and it is a good tool in experiments as those made by Michelson and others. However, to obtain images of distant bodies a much more powerful technique consists of formulating interferometry in terms of complex signals, expressed by means of the *analytic signal* (see section 1.3). Now the two signals coming from the source are $b_1(t) = \mathrm{A} \exp j2\pi f_0(t - \Delta r/c)$ and $b_2(t) = \mathrm{A} \exp j2\pi f_0 t$ where again A is the random amplitude, f_0 the frequency and c the velocity of light. The complex correlation between these tho signals is

$$\frac{1}{2}\left\langle b_1(t) b_2^*(t)\right\rangle = \frac{\left\langle \mathrm{A}^2\right\rangle}{2} e^{-j2\pi f_0 \Delta r/c} \tag{3.4}$$

where the factor $1/2$ is included for consistency with (3.1). This complex correlation has constant amplitude and linear phase variation with respect to the path length difference Δr. Fringes are only observed in either the real or imaginary part.

A finite size source is again modeled as multiple elementary radiators, so that the correlation is the sum of all differential contributors. With the same definitions of u and ξ, now (3.2) becomes

$$\frac{1}{2}\left\langle b_1(t) b_2^*(t)\right\rangle = V(u) = \int_{-\infty}^{\infty} I(\xi) e^{-j2\pi\xi u} d\xi \tag{3.5}$$

where the intensity function $I(\xi)$ is understood to be zero outside the source extent $\pm\Delta\xi/2$. The above equation is easily recognized as a Fourier transform (A.37) where u and ξ play the role of frequency and time, respectively. This result is the starting point of the so-called aperture synthesis technique, devoted to make images of distant bodies derived from correlation measurements between signals collected by sensors located at different relative spacings. The left-hand side of (3.5) is named "visibility" and is complex in general. The real part is measured by real signal correlation as in (3.1) and the imaginary part in the same way but adding a 90° phase shift in one of the signals. Fourier inversion of the visibility, measured as a function of u, results in the source intensity image $I(\xi)$.

For example, consider a source that is constant in a small range of director cosines $\Delta\xi$ and zero outside it (figure 3.2). The intensity is thus a rectangular function and therefore its Fourier transform, the visibility, is a sinc function (see Table A.2). They are both depicted in the figure. Note that the visibility vanishes for a normalized antenna spacing equal to $u = \pm 1/\Delta\xi$, which is precisely the feature that Michelson used to measure the size of Betelgeuse. The intensity image $I(\xi)$ can be estimated by performing an inverse discrete Fourier transform from measured $V(u)$ samples taken at selected antenna spacings.

A mathematical formulation derived by van Cittert (1934) and Zernike (1938) was fundamental in using two-dimensional interferometric measurements for obtaining high-resolution images of distant objects. It is based on describing the radiated signal by a *complex* scalar field (by means of the analytic signal) that is a function of the spatial position. The "mutual intensity" is defined as the cross-correlation of the field at two different spatial positions and the "intensity" as the autocorrelation at a given point. The so-called van Cittert-Zernike theorem states that the mutual intensity, as a function of sensor separation, is equal to the two-dimensional Fourier Transform of the intensity distribution across the

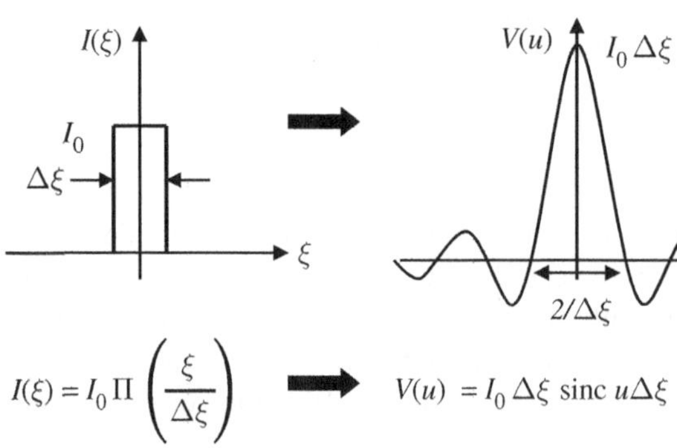

Figure 3.2 Example of the Fourier relationship in a 1-D interferometer imaging a small source at boresight.

source [2]. This theorem extends the simple equation (3.5) to a two-dimensional system and opens the door to get improved images with a consistent formulation.

The van Cittert-Zernike theorem, originally derived for optical wavelengths, was soon applied to Radio Frequency systems, in particular for radioastronomy applications [3]. In this case, optical sensors (telescopes) are substituted by directive antennas so that distant stars can be characterized at the frequency bands of interest by choosing the appropriate receivers' technology. Combining interferometric measurements at different antenna spacings and using Fourier inversion, high-resolution images of celestial bodies have been and are still obtained regularly. An interesting example is the Very Long Baseline Interferometry (VLBI) where two antennas forming a baseline are located at very distant sites on the Earth surface and the Earth rotation is used to change their relative orientation with respect to the source.

Use of interferometric radiometry for the remote sensing of the Earth, either from aircraft or satellite, is more recent. It was initially proposed in [4, 5], where the technique was reformulated using the heritage of Radioastronomy but taking into account the specificities of this application, basically the fact that the source extent is usually very large with respect to the antenna beamwidth. System performance, error analysis, figures of merit, calibration and imaging techniques were comprehensively presented in [6]. A systematic formulation of the visibility equation for Earth observation was rigorously derived in [7] and later extended to the polarimetric case in [8]. These last three references are related to the most emblematic instrument of this kind for Earth observation, the Microwave Interferometric Radiometer with Aperture Synthesis (MIRAS), which is the single payload of the Soil Moisture and Ocean Salinity (SMOS) mission [9] of the European Space Agency (ESA). A good historical review of microwave interferometric radiometry for Earth observation can be found in [10].

3.1.1 The Proposed Formulation

In microwave radiometry the radiated field is wide band. Its power distribution over frequency is characterized by the "brightness temperature" (section 2.1.1), which is in general a smooth function of frequency (resonant peaks exist only for some specific molecules). The collected signal is thus only shaped by the receiver frequency response, which is tailored to meet the required band of interest depending on the specific application. Receiver output signals do have limited bandwidth and therefore they are conveniently described by the analytic signals, but the emitted field is not so well described this way. The complex correlation between output signals is still defined as in (3.4) and the interferometric technique remains valid, but it must by reformulated to account for this conceptual difference. To this end, the present chapter derives the visibility equation as the integral relation between visibility and brightness temperature, both in units of Kelvin, using a *frequency domain description* of the incoming signals based on the concept of "phasor", duly formulated in chapter 1 (section 1.4). Analytic signals are used only after the filtering inherent to the receivers' operation. Under certain restrictive conditions, this visibility equation reduces to a two-dimensional Fourier transform and simplifies the inversion processing.

3.2 A Single Baseline

While a total power radiometer measures the *mean power* of the signal collected by a *single antenna*, the basic measurement of an interferometric radiometer is the *cross-correlation* between the output signals of *two antennas* pointing to the same source of radiation. In interferometric terminology, a set of two antennas is commonly referred to as *a baseline*.

3.2.1 Visibility

The antenna temperature T_a is defined in section 2.3 as the equivalent noise temperature of a receiving antenna modeled as a noisy one-port network. In particular, (2.45) states that $\overline{|b'_a|^2} = kT_a$ where b'_a is the signal collected by the antenna precisely defined in (1.262). In a similar way, for two receiving antennas, the "*antenna visibility*"[1] $\mathcal{V}_{12}$ is here defined as

$$k\mathcal{V}_{12} = \overline{b'_{a1} b'^{*}_{a2}} \quad \text{antenna visibility } [K] \tag{3.6}$$

where b'_{a1} and b'_{a2} are, respectively, the signals collected by each antenna (1.262) and $\overline{b'_{a1} b'^{*}_{a2}}$ the cross-correlation between them. To help finding expressions for the antenna visibility, figure 3.3 shows two antennas and an extended source of thermal radiation.

It is identical to figure 2.5 but adding a second receiving antenna. The normalized electric field (1.270) arriving at each antenna as well as both antenna patterns (1.238) and output signals b'_{a1} and b'_{a2} are duly indicated. Following the methodology of section 2.3.1, these ones are expressed as the integral of all contributions of the electric field emitted by elementary sources and arriving at the antennas from different directions of space (2.44). The visibility follows by substituting both expressions into (3.6) in the same way it is carried out in (2.46) for the antenna temperature.

1 The term visibility will also be used when there is no confusion from the context (see section 3.3.1).

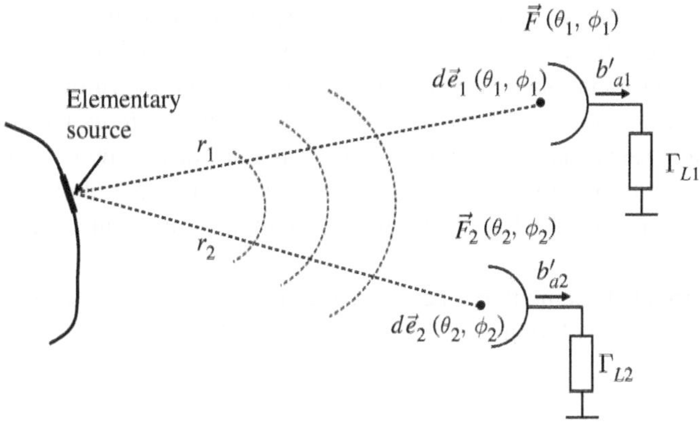

Figure 3.3 Two antennas collecting energy from an extended source of radiation.

The general case, with different antenna orientations and arbitrary separation, is difficult to handle so a simplifying approximation is here proposed. Both antennas are assumed to be located close to each other with respect to the distance to the source and a *common coordinate system* centered near both antennas is defined. Using the spherical coordinates (r, θ, ϕ) defined in appendix (A.4), the following approximations are considered

$$\begin{aligned} \theta_1 &= \theta_2 = \theta \\ \phi_1 &= \phi_2 = \phi \\ d\vec{e}_i(\theta, \phi) &= d\vec{e}(\theta, \phi)e^{-jk(r_i - r)} \quad (i = 1, 2) \end{aligned} \tag{3.7}$$

where $d\vec{e}(\theta, \phi)$ is the normalized electric field arriving at the *center of coordinates* from an elementary source at (r, θ, ϕ) and r_i the distance from this one to the corresponding antenna. In writing this approximation, the electric field radiated by an elementary source is assumed to be a spherical wave (1.236) and the ratio of distances from the source to each antenna is approximated to one. These approximations hold only for large distances to the source and are perfectly reasonable in, for example, instruments sensing the Earth from space or in radioastronomy. In applications where this distance does not allow endorsing them, the formulation provided below fails to be valid. This case, sometimes denoted as "near field interferometer," is out of the scope of this book.

Introducing (3.7) into (2.44), the output signal of each antenna becomes

$$b'_{ai} = \frac{1}{\sqrt{\Omega_{ai}}} \iint \vec{F}_i(\theta, \phi) \odot d\vec{e}(\theta, \phi)e^{-jk(r_i - r)} \quad (i = 1, 2) \tag{3.8}$$

where the integral extends to all elementary sources.

The antenna pattern $\vec{F}_i(\theta, \phi)$ characterizes the corresponding antenna embedded in the whole mechanical structure and includes the effect of the other antenna as well as additional ones that may be present in a given instrument. It is referenced to a local coordinate frame *centered at the antenna position* which, according to section 1.6.6, may be taken nominally as the phase center. This selection ensures a pattern with small phase variation, but a different choice is preferred to simplify imaging. Additional discussion about this important topic is provided in section 4.4.1.

Assuming uncorrelated elementary radiators, the antenna visibility follows immediately by introducing (3.8) into (3.6)

$$k\mathcal{V}_{12} = \frac{1}{\sqrt{\Omega_{a1}\Omega_{a2}}}$$
$$\cdots \iint \overline{\left(\vec{F}_1(\theta,\phi) \odot d\vec{e}(\theta,\phi)\right)\left(\vec{F}_2^*(\theta,\phi) \odot d\vec{e}^*(\theta,\phi)\right)e^{-jk(r_1-r_2)}} \quad (3.9)$$

where, again, the integral extends to all sources. The reader can immediately perceive that this equation is just a generalization of (2.46) when two antennas are considered. Not surprisingly, it converges into it when they both collapse into a single one. Following the same procedure carried out from (2.46), both electric field and antenna patterns are expressed as in (2.47) in terms of two orthogonal polarizations in a common reference frame

$$\begin{aligned} \vec{F}_i(\theta,\phi) &= F_{i_p}(\theta,\phi)\hat{p} + F_{i_q}(\theta,\phi)\hat{q} \quad (i = 1, 2) \\ d\vec{e}(\theta,\phi) &= de_p(\theta,\phi)\hat{p} + de_q(\theta,\phi)\hat{q} \end{aligned} \quad (3.10)$$

where $\hat{p}$ and $\hat{q}$ are arbitrary orthogonal unit vectors contained in a plane normal to the radial direction $\hat{r}$.

3.2.1.1 Unpolarized Field

It is not uncommon that a source radiates an uncorrelated field, so with brightness temperatures satisfying (2.20). It is the case of nadir view of most Earth surfaces (as long as there are no heavily marked corrugations) or radiation coming from distant stars, typical in radioastronomy. Introducing (3.10) into (3.9) and using (2.19) with (2.20), the antenna visibility of an unpolarized source becomes

$$\mathcal{V}_{12} = \frac{1}{\sqrt{\Omega_{a1}\Omega_{a2}}} \iint_{4\pi} T_B(\theta,\phi)\left(\vec{F}_1(\theta,\phi) \cdot \vec{F}_2(\theta,\phi)\right) e^{-jk(r_1-r_2)} d\Omega \quad (3.11)$$

where $T_B(\theta, \phi)$ is the brightness temperature, now polarization independent, of the source and the centered dot denotes Hermitian dot product (A.1). For example, if both antennas have identical polarization vectors, then $\vec{F}_1(\theta,\phi) \cdot \vec{F}_2(\theta,\phi) = F_1(\theta,\phi)F_2^*(\theta,\phi)$. For linearly polarized antennas, in general $\vec{F}_1(\theta,\phi) \cdot \vec{F}_2(\theta,\phi) = F_1(\theta,\phi)F_2^*(\theta,\phi)\cos\alpha$ where α is the rotation angle between the polarization vectors of both antennas. Eventually, if both antennas are perpendicular, the visibility vanishes.

3.2.1.2 Flat Target

An interesting special case is a flat target, defined in section 2.3.2 as an unpolarized thermal source with brightness temperature independent of the direction of observation. From (3.11), the antenna visibility of a flat target at temperature T_{FT} is then

$$\mathcal{V}_{12} = T_{FT}\, FTR_{12} \quad \text{(Flat target)} \tag{3.12}$$

where FTR_{12} is the so-called "Flat Target Response" [11], sometimes also denoted as "1K visibility,"

$$FTR_{12} = \frac{1}{\sqrt{\Omega_{a1}\Omega_{a2}}} \iint_{4\pi} \left(\vec{F}_1(\theta,\phi) \cdot \vec{F}_2(\theta,\phi)\right) e^{-jk(r_1-r_2)} d\Omega \tag{3.13}$$

This expression will be used later to deduce a general property of antennas.

The particular case of a single antenna (1 or 2) is included here for completeness. Taking into account that now $r_1 = r_2$, applying property (1.243), it follows that the single antenna flat target response is always one.

$$FTR_1 = FTR_2 = 1 \tag{3.14}$$

3.2.2 Single Polarization

Assume that both antennas have, in all directions, identical polarization vectors. From (1.238), substituting $\vec{F}_i = F_i \hat{e}_a$ into (3.9) and taking into account (2.19) to relate the normalized electric field to brightness temperature,

$$\mathcal{V}_{12} = \frac{1}{\sqrt{\Omega_{a1}\Omega_{a2}}} \iint_{4\pi} T_B(\theta,\phi) F_1(\theta,\phi) F_2^*(\theta,\phi) e^{-jk(r_1-r_2)} d\Omega \tag{3.15}$$

where T_B is the brightness temperature *at the antenna polarization basis*. The above equation provides the integral relation between brightness temperature and antenna visibility (3.6) for single polarization. As expected, the antenna temperature (2.50) is the particular case of this expression for single antenna.

3.2.2.1 Point Source

Following the same procedure as in section 2.3.3, the visibility (3.15) of a point source is immediate

$$\mathcal{V}_{12} = \frac{\Omega_s}{\sqrt{\Omega_{a1}\Omega_{a2}}} T_s F_1(\theta_s, \phi_s) F_2^*(\theta_s, \phi_s) e^{-jk(r_1-r_2)} \tag{3.16}$$

which obviously becomes (2.53) when both antennas collapse into one. Furthermore, using (2.53), the above result can be written in terms of the corresponding antenna temperatures T_{a1} and T_{a2}

$$\mathcal{V}_{12} = \sqrt{T_{a1}T_{a2}}\, e^{j[\alpha_1(\theta_s,\phi_s)-\alpha_2(\theta_s,\phi_s)]}\, e^{-jk(r_1-r_2)} \tag{3.17}$$

where $\alpha_i(\theta_s, \phi_s)(i = 1, 2)$ are the phases of the antenna patterns in the direction of the source. As seen, the amplitude of the point source visibility is constant, independent of antenna separation and equal to the geometric mean of antenna temperatures. The above expression can be further simplified by assuming the same antenna temperature for both antennas and neglecting the antenna pattern phase difference. In this case, the antenna visibility of a point source at boresight becomes identical to the antenna temperature.

3.2.3 Polarimetric Radiometry: Ideal Case

Assume now that one antenna has polarization p and the other q consistently orthogonal at all spatial directions. For wide beamwidth antennas, this is not possible, but it can be approximately achieved by narrow beams with, for example, two identical linearly polarized antennas rotated 90° one respect to the other. Anyway, it is an interesting academic exercise before treating later the most general case. Introducing $\vec{F}_1 = F_1^p\hat{p}$ and $\vec{F}_2 = F_2^q\hat{q}$ into (3.9), where the superscripts indicate the nominal polarization of the antennas, and using (2.19), the visibility becomes

$$\mathcal{V}_{12}^{pq} = \frac{1}{\sqrt{\Omega_{a1}^p\Omega_{a2}^q}}\iint_{4\pi} T_{B_{pq}}(\theta,\phi)F_1^p(\theta,\phi)F_2^{q*}(\theta,\phi)e^{-jk(r_1-r_2)}d\Omega \tag{3.18}$$

where $T_{B_{pq}}(\theta, \phi)$ is the polarimetric brightness temperature and, consistently, the superscript pq in the visibility makes explicit that it is related to the cross correlation between the output signal of the "p-antenna" and that of the "q-antenna." The antenna temperature of each one is (2.50) with the brightness temperature replaced by $T_{B_p}(\theta, \phi)$ and $T_{B_q}(\theta, \phi)$, respectively.

3.2.3.1 Dual-Polarization Antenna

The above equation also applies to a dual-polarization antenna consisting of a *single aperture* and an "orthomode transducer" (OMT), a device that provides two output signals sensitive each one to the field component at polarization p or q (see figure 3.4). The antenna pattern for port p is now $F^p\hat{p}$ and the one for port q $F^q\hat{q}$. Since in this case $r_1 = r_2$ (3.18) reduces to

$$T_a^{pq} = \frac{1}{\sqrt{\Omega_a^p\Omega_a^q}}\iint_{4\pi} T_{B_{pq}}(\theta,\phi)F^p(\theta,\phi)F^{q*}(\theta,\phi)d\Omega \tag{3.19}$$

where the left-hand side has been renamed to T_a^{pq} to indicate that it is no longer an antenna visibility but a "polarimetric antenna temperature."

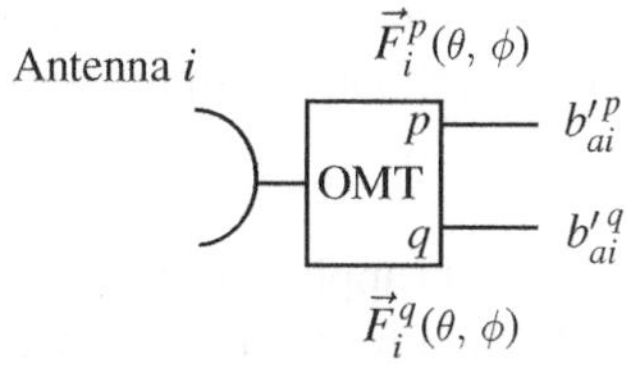

Figure 3.4 A dual polarization antenna using an orthomode transducer (OMT). Output signals are defined as in (1.262).

A polarimetric radiometer is based on this equation in the same way as a total power radiometer relies on (2.50). Assuming a highly directive antenna pointing to an extended source, an approximation similar to (2.55) can be applied leading to

$$T_a^{pq} = T_{B_{pq}}(\theta_0, \phi_0)\eta_b^{pq} + T_{SL} \tag{3.20}$$

where T_{SL} is the contribution of the side (including back) lobes and η_b^{pq} the main beam efficiency defined for the two patterns at orthogonal polarization

$$\eta_b^{pq} = \frac{1}{\sqrt{\Omega_{a1}\Omega_{a2}}} \iint_{\text{MB}} F^p(\theta, \phi) F^{q*}(\theta, \phi) d\Omega \tag{3.21}$$

with MB standing for main beam, consistently defined to be applicable to the product of both patterns. If both, antenna beam efficiency and side lobe contribution, are known, either from estimation or measurement, they can be used to infer the polarimetric brightness temperature $T_{B_{pq}}$ out of the correlation measurement implicit in T_a^{pq}.

A polarimetric radiometer can also be interferometric by using antennas at different locations. The polarimetric antenna visibility for a single baseline is (3.18), and the polarimetric antenna temperature of a single antenna is (3.19). From them, imaging techniques are applied to retrieve the polarimetric brightness temperature with the methods described in section 4.4.

3.2.4 Full Polarimetric Case

The most general case consists of two dual polarization antennas, each one including an OMT (figure 3.4) to get two different output signals, one for each polarization. The superscripts p and q in the figure are labels to differentiate both ports. *Nominally*, the one labeled p is mainly sensitive to the p component of the impinging electric field with some contribution from the other. And similarly for the q port. Nevertheless, in the formulation that follows, there is *no restriction* regarding this association between port and polarization. The contribution of the q component of the electric field to the signal at port p needs not to be minor.

For any combination of the four output signals from both antennas, the antenna visibility (3.6) is now conveniently denoted as

$$\overline{b'^p_{a1} b'^{q*}_{a2}} = k\, \mathcal{V}_{12}^{pq} \quad \text{Polarimetric Visibility} \tag{3.22}$$

which is fully compatible with (3.18). As it is apparent from the above definition, the polarimetric visibility satisfies the following *conjugation property*

$$\mathcal{V}_{12}^{pq} = \left(\mathcal{V}_{21}^{qp}\right)^* \tag{3.23}$$

obviously applicable also to the single polarization case, $\mathcal{V}_{12} = \mathcal{V}_{21}^*$

Each output port has associated a *different* antenna pattern that can be decomposed into two components as in (3.10)

$$\begin{aligned} \vec{F}_i^p(\theta,\phi) &= F_{i_p}^p(\theta,\phi)\hat{p} + F_{i_q}^p(\theta,\phi)\hat{q} \\ \vec{F}_i^q(\theta,\phi) &= F_{i_p}^q(\theta,\phi)\hat{p} + F_{i_q}^q(\theta,\phi)\hat{q} \end{aligned} \quad (i = 1, 2) \tag{3.24}$$

Antennas are said to have infinite cross-coupling isolation if the cross-terms above $F_{i_q}^p(\theta,\phi)$ and $F_{i_p}^q(\theta,\phi)$ vanish, and this is the situation in which (3.18) holds. It is not considered now.

Antenna polarization is determined by unit vectors $\hat{p}$ and $\hat{q}$, which can be any of the ones defined in section 1.6.3 or others that one may deem to be appropriate to the particular situation. It is often convenient to use the co-polar and cross-polar patterns (1.254) directly from measurements, but then care must be taken to apply appropriate sign changes to the above equations. The reader must be aware that $\hat{p}$ and $\hat{q}$ are defined in the *common system* of coordinates defined for both antennas, which may be different from any "local system" that could be useful to define the nominal polarization of each antenna individually, and that can vary from one to another.

Introducing now the definitions (3.22) into (3.9) and using also (3.24) and (2.19), the following general expression for the polarimetric visibility is found

$$\begin{aligned} \mathcal{V}_{12}^{pq} = \frac{1}{\sqrt{\Omega_{a1}^p \Omega_{a2}^q}} \iint_{4\pi} \Big[& F_{1_p}^p(\theta,\phi) F_{2_p}^{q*}(\theta,\phi)\, T_{B_p}(\theta,\phi) \\ & + F_{1_q}^p(\theta,\phi) F_{2_q}^{q*}(\theta,\phi)\, T_{B_q}(\theta,\phi) \\ & + F_{1_p}^p(\theta,\phi) F_{2_q}^{q*}(\theta,\phi)\, T_{B_{pq}}(\theta,\phi) \\ & + F_{1_q}^p(\theta,\phi) F_{2_p}^{q*}(\theta,\phi)\, T_{B_{qp}}(\theta,\phi) \Big] e^{-jk(r_1 - r_2)} d\Omega \end{aligned} \tag{3.25}$$

where the antenna equivalent solid angles (1.243) for each pattern are

$$\Omega_{ai}^p = \iint_{4\pi} \left[|F_{i_p}^p(\theta,\phi)|^2 + |F_{i_q}^p(\theta,\phi)|^2 \right] d\Omega \quad (i = 1, 2) \tag{3.26}$$

and a similar equation for Ω_{ai}^q. The reader may want to check that (3.25) is compliant with the conjugation property (3.23). As expected, if the cross-terms of (3.24) vanish, then it simplifies to either (3.15) for single polarization (superscripts *pp* or *qq*) or (3.18) for the polarimetric case *pq*. Also, for an unpolarized field (2.20), it reduces to (3.11). Equation (3.25) was initially derived in [12].

3.2.4.1 Individual Antennas

The antenna temperature for each of the two ports of figure 3.4 is straightforwardly obtained from (3.25) just by making $r_1 = r_2$ and using patterns from one single antenna, so for superscripts pp for $T^p_{a_i}$ or qq for $T^q_{a_i}$ ($i = 1, 2$). The resultant expressions coincide with (2.48) with the appropriate inclusion of the antenna number so they are not repeated here.

The polarimetric antenna temperature (3.19) for each antenna is also derived from (3.25) with $r_1 = r_2$ but now with different superscripts,

$$T^{pq}_{a_i} = \frac{1}{\sqrt{\Omega^p_{ai}\Omega^q_{ai}}} \iint_{4\pi} \left[F^p_{i_p}(\theta,\phi) F^{q*}_{i_p}(\theta,\phi)\, T_{B_p}\theta,\phi) \right.$$
$$+ F^p_{i_q}(\theta,\phi) F^{q*}_{i_q}(\theta,\phi)\, T_{B_q}(\theta,\phi)$$
$$+ F^p_{i_p}(\theta,\phi) F^{q*}_{i_q}(\theta,\phi)\, T_{B_{pq}}(\theta,\phi)$$
$$\left. + F^p_{i_q}(\theta,\phi) F^{q*}_{i_p}(\theta,\phi)\, T_{B_{qp}}(\theta,\phi) \right] d\Omega \tag{3.27}$$

with $i = 1, 2$. An interferometric polarimetric radiometer measures, for each element, the total power and polarimetric antenna temperatures, and for each baseline the polarimetric visibilities. The whole set of brightness temperatures, at both polarizations and polarimetric, are then obtained through mathematical inversion (see section 4.4.6).

If the above equation is now used instead of (3.19) for a polarimetric radiometer with a *single* narrow beam antenna, equation (3.20) becomes

$$\begin{bmatrix} T^p_a \\ T^q_a \\ T^{pq}_a \\ T^{qp}_a \end{bmatrix} = \begin{bmatrix} \eta^{pp}_{pp} & \eta^{pp}_{qq} & \eta^{pp}_{pq} & \eta^{pp}_{qp} \\ \eta^{qq}_{pp} & \eta^{qq}_{qq} & \eta^{qq}_{pq} & \eta^{qq}_{qp} \\ \eta^{pq}_{pp} & \eta^{pq}_{qq} & \eta^{pq}_{pq} & \eta^{pq}_{qp} \\ \eta^{qp}_{pp} & \eta^{qp}_{qq} & \eta^{qp}_{pq} & \eta^{qp}_{qp} \end{bmatrix} \begin{bmatrix} T_{B_p}(\theta_0,\phi_0) \\ T_{B_q}(\theta_0,\phi_0) \\ T_{B_{pq}}(\theta_0,\phi_0) \\ T_{B_{qp}}(\theta_0,\phi_0) \end{bmatrix} + \begin{bmatrix} T^p_{SL} \\ T^q_{SL} \\ T^{pq}_{SL} \\ T^{qp}_{SL} \end{bmatrix} \tag{3.28}$$

where the last column vector is the contribution from the side lobes and the η parameters are the beam efficiencies defined as

$$\eta^{pq}_{\alpha\beta} = \frac{1}{\sqrt{\Omega^p_a\Omega^q_a}} \iint_{\mathrm{MB}} F^p_\alpha(\theta,\phi) F^{q*}_\beta(\theta,\phi) d\Omega \tag{3.29}$$

with α and β conveniently substituted by p or q. Note that the antenna number i is not included since a single one is used in this case. Dual polarization and polarimetric brightness temperatures can thus be retrieved by a simple inversion as long as the matrix of main beam efficiencies and the side lobe contributions are known.

3.2.5 Receivers Interaction

If the two antennas of an interferometer are close enough to each other, there may exist a nonnegligible interaction between both. Noise generated by one receiver is captured by the antenna of the other and the visibility measurement is affected. This is particularly important in instruments designed for Earth observation, where short baselines are required to reduce aliasing (see section 4.3.3).

Assume that both antennas are connected to corresponding microwave receivers, as shown in figure 3.5. Mutual interaction is included by modeling the system formed by the antennas and the whole space around them as a linear two-port network characterized by an S-parameter matrix (1.206), so the following relations hold

$$\begin{aligned} a_1^{(1)} &= S_{11}b_1^{(1)} + S_{12}b_1^{(2)} + b_{a1} \\ a_1^{(2)} &= S_{21}b_1^{(1)} + S_{22}b_1^{(2)} + b_{a2} \end{aligned} \tag{3.30}$$

where b_{a1} and b_{a2} are the intrinsic noise waves at antenna ports, generated in this case by the distant source. In this model, the antennas themselves are assumed lossless, so they do not contribute to thermal noise. As indicated in section 1.6.11, antenna loss can be assigned to a lossy network in cascade, which can be formally embedded in the corresponding receiver so no generality is lost. Due to the reciprocity theorem (section 1.6.9), forward and backward transmission parameters are always identical $S_{12} = S_{21}$ although different symbols will be kept for clarity. In some situations, for example, if the antennas are very distant from each other, they can be neglected and these equations become decoupled. The full equations are here considered.

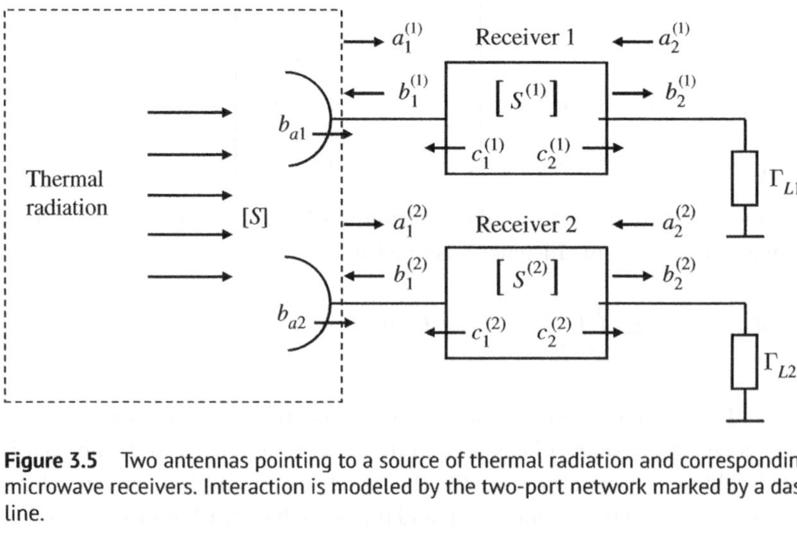

Figure 3.5 Two antennas pointing to a source of thermal radiation and corresponding microwave receivers. Interaction is modeled by the two-port network marked by a dashed line.

The first difficulty encountered in this setup is establishing a suitable definition for both antenna temperature and visibility. One consistent with (2.45) and (3.6) is

$$kT_{ai} = \frac{\overline{|b_{ai}|^2}}{1 - |\Gamma_{ai}|^2} \quad (i = 1, 2) \tag{3.31}$$

$$k\mathcal{V}_{12} = \frac{\overline{b_{a1} b_{a2}^*}}{\sqrt{1 - |\Gamma_{a1}|^2}\sqrt{1 - |\Gamma_{a2}|^2}} \tag{3.32}$$

where Γ_{ai} is the input reflection coefficient of the corresponding antenna (not shown in the figure to avoid cluttering), which in practice can be approximated by S_{ii} even for relatively close antennas [see (1.185) or (1.188)]. An important conceptual difference of (3.31) with respect to (2.45) is that it can no longer be interpreted as "available power" since this one depends on the noise temperature of the other receiver [see (3.41)].

3.2.5.1 Flat Target

If both antennas are enclosed in an anechoic chamber at a constant temperature T_{FT} (flat target) the intrinsic waves b_{a1} and b_{a2} must satisfy the Bosma theorem (1.233) so,

$$\begin{aligned} T_{a1} &= T_{FT}\frac{1 - |S_{11}|^2 - |S_{12}|^2}{1 - |S_{11}|^2} \\ T_{a2} &= T_{FT}\frac{1 - |S_{22}|^2 - |S_{21}|^2}{1 - |S_{22}|^2} \qquad \text{(Flat target)} \\ \mathcal{V}_{12} &= \frac{-T_{FT}(S_{11}S_{21}^* + S_{12}S_{22}^*)}{\sqrt{1 - |S_{11}|^2}\sqrt{1 - |S_{22}|^2}} \end{aligned} \tag{3.33}$$

where the antenna reflection coefficients have been substituted by S_{ii}. The slight inconsistency of the first two equations with (2.52) is due to the already mentioned fact that the available power of each antenna has a contribution from the power leaked from the other one. This effect will be normally neglected. On the other hand, by substituting (3.12) and (3.13) in the last equation, the following general expression is found:

$$\begin{aligned} &\frac{1}{\sqrt{\Omega_{a1}\Omega_{a2}}}\iint_{4\pi}\left(\vec{F}_1(\theta,\phi)\cdot\vec{F}_2(\theta,\phi)\right) e^{-jk(r_1 - r_2)} d\Omega \\ &= \frac{-(S_{11}S_{21}^* + S_{12}S_{22}^*)}{\sqrt{1 - |S_{11}|^2}\sqrt{1 - |S_{22}|^2}} \end{aligned} \tag{3.34}$$

This is a general property of antenna patterns valid for any situation. It ultimately relies on energy conservation, and it is valid for any pair of antennas independently of their design. It was first derived in [13] and [14] using the convention of section 1.6.10 for antenna parameters. The expression provided in these two references is compatible with (3.34) when (1.284) is substituted.

3.2.5.2 System Visibility

The second issue to be decided in the setup of figure 3.5 is the precise definition of the signals involved in the cross-correlation measurement. This is here done by a generalization of the autocorrelation (power) defined in an individual receiver. Neglecting the contribution of the loads' noise, the mean output power of each receiver is (1.222)

$$P_i = \overline{|b_2^{(i)}|^2}(1 - |\Gamma_{Li}|^2) \quad (i = 1, 2) \tag{3.35}$$

where Γ_{Li} is the load reflection coefficient of each receiver (see figure 3.5). Consistently, the output cross-correlation (or cross-power) of two receivers is here defined as

$$P_{12} = \overline{b_2^{(1)} b_2^{(2)*}} \sqrt{(1 - |\Gamma_{L1}|^2)(1 - |\Gamma_{L2}|^2)} \tag{3.36}$$

which reduces to (3.35) for a single receiver.

Receivers are modeled as active two-port networks characterized by S-parameter matrices $\boldsymbol{S}^{(i)}$. From (1.206) and assuming negligible reverse gain $S_{12}^{(i)} \approx 0$, the receivers' outward waves in both ports become

$$\begin{aligned} b_1^{(i)} &= S_{11}^{(i)} a_1^{(i)} + c_1^{(i)} \\ b_2^{(i)} &= S_{21}^{(i)} a_1^{(i)} + S_{22}^{(i)} a_2^{(i)} + c_2^{(i)} \end{aligned} \quad (i = 1, 2) \tag{3.37}$$

Since the noise generated by the receiver load is neglected, waves $a_2^{(i)}$ above are directly equal to $\Gamma_{Li} b_2^{(i)}$ so the second equation becomes

$$b_2^{(i)} = \Lambda_i' \left(S_{21}^{(i)} a_1^{(i)} + c_2^{(i)} \right) \quad (i = 1, 2) \tag{3.38}$$

where $\Lambda_i' = 1/(1 - S_{22}^{(i)} \Gamma_{Li})$ is a parameter similar to that defined in (1.209), useful to simplify the final expressions. The outward waves $b_2^{(i)}$ are obtained by introducing (3.30) into (3.38) and operating as in section 1.5.7. The procedure is quite straightforward, and the result can be expressed elegantly in matrix form [15]. Here, instead, the equations are simplified by assuming that

$$S_{11}^{(i)} = 0 \; ; \; \mathcal{T}_{12}^{(i)} = 0 \quad (i = 1, 2) \tag{3.39}$$

where $\mathcal{T}_{12}^{(i)}$ is defined in (1.213), along with $\mathcal{T}_1^{(i)}$ and $\mathcal{T}_2^{(i)}$ used later. The above approximations are reasonably fulfilled in most systems, especially if the receivers include a good isolator at their input, and they are useful to simplify expressions for which the exact versions could obscure a suitable interpretation. Later, their impact will be illustrated by an example (see figure 3.6).

The first equation (3.37) now reduces to $b_1^{(i)} = c_1^{(i)}$ and the power and cross-correlation of the outward waves (3.38) becomes

$$\begin{aligned}
\overline{|b_2^{(1)}|^2} &= |\Lambda_1'|^2 |S_{21}^{(1)}|^2 \left(\overline{|b_{a1}|^2} + |S_{11}|^2 k\mathcal{T}_1^{(1)} + |S_{12}|^2 k\mathcal{T}_1^{(2)} + k\mathcal{T}_2^{(1)} \right) \\
\overline{|b_2^{(2)}|^2} &= |\Lambda_2'|^2 |S_{21}^{(2)}|^2 \left(\overline{|b_{a2}|^2} + |S_{22}|^2 k\mathcal{T}_1^{(2)} + |S_{21}|^2 k\mathcal{T}_1^{(1)} + k\mathcal{T}_2^{(2)} \right) \\
\overline{b_2^{(1)} b_2^{(2)*}} &= \Lambda_1' \Lambda_2'^* S_{21}^{(1)} S_{21}^{(2)*} \left(\overline{b_{a1} b_{a2}^*} + S_{11} S_{21}^* k\mathcal{T}_1^{(1)} + S_{12} S_{22}^* k\mathcal{T}_1^{(2)} \right)
\end{aligned} \tag{3.40}$$

These equations (with $\Gamma_{Li} = 0$ and hence $\Lambda_i' = 1$) were originally derived in [14], where the term $\overline{|b_{ai}|^2}$ was used to re-define the antenna temperature and $\overline{b_{a1} b_{a2}^*}$ the visibility. Later in [15], this last one was renamed antenna visibility. However, in order to gain compatibility with the case of a single antenna, these definitions are *not supported* here. Instead, (3.40) is introduced into (3.35) and (3.36) and then rewritten in terms of antenna temperature (3.31) and visibility (3.32), receiver noise temperature (1.214), now designated by T_R as in (1.219), and transducer power gain G_T (1.186). Taking into account that $\Gamma_{out}^{(i)} = S_{22}^{(i)}$ (due to having negligible $S_{12}^{(i)}$) and that (3.39) implies $\Gamma_{ai} = S_{ii}$, for receiver number 1 (3.35) becomes

$$P_1 = G_{T1} \left(T_{a1} + T_{R1} + \frac{|S_{12}|^2 \mathcal{T}_1^{(2)}}{1 - |S_{11}|^2} \right) = kG_{T1} T_{sys1} \tag{3.41}$$

which is equivalent to (1.221), applicable to an isolated receiver, but with the added contribution of the noise coupled from the other. The last term inside the parenthesis is recognized as the available power at port 1 due to the noise generated by receiver 2. In most practical situations, it can be neglected, although it represents a small increase in system temperature with respect to that of a single receiver (2.59) due to noise coupling.

The cross-power (3.36) can be written in a similar way from the third equation of (3.40) by introducing the receivers' frequency responses defined in (1.226) based on the corresponding transducer power gains

$$H_i = \sqrt{G_{Ti}} e^{\psi_{21}^{(i)}} \quad (i = 1, 2) \tag{3.42}$$

where $\psi_{21}^{(i)}$ is the phase of the frequency response. Then

$$P_{12} = kH_1 H_2^* \mathbb{V}_{12} \tag{3.43}$$

in which $\mathbb{V}_{12}$, here designated as *system visibility*, is

$$\mathbb{V}_{12} = \mathcal{V}_{12} + \frac{S_{11} S_{21}^* \mathcal{T}_1^{(1)} + S_{12} S_{22}^* \mathcal{T}_1^{(2)}}{\sqrt{1 - |S_{11}|^2} \sqrt{1 - |S_{22}|^2}} \tag{3.44}$$

where $S_{ii} = \Gamma_{ai}$ due to the assumption (3.39). The system visibility of a flat target is obtained by substituting $\mathcal{V}_{12}$ by (3.33), so it vanishes when $\mathcal{T}_1^{(1)} = \mathcal{T}_1^{(2)} = T_{FT}$.

The system visibility is a concept analog to the system temperature, as it includes the contribution from both the external source and the receivers' noise. To simplify the notation, the subscript *sys*, common to denote system temperature, is not used in the system visibility, but it is distinguished from the antenna visibility by a different font.

> **Remark**: From the definitions (3.35) and (3.36) introduced into (3.41) and (3.43), respectively; and from (3.31) and (3.32) the following conventions can be established:
>
> $$\begin{aligned} \mathbb{V}_{ii} &= T_{sysi} \quad (i = 1, 2) \\ \mathcal{V}_{ii} &= T_{ai} \quad (i = 1, 2) \end{aligned} \tag{3.45}$$
>
> which justifies their denomination as system and antenna visibilities, respectively. However, (3.44) only applies to two *different* receivers since it fails to account for the receiver internal noise if only one is included.

3.2.5.3 A Simulation

As stated, equation (3.44) holds only if the approximation (3.39) is valid and the reverse S-parameters of the receivers $S_{12}^{(i)}$ are neglected. It is interesting then to compare (3.44) with the system visibility deduced by (3.43) with P_{12} obtained from (3.36) by a complete circuit simulation. The result of this exercise for a particular case is presented here. The interested reader may take this idea to simulate other systems that he/she has designed.

Receivers in this example use the components described in reference [15], corresponding to actual units mounted in the SMOS MIRAS instrument. Figure 3.3 of this reference shows the S-parameters and temperatures (1.213) of a low noise amplifier and figure 3.4 those of an isolator. Two different setups are analyzed based on the block diagram of figure 3.5. The first one assumes the amplifiers' input ports directly connected to the antennas, and the second one includes isolators in between. Slightly different subsystems are included in each branch and the loads Γ_{Li} are set to arbitrary different values (0.13 and 0.15, respectively), constant in frequency. Finally, the equivalent two-port network characterizing the antennas and the thermal radiation are defined by

$$\begin{aligned} &S_{11} = 0.1\angle 10^\circ \quad S_{22} = 0.15\angle 20^\circ \quad S_{21} = 0.013\angle -30^\circ \\ &\overline{|b_{a1}|^2}/k = 210K \quad \overline{|b_{a2}|^2}/k = 212K \quad \overline{b_{a1}b_{a2}^*}/k = 5\angle 60^\circ K \end{aligned} \tag{3.46}$$

which are reasonable values for a typical baseline pointing to an extended source.

The cross-correlation between the output waves $b_2^{(i)}$ is computed using the methodology of [16], and the result is introduced into (3.36) an then into (3.43) to get the system visibility, which is compared to (3.44). The receivers' frequency responses are computed according to (1.226). The result is provided in figure 3.6

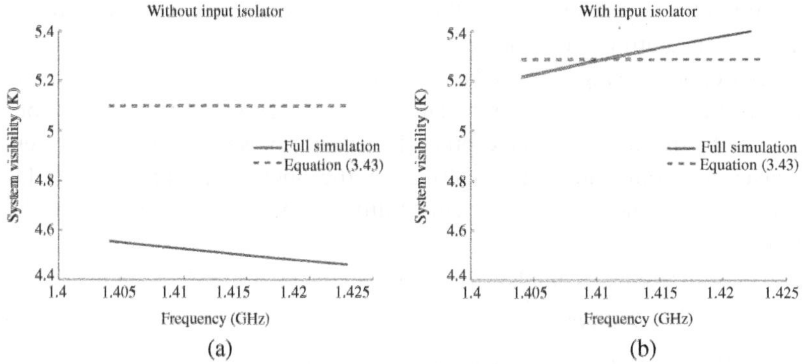

Figure 3.6 Simulation of the system visibility (amplitude). (a) Antenna directly connected to amplifiers. (b) Intercalating isolators.

as a function of frequency for the SMOS nominal band. The plot on figure 3.6(a) shows the case of direct connection between antennas and amplifiers while that of figure 3.6(b) when isolators are included. As seen, there is a small frequency-dependent error in the system temperature, which is highly reduced when the receivers have input isolators. Needless to say, if an ideal isolator is used, both curves overlap, as expected. The different levels of system visibility in figure 3.6(a) and (b) plots are due to the fact that, when an isolator is used the equivalent temperature of the intrinsic outward wave at input becomes the physical temperature $\mathcal{T}_1^{(i)} \approx T_{ri}$, set to 298 and 315 K for $i = 1, 2$, respectively, in this example.

3.2.6 The "$-T_r$" Term

The so-called $-T_r$ term is a way to include antennas cross-coupling and matching into the system visibility by introducing (3.34) into (3.44). If receivers include isolators at their input port, both kept at the same physical temperature T_r, then $\mathcal{T}_1^{(1)} = \mathcal{T}_1^{(2)} = T_r$ and (3.44) becomes

$$\mathbb{V}_{12} = \mathcal{V}_{12} - \frac{T_r}{\sqrt{\Omega_{a1}\Omega_{a2}}} \iint_{4\pi} (\vec{F}_1(\theta,\phi) \cdot \vec{F}_2(\theta,\phi))\, e^{-jk(r_1-r_2)} d\Omega \tag{3.47}$$

Substituting now the single polarization expression (3.15),

$$\mathbb{V}_{12} = \frac{1}{\sqrt{\Omega_{a1}\Omega_{a2}}} \iint_{4\pi} (T_B(\theta,\phi) - T_r) F_1(\theta,\phi) F_2^*(\theta,\phi) e^{-jk(r_1-r_2)} d\Omega \tag{3.48}$$

which in some papers is referred to as "Corbella equation" and was experimentally validated in [17]. As seen, the net effect of the antenna coupling and matching can be accounted for by just subtracting the receivers physical temperature from the brightness temperature of the source.

The restriction of using input isolators is needed to ensure that (3.44) correctly represents the system visibility, defined in terms of the cross-correlation of output signals (see figure 3.6). However, even in the absence of isolators, if the receivers themselves are compliant with (3.39), then the above expression still holds provided the equivalent backward temperature at input is used instead of the physical one.

The $-T_r$ term should be applied, in general, to the polarimetric visibility (3.25),

$$\mathbb{V}_{12}^{pq} = \mathcal{V}_{12}^{pq} - T_r\, FTR_{12}^{pq} \tag{3.49}$$

where FTR_{12}^{pq} is the flat target response (3.13) for antennas p and q,

$$FTR_{12}^{pq} = \frac{1}{\sqrt{\Omega_{a1}^{p}\Omega_{a2}^{q}}} \iint_{4\pi} \left(\vec{F}_1^{p}(\theta,\phi)\cdot\vec{F}_2^{q}(\theta,\phi)\right) e^{-jk(r_1-r_2)} d\Omega \tag{3.50}$$

The polarimetric antenna temperature is obtained from $T_{a_i}^{pq}$ (3.27) in the same way, which means making $r_1 = r_2$ in the above equation. In this case, however, both superscripts must be *different* in agreement with the remark given after equation (3.44).

The reader may be willing to expand (3.49) taking into account the definition (A.1) and notice that the receiver temperature T_r is only subtracted from the single polarization brightness temperatures but not from the polarimetric ones. The resultant expressions are not given here to avoid excess repetition of long equations. As expected, all of them vanish for a flat target (2.51) at temperature T_r.

> **Comment**: In systems where the antennas are very well matched and weakly coupled, the right-hand side of (3.34) becomes negligible and the $-T_r$ term can be neglected. For example, in radioastronomy, where antennas are usually far away from each other and mutual coupling is negligible, it is never included. Otherwise, for instruments with close antennas, it must be removed by a calibration procedure based on measuring a flat target (see section 6.2.4).

3.3 The Visibility Equation

The fundamental operation of an interferometric radiometer consists of inverting the so-called "visibility equation." It relates the brightness temperature of the source to the *measurable* visibility, retrieved from the complex correlation measurement of the *analytic signals* corresponding to the receivers' output waveforms. This section is devoted to precisely formulate this equation while its inversion is covered in chapter 4.

3.3.1 Complex Correlation

The real-time-domain signals that are present at the receivers' output ports of figure 3.5 are zero mean normal stochastic processes, generally designated in what follows by $b_i(t)$ with i the receiver number (1 or 2). This is compatible with (2.60) in total power radiometers and is based on modeling the receivers as in section 1.5.9. Following (1.223), the phasor of $b_i(t)$ is[2] $b_i = b_2^{(i)}\sqrt{1-|\Gamma_{Li}|^2}$, with output power $\overline{|b_i|^2}$ consistent with (3.35). Similarly, the cross-power P_{12} (3.36) is the cross-correlation between phasors $\overline{b_1 b_2^*}$. Therefore, using (3.43),

$$\overline{b_1(f)b_2^*(f)} = kH_1(f)H_2^*(f)\mathbb{V}_{12}(f) \qquad [W/Hz] \tag{3.51}$$

where the explicit dependence on frequency has been included.

Cross-correlation between phasors cannot be directly measured. The actual measurement is instead the cross-correlation between the corresponding *analytic signals* (see section 1.3). Using the second equation (1.161),

$$\frac{1}{2}\left\langle b_1(t)b_2^*(t)\right\rangle = k\int_0^{\infty} H_1(f)H_2^*(f)\,\mathbb{V}_{12}(f)df \qquad [W] \tag{3.52}$$

where $b_i(t)$ is the analytic signal corresponding to the phasor $b_i(f)$. This convention, where explicit time or frequency dependency is used to differentiate both concepts, was already introduced in sections 1.3 and 1.4.

The companion equation for a total power radiometer (2.60) is simplified in section 2.4.1 by assuming the system temperature independent of frequency within the receiver's bandwidth. In (3.52), the system visibility cannot be taken outside the integral since it includes the complex exponential that depends on frequency. Introducing (3.48) into (3.52), interchanging the order of integration and assuming brightness temperature and antenna patterns constant within the bandwidth,

$$\frac{1}{2}\left\langle b_1(t)b_2^*(t)\right\rangle = \frac{k}{\sqrt{\Omega_{a1}\Omega_{a2}}}\iint_{4\pi}(T_B(\theta,\phi)-T_r)F_1(\theta,\phi)F_2^*(\theta,\phi)$$

$$\cdots\int_0^{\infty} H_1(f)H_2^*(f)e^{-jk(r_1-r_2)}df\, d\Omega \tag{3.53}$$

The above expression can be written in a more compact form using the so called "*fringe washing function*" [see also section 3.3.2]

$$\tilde{r}_{12}(t) = \frac{e^{-j2\pi f_0 t}}{\sqrt{B_1B_2}}\int_0^{\infty}\bar{H}_1(f)\bar{H}_2^*(f)e^{j2\pi f t}df \tag{3.54}$$

where f_0 is the center frequency, the bar over the frequency responses means normalization with respect to their amplitude at f_0, $\bar{H}_i(f) = H_i(f)/|H_i(f_0)|$ and

2 The subscript L is dropped for simplicity.

B_i is the noise equivalent bandwidth already defined in (2.62)

$$B_i = \int_0^{\infty} |\bar{H}_i(f)|^2 df \tag{3.55}$$

As already pointed out in section 2.4.1, the choice of center frequency f_0 is arbitrary, but it is convenient to use the *nominal* frequency at which the receivers are designed, independently of the symmetry of their band with respect to it. In any case, the same must be used for both receivers.

The fringe washing function at the origin is a calibration parameter that depends on the dissimilarity between receivers. It is then useful to introduce the *normalized* version of the function

$$\bar{\tilde{r}}_{12}(t) = \frac{\tilde{r}_{12}(t)}{\tilde{r}_{12}(0)} \tag{3.56}$$

With all these definitions and denoting $\mathrm{G}_i = \mathrm{G}_{Ti}(f_0)$, equation (3.53) becomes

$$\frac{1}{2}\left\langle b_1(t) b_2^*(t)\right\rangle = k\sqrt{B_1 B_2}\sqrt{\mathrm{G}_1 \mathrm{G}_2}\,\tilde{r}_{12}(0) V_{12} \tag{3.57}$$

where V_{12} is now the "measurable" system visibility

$$V_{12} = \frac{1}{\sqrt{\Omega_{a1}\Omega_{a2}}} \iint_{4\pi} (T_B(\theta, \phi) - T_r) F_1(\theta, \phi) F_2^*(\theta, \phi) \cdots \bar{\tilde{r}}_{12}\left(-{(r_1 - r_2)}/{c}\right) e^{-jk_0(r_1 - r_2)} d\Omega \tag{3.58}$$

in which k_0 is the wave number at the receivers' center frequency: $k_0 = 2\pi f_0 / c$.

This *measurable* visibility differs from (3.48) in the presence of the normalized fringe washing function inside the integral. As explained in section 3.3.2, for similar filters this function has a smooth behavior, with a maximum at the origin and zero at the inverse of the signal bandwidth. Therefore, if the source extent is limited in such a way that, in all the field of view, $|r_1 - r_2|/c \ll 1/B$, then the normalized fringe washing function inside (3.58) can be approximated to unity and then

$$V_{12} \approx \mathbb{V}_{12}(f_0) \quad \text{(small source)} \tag{3.59}$$

where $\mathbb{V}_{12}(f)$ is the system visibility (3.48). If, additionally, the receivers interaction, and hence the $-T_r$ term, is neglected, then the measurable visibility reduces to the antenna visibility (3.15) $V_{12} \approx \mathcal{V}_{12}(f_0)$. In general, this approximation is not contemplated in the remaining of this book.

To improve text readability, the measurable system visibility V_{12} is, from now on, referred to as simply "visibility," except when needed to differentiate with $\mathcal{V}_{12}$ (3.6) or $\mathbb{V}_{12}$ (3.44). Note that these two visibilities are function of frequency while V_{12} is a constant since the complex exponential of (3.58) is evaluated at the center frequency.

Needless to say, since the fringe washing function is an effect of the receivers' limited bandwidth, it must also be included in the polarimetric visibility (3.25) or polarimetric antenna temperature (3.27).

For completeness, the mean power out of each receiver (2.61) is here rewritten in terms of the analytic signals in order to provide an equation similar to (3.57):

$$\frac{1}{2}\left\langle |b_i(t)|^2 \right\rangle = kB_i G_i T_{sysi} \quad (i = 1, 2) \tag{3.60}$$

where T_{sysi} is the system temperature (2.59) of the corresponding receiver. By comparing with (3.57), it is apparent that the system temperature coincides with the visibility when both antennas collapse into one ($\tilde{r}_{ii}(0) = 1$). This result is consistent with the remark (3.45).

3.3.2 The Fringe Washing Function

Given the frequency responses of two filters, the fringe washing function with respect to an *arbitrary* reference frequency f_0 is (3.54). In terms of the inverse Fourier transform (A.38), it can be written as

$$\begin{aligned}\tilde{r}_{12}(t) &= \frac{e^{-j2\pi f_0 t}}{\sqrt{B_1 B_2}} \mathcal{F}^{-1}\left[\bar{H}_1(f)\bar{H}_2^*(f)\mathrm{u}(f)\right] \\ &= \frac{1}{\sqrt{B_1 B_2}} \mathcal{F}^{-1}\left[\bar{H}_1(f+f_0)\bar{H}_2^*(f+f_0)\mathrm{u}(f+f_0)\right]\end{aligned} \tag{3.61}$$

where $\mathrm{u}(\cdot)$ is the unit step function (A.31) and the bar over the frequency response means normalization with respect to their absolute value at f_0. The second expression above makes use of the frequency translation property of the Fourier transform.

The fringe washing function is thus the inverse Fourier transform of the conjugated product of the receivers' frequency responses considering only the positive part of the spectrum and shifted to the origin by an amount equal to f_0. Note that, taking a look at figure 1.4, the fringe washing function can be interpreted as a complex envelope and this is the reason why a tilde is included in its symbol [see comment after equation (2.65)].

For a single filter the fringe washing function is defined by collapsing in (3.54) both frequency responses into one:

$$\tilde{r}(t) = \frac{e^{-j2\pi f_0 t}}{B} \int_0^\infty |\bar{H}(f)|^2 e^{j2\pi f t} df \tag{3.62}$$

from which it turns out that $\tilde{r}(0) = 1$. This function is the same as (2.65), used in section 2.4.1 in the context of total power radiometers.

The fringe washing function is in general complex and has the following properties, most of them inherited from those of the Fourier transform (see Table A.1 for reference):

1. Swapping the filter order implies both conjugation and sign change $\tilde{r}_{21}(t) = \tilde{r}_{12}^*(-t)$.

2. If the product $\bar{H}_1(f)\bar{H}_2^*(f)$ is real, then $\tilde{r}_{12}(-t) = \tilde{r}_{12}^*(t)$, so the real part is even and the imaginary part is odd. This happens either if both receivers have identical frequency responses or if a single receiver is considered (3.62).
3. If both frequency responses are symmetrical around the reference frequency (that is, if $\bar{H}_i(f_0 + \Delta f) = \bar{H}_i(f_0 - \Delta f)$), then the fringe washing function is real.
4. Inserting time delays in the receivers is equivalent to introducing linear phase shifts in the corresponding frequency responses. For delays t_1 and t_2

$$H_1'(f) = H_1(f)e^{-j2\pi f t_1} \; ; \; H_2'(f) = H_2(f)e^{-j2\pi f t_2} \tag{3.63}$$

   Using these expressions in (3.61) along with the time delay property of the Fourier transform results in the fringe washing function corresponding to the filters with added time delay

$$\tilde{r}_{12}'(t) = \tilde{r}_{12}(t - \Delta t)e^{-j2\pi f_0 \Delta t} \tag{3.64}$$

   where $\tilde{r}_{12}(t)$ is the fringe washing function corresponding to the original filters and $\Delta t = t_1 - t_2$ is the added time delay difference between both. Note that adding the same delay to both ($t_1 = t_2$) does not change the fringe washing function. Adding a delay only to receiver 1 means $\Delta t = t_1$ while adding it only to receiver 2 means $\Delta t = -t_2$.
5. Changing the reference frequency of the fringe washing function adds a linear term to the phase with no change in amplitude:

$$\tilde{r}_{12}(t; f_c) = \tilde{r}_{12}(t; f_0)e^{-j2\pi(f_c - f_0)t} \tag{3.65}$$

6. If the product $\bar{H}_1(f)\bar{H}_2^*(f)$ is approximated by a rectangular function of width B centered at f_0, then the fringe washing function referenced to this frequency is

$$\tilde{r}_{12}(t; f_0) = \frac{B}{\sqrt{B_1 B_2}} \operatorname{sinc} Bt \approx \operatorname{sinc} Bt \tag{3.66}$$

   where the last expression holds for identical filters $B_1 = B_2 = B$. For filters centered on another frequency, (3.65) applies.
7. Independently of the reference frequency, the fringe washing function at the origin is

$$\tilde{r}_{12}(0) = \frac{1}{\sqrt{B_1 B_2}} \int_0^\infty \bar{H}_1(f)\bar{H}_2^*(f)df \tag{3.67}$$

   which is complex in general and normally with amplitude close to unity (if filters are similar enough). For two *identical* filters, $\tilde{r}_{12}(0) = 1$.

The fringe washing function can be estimated from measured or simulated frequency responses. A good approximation is the last term of (3.66), strictly corresponding to two identical rectangular filters of bandwidth B. In the visibility equation (3.58) it is evaluated at $t = -\Delta r/c$, so it produces a "decorrelation" of the signals due to delay in their arrival time to each antenna. It can be ignored for instruments imaging small sources or if the bandwidth is small, but otherwise it has to be taken into account in the processing.

The name "fringe washing" makes reference to the effect that this function has on the correlation measurement of a point source. Proceeding with (3.58) in a similar way as in the derivation of (3.17) and neglecting the "$-T_r$" term, the measurable visibility of a point source becomes

$$V_{12} = \frac{\Omega_s}{\Omega_a} T_s \tilde{r}_{12} \left(-{}^{r_1-r_2}/_{c}\right) e^{-jk_0(r_1-r_2)} \tag{3.68}$$

Using now the approximation (3.66), the real part of the measurable visibility, which is the one having "fringes," becomes

$$\Re e[V_{12}] = \frac{\Omega_s}{\Omega_a} T_s \operatorname{sinc}(B\Delta t) \cos k_0(r_1 - r_2) \tag{3.69}$$

where $\Delta t = (r_1 - t_2)/c$ is the time delay between the signals arriving at both sensors. This equation is the same as (3.3) deduced in the historic introduction of this chapter. The fringe washing function is effectively reducing the fringes' amplitude and hence its name. When $B = 1/\Delta t$ the fringes disappear whereas they have constant amplitude for $B = 0$ (sinusoidal signals)

3.3.3 Director Cosines

Since a common system of coordinates is implicit in all equations, the distance from each antenna to a source element located at coordinates (x, y, z) is

$$r_i = \sqrt{(x - x_i)^2 + (y - y_i)^2 + (z - z_i)^2} \quad (i = 1, 2) \tag{3.70}$$

where (x_i, y_i, z_i) are the antenna coordinates. Assuming the source at a distance r large with respect to the antennas coordinates, this equation can be approximated using Taylor linear expansion as

$$r_i \approx r - \left(\frac{x}{r}x_i + \frac{y}{r}y_i + \frac{z}{r}z_i\right) = r - (\xi x_i + \eta y_i + \gamma z_i) \tag{3.71}$$

where $\xi = x/r$, $\eta = y/r$, and $\gamma = z/r$ are the *director cosines* of the source location with respect to the instrument coordinate system. In terms of elevation and azimuth angles (A.17), they are

$$\xi = \sin\theta\cos\phi \quad \eta = \sin\theta\sin\phi \quad \gamma = \cos\theta \tag{3.72}$$

Using the approximation (3.71) and assuming that both antennas are placed on the (x, y) plane ($z_1 = z_2 = 0$), the complex exponential of the visibility equation (3.58) can be approximated by

$$k_0(r_1 - r_2) \approx 2\pi(\xi u + \eta v) \tag{3.73}$$

where u and v are the antenna differential coordinates normalized to the wavelength:

$$u = \frac{x_2 - x_1}{\lambda_0} \qquad v = \frac{y_2 - y_1}{\lambda_0} \tag{3.74}$$

where $\lambda_0 = c/f_0$ is the wavelength at the center frequency. *Important*: note that the order of the receiver numbers in the definition of u and v is swapped with respect to the distance difference.

The visibility equation can now be rewritten in terms of these variables after applying the corresponding coordinate transformation. Using the Jacobian of (3.72),

$$d\xi d\eta = \left\| \begin{matrix} \frac{\partial \xi}{\partial \theta} & \frac{\partial \xi}{\partial \phi} \\ \frac{\partial \eta}{\partial \theta} & \frac{\partial \eta}{\partial \phi} \end{matrix} \right\| d\theta d\phi = \sin\theta \cos\theta d\theta d\phi \tag{3.75}$$

the differential solid angle (A.20) becomes

$$d\Omega = \frac{d\xi d\eta}{\sqrt{1 - \xi^2 - \eta^2}} \tag{3.76}$$

where the term $\cos\theta = \sqrt{1 - \xi^2 - \eta^2}$ is often called "obliquity factor." Introducing this change of variables into (3.58) results in

$$V_{12} = \frac{1}{\sqrt{\Omega_{a1}\Omega_{a2}}} \iint_{\xi^2+\eta^2<1} \frac{T_B(\xi,\eta) - T_r}{\sqrt{1-\xi^2-\eta^2}} F_1(\xi,\eta) F_2^*(\xi,\eta)$$

$$\tilde{\bar{r}}_{12}\left(-{}^{\xi u + \eta v}/_{f_0}\right) e^{-j2\pi(\xi u + \eta v)} d\xi d\eta \tag{3.77}$$

where the integration area $\xi^2 + \eta^2 < 1$ is usually referred to as "unit circle." Proceeding similarly for the antenna temperature (2.50)

$$T_{a_i} = \frac{1}{\Omega_{ai}} \iint_{\xi^2+\eta^2<1} \frac{T_B(\xi,\eta)|F_i(\xi,\eta)|^2}{\sqrt{1-\xi^2-\eta^2}} d\xi d\eta \quad (i = 1, 2) \tag{3.78}$$

These are the equations that must be solved to retrieve the brightness temperature out of the measurements of antenna temperature and visibility (see section 4.4). Needless to say, the change of variables applies also to the polarimetric case (3.49), including the flat target response (3.50).

Remark: The integration region $\xi^2 + \eta^2 \leq 1$ corresponds actually to two zones: the top ($z > 0$) and the bottom ($z < 0$) hemispheres. Therefore all the integrals in (ξ, η) have to be understood as the sum of two, one for each hemisphere. If the orientation and design of the antennas are such that the patterns are negligible in the back hemisphere, then the corresponding term can be neglected. Before applying these equations it is important to choose the coordinates such that this approximation holds.

3.3.4 Fourier Relation

If both antenna patterns happen to be *identical* (which, strictly speaking, is not possible due to the asymmetrical interaction of each antenna with its neighbor), then (3.77) becomes independent of the particular unit and depends only on the variables (u, v). If, furthermore, the $-T_r$ term is neglected or canceled out and the differential path decorrelation is neglected (fringe washing function approximated to unity), then (3.77) simplifies to

$$V(u, v) = \iint_{\xi^2+\eta^2<1} \frac{T_B(\xi, \eta)t(\xi, \eta)}{\Omega_a\sqrt{1-\xi^2-\eta^2}} e^{-j2\pi(\xi u+\eta v)} d\xi d\eta \tag{3.79}$$

where $t(\xi, \eta) = |F(\xi, \eta)|^2$ is the – now unique – antenna power pattern (1.239) and Ω_a the antenna equivalent solid angle (1.243). Subscripts 1 and 2 are now unneeded and have not included.

The above equation is recognized as a two-dimensional Fourier transform (A.46) with the convention that the brightness temperature outside the unit circle $\xi^2 + \eta^2 < 1$ is zero. It can thus be written as

$$V(u, v) = \mathscr{F}\left[T'_B(\xi, \eta)\right] \tag{3.80}$$

where $T'_B(\xi, \eta)$ is the "modified brightness temperature"

$$T'_B(\xi, \eta) = \frac{T_B(\xi, \eta)t(\xi, \eta)}{\Omega_a\sqrt{1-\xi^2-\eta^2}} \quad (\xi^2 + \eta^2 < 1) \tag{3.81}$$

The antenna temperature (3.78) is in this case the same for both antennas and equal to the visibility at the origin $V(0, 0) = T_a$.

Equation (3.80) is the Fourier relationship predicted by the van Cittert-Zernike theorem between mutual intensity and intensity. The present derivation points out all the approximations and conditions required for this equation to hold. For a small source around boresight the obliquity factor can be neglected and, provided the antenna beamwidth is much wider than the source extent, the antenna pattern may be considered constant. With these added conditions, the modified brightness temperature (3.81) becomes equal to the brightness temperature divided by the antenna solid angle $T'_B(\xi, \eta) \approx T_B(\xi, \eta)/\Omega_a$.

3.4 Correlation Measurement

Equation (3.57) states that the cross-correlation between the analytic signal representation of the receiver output waveforms is proportional to the measurable visibility (3.58). This is the basic measurement of an interferometric radiometer and also applies to a polarimetric radiometer just by substituting the visibility by the polarimetric antenna temperature (3.27). For convenience, (3.57) is here rewritten as

$$\Gamma_{12} \triangleq \frac{1}{2}\left\langle b_1(t)b_2^*(t)\right\rangle = G_{12}V_{12} \tag{3.82}$$

where G_{12} is a complex constant interpreted as "correlator gain" in units of W/K:

$$G_{12} = k\sqrt{B_1B_2}\sqrt{\mathrm{G}_1\mathrm{G}_2}\,\tilde{r}_{12}(0) \tag{3.83}$$

Similarly, the total output power of each receiver (3.60) is

$$\mathcal{P}_i \triangleq \frac{1}{2}\left\langle |b_i(t)|^2\right\rangle = G_iT_{sysi} \qquad (i = 1, 2) \tag{3.84}$$

where G_i is the radiometric gain

$$G_i = kG_iB_i \qquad (i = 1, 2) \tag{3.85}$$

equivalent to (2.73) with $C_d = 1$. Now this constant is not included since the forthcoming analysis is done in terms of mean power and cross-correlation. The hardware-related constants that relate them to the corresponding measurements can be included at a later stage for specific implementations. It is useful to point out here that the above definitions imply that $G_{12} = \sqrt{G_1G_2}\tilde{r}_{12}(0)$.

3.4.1 Sensitivity

Figure 3.7 shows a conceptual block diagram of the complex correlation measurement. Both receivers include high gain amplifiers and band-pass filters

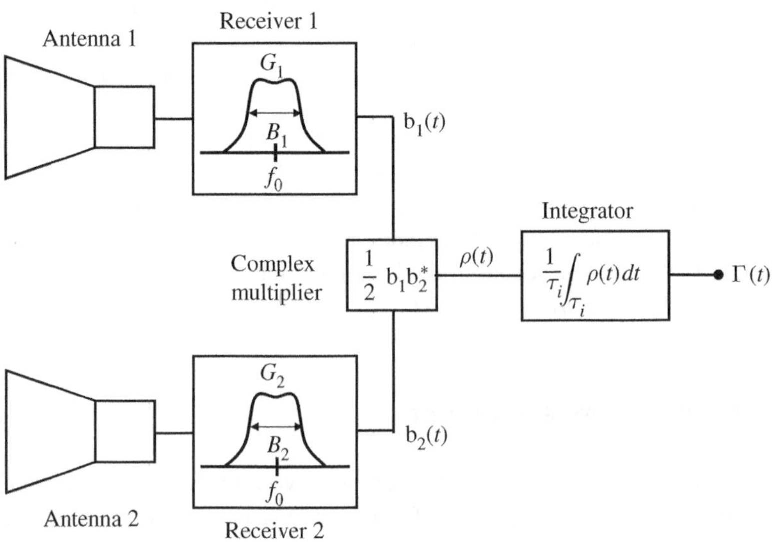

Figure 3.7 Conceptual block diagram of cross-correlation measurement.

with center frequency f_0 and may or not include frequency conversions (see section 5.1). The frequency responses are defined consistently by (3.42) so in each case the amplitude squared is the transducer power gain. More details about the receivers' output signals $b_1(t)$ and $b_2(t)$, including their spectral densities and autocorrelation functions can be consulted in section 2.4.1 in the context of total power radiometers [in particular, see (2.67) and (2.68), respectively]. As indicated in figure 3.7, cross-correlation is measured by an ideal complex signal multiplier followed by an integrator with limited integration time τ_i. Details about the practical implementation of the complex multiplier are given later.

For a given receiver pair, the correlator gain (3.83) is estimated by a calibration process (see section 6.2.1) and then the visibility is deduced *at each measurement epoch* from (3.82)

$$V_{12}(t) = \frac{\Gamma(t)}{G_{12}} \tag{3.86}$$

where $\Gamma(t)$ is the integrator output signal, a random process with mean Γ_{12}. The uncertainty in the visibility measurement, or sensitivity, is defined – similar to (2.75) for the antenna temperature – as the standard deviation of $V_{12}(t)$:

$$\Delta V_{12} = \frac{\sigma_\Gamma}{|G_{12}|} \tag{3.87}$$

where σ_Γ is the standard deviation of $\Gamma(t)$. This signal is the result of averaging during an integration time τ_i the "instantaneous cross-correlation"[3]

$$\rho(t) = \frac{1}{2} b_1(t) b_2^*(t) \tag{3.88}$$

which, obviously, has the same mean as that of $\Gamma(t)$,

$$\langle \rho(t) \rangle = \langle \Gamma(t) \rangle = \Gamma_{12} = G_{12} V_{12} \tag{3.89}$$

Knowing that the analytic signals of jointly normal processes are circular (see remark at the end of section 1.3.3), the autocorrelation function of $\rho(t)$ is directly given by (1.63) when $z(t)$ and $w(t)$ are identified with $b_1(t)$ and $b_2(t)$, respectively, and the factor 1/2 is included. Defining, furthermore, $\rho'(t) = \rho(t) - \langle \rho(t) \rangle$ and using (1.49),

$$R_{\rho'}(\tau) = \frac{1}{4} R_{b_1}(\tau) R_{b_2}^*(\tau) \tag{3.90}$$

This equation is the generalization of (2.102) for two different signals (with $C_d = 1$), so the procedure below is essentially the same as the one carried out from this equation. The standard deviation of $\rho(t)$ is immediate from $R_{\rho'}(0)$ using (3.84)

$$\sigma_\rho = \sqrt{\mathcal{P}_1 \mathcal{P}_2} \tag{3.91}$$

---

3 For simplicity renamed to instantaneous correlation from now on.

which could also have been obtained directly from (1.64). Following the methodology of section 1.2.6, the variance of $\Gamma(t)$ is approximately equal to the power spectral density of $\rho'(t)$ at $f = 0$ divided by the integration time. The Fourier transform of (3.90) is straightforward using the multiplication and conjugation properties (see Table A.1)

$$S_{\rho'}(f) = \frac{1}{4} S_{b_1}(f) * S^*_{b_2}(-f) = \frac{1}{4} \int_{-\infty}^{\infty} S_{b_1}(f') S^*_{b_2}(-f + f')\, df' \tag{3.92}$$

and then the variance of the output signal follows from (1.111)

$$\sigma_\Gamma^2 \approx \frac{S_{\rho'}(0)}{\tau_i} = \frac{1}{4\tau_i} \int_0^{\infty} S_{b_1}(f) S^*_{b_2}(f)\, df \tag{3.93}$$

where the dummy variable f' has been substituted by f. Introducing now the spectral densities of the receiver output signals (2.67),

$$\sigma_\Gamma = \frac{\sqrt{P_1 P_2}}{\sqrt{B\tau_i}} \alpha_F \tag{3.94}$$

where the following definitions have been used

$$\begin{aligned} B &= \sqrt{B_1 B_2} \\ \alpha_F^2 &= \frac{1}{B} \int_0^{\infty} |\overline{H}_1(f)|^2 |\overline{H}_2(f)|^2\, df \end{aligned} \tag{3.95}$$

For two identical filters, $B_1 = B_2 = B$ and α_F becomes identical to (2.80), defined for a single filter. If, furthermore, the filters have ideal rectangular shape, then $\alpha_F = 1$.

Substituting now (3.94) into (3.87), taking into account that $G_{12} = \sqrt{G_1 G_2} \tilde{r}_{12}(0)$ and using (3.84) to express the total output powers as function of system temperatures, the visibility uncertainty results in

$$\Delta V_{12} = \frac{\sqrt{T_{sys1} T_{sys2}}}{\sqrt{B\tau_i}} \frac{\alpha_F}{|\tilde{r}_{12}(0)|} \tag{3.96}$$

where both the fringe washing function at the origin (3.67) and the constant α_F (3.95) depend only on the filters' shape. If they are both identical, then $\tilde{r}_{12}(0) = 1$ and the above result becomes *exactly the same* as the one found for the measurement uncertainty of a total power radiometer (2.82) provided the system temperature is taken as the geometric mean of the system temperatures. It is interesting to point out that the uncertainty of the visibility measurement *does not depend on the visibility itself* but only on the total signal power, which is not to be expected a priori.

3.4.1.1 Power Spectral Density of the Instantaneous Correlation

The autocorrelation function of $\rho'(t) = \rho(t) - \langle\rho(t)\rangle$ is easily deduced by including (2.68), particularized for each receiver, into (3.90)

$$R_{\rho'}(\tau) = \mathcal{P}_1\mathcal{P}_2\tilde{r}_1(\tau)\tilde{r}_2^*(\tau) \tag{3.97}$$

where $\tilde{r}_1(\tau)$ and $\tilde{r}_2(\tau)$ are the single-filter fringe washing functions (3.62) [or (2.65)].

The power spectral density is the Fourier transform of this expression. For example, if both filters are identical and have ideal rectangular shape then $\tilde{r}_1(\tau) = \tilde{r}_2(\tau) = \text{sinc}\, B\tau_i$ and the spectrum follows by using Table A.2,

$$S_{\rho'}(f) = \frac{\mathcal{P}_1\mathcal{P}_2}{B}\Lambda\left(\frac{f}{B}\right) \quad \text{(Identical rectangular filters)} \tag{3.98}$$

which, as seen, is limited to frequencies lower than B. The reader may take a glance to the spectrum at the right of figure 2.12 to get a graphical visualization. The equation above is the triangular low frequency band. More realistic and different filters have different spectrum shapes but always limited to a low frequency band up to about the bandwidth of the signal B.

In any case, the spectrum of $\rho'(t)$ can be used to infer the dependence of σ_Γ on the product $B\tau_i$ by using (2.83) properly rewritten for this case (changing v by Γ and p′ by ρ'). If both filters are identical to each other, then, obviously the result becomes *exactly the same* as that shown in the plot at the right of figure 2.11. For slightly different filters, the plot may change, but the conclusions remain the same.

3.4.2 Four Signal Multipliers

A complex correlator is in practice implemented by real signal multipliers after separating the analytic signals into their real and imaginary parts by means of quadrature filters (1.123). From (3.82) and making use of (1.143) and (1.144),

$$\begin{aligned}\mathfrak{Re}[\Gamma_{12}] &= \frac{1}{2}\left(\langle \mathrm{b}_1(t)\mathrm{b}_2(t)\rangle + \left\langle \hat{\mathrm{b}}_1(t)\hat{\mathrm{b}}_2(t)\right\rangle\right)\\ &= \langle \mathrm{b}_1(t)\mathrm{b}_2(t)\rangle = \left\langle \hat{\mathrm{b}}_1(t)\hat{\mathrm{b}}_2(t)\right\rangle\end{aligned} \tag{3.99}$$

and

$$\begin{aligned}\mathfrak{Im}[\Gamma_{12}] &= \frac{1}{2}\left(\left\langle \hat{\mathrm{b}}_1(t)\mathrm{b}_2(t)\right\rangle - \left\langle \mathrm{b}_1(t)\hat{\mathrm{b}}_2(t)\right\rangle\right)\\ &= \left\langle \hat{\mathrm{b}}_1(t)\mathrm{b}_2(t)\right\rangle = -\left\langle \mathrm{b}_1(t)\hat{\mathrm{b}}_2(t)\right\rangle\end{aligned} \tag{3.100}$$

where $\mathrm{b}_1(t)$ and $\mathrm{b}_2(t)$ are the real time-domain signals at the receivers' output ports and the hat stands for Hilbert transform (1.122). These same relations, with finite integration time replacing the averaging, are applied to get the real and imaginary parts of the output signal $\Gamma(t)$ in figure 3.7. Any version of them provides,

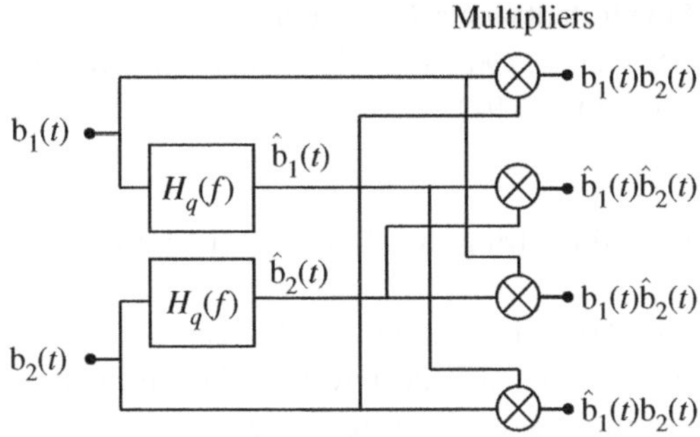

Figure 3.8 Implementation of a complex multiplier.

obviously, the same mean value Γ_{12} but also, not so obvious, *the same standard deviation* σ_Γ. This result becomes apparent in the ensuing analysis.

A conceptual block diagram of (3.99) and (3.100) is sketched in figure 3.8. The mean of either one of the two output products on top or of the average of both gives the real part (3.99). The imaginary part (3.100) is similarly obtained from the two bottom products. The Hilbert filters can be implemented, for example, through in phase quadrature mixers, covered in section 5.2.

Although in figure 3.7 the integrator is applied at the end to the complex instantaneous correlation, linearity allows applying it separately to the real and imaginary parts and then combine the corresponding outputs. It can even be applied to each individual product of figure 3.8 and then combine the outputs using (3.99) and (3.100).

If the instantaneous correlation (3.88) is defined on the basis of the full expressions (3.99) and (3.100), the real and imaginary parts become

$$\begin{aligned} \mathfrak{Re}[\rho(t)] &= \rho_r(t) = \frac{1}{2}\left(\mathrm{b}_1(t)\mathrm{b}_2(t) + \hat{\mathrm{b}}_1(t)\hat{\mathrm{b}}_2(t)\right) \\ \mathfrak{Im}[\rho(t)] &= \rho_i(t) = \frac{1}{2}\left(\hat{\mathrm{b}}_1(t)\mathrm{b}_2(t) - \mathrm{b}_1(t)\hat{\mathrm{b}}_2(t)\right) \end{aligned} \tag{3.101}$$

for which the corresponding variances are deduced from (1.76) after including the factor 1/2 and expressing the result in terms of power $P_i = \sigma^2_{\mathrm{b}_i} = \sigma^2_{\hat{\mathrm{b}}_i}$ $(i = 1, 2)$ and cross-correlation $\langle\rho(t)\rangle = \Gamma_{12}$:

$$\begin{aligned} \sigma^2_{\rho_r} &= \frac{1}{2}P_1P_2 + \frac{1}{2}\mathfrak{Re}[\Gamma_{12}]^2 - \frac{1}{2}\mathfrak{Im}[\Gamma_{12}]^2 \\ \sigma^2_{\rho_i} &= \frac{1}{2}P_1P_2 + \frac{1}{2}\mathfrak{Im}[\Gamma_{12}]^2 - \frac{1}{2}\mathfrak{Re}[\Gamma_{12}]^2 \end{aligned} \tag{3.102}$$

This result is obviously consistent with (3.91) when the complex instantaneous correlation $\rho(t) = \rho_r(t) + j\rho_i(t)$ is formed. The terms that depend on the cross-correlation cancel out in the sum.

3.4.3 Two Signal Multipliers

Using only two signal multipliers instead of four is advantageous to reduce hardware complexity and has some interesting theoretical implications regarding the measurement uncertainty. Using for example the first options[4] of the second rows of (3.99) and (3.100)

$$\begin{aligned} \rho_r(t) &= \mathrm{b}_1(t)\mathrm{b}_2(t) \\ \rho_i(t) &= \hat{\mathrm{b}}_1(t)\mathrm{b}_2(t) \end{aligned} \tag{3.103}$$

and then, using (1.66) with appropriate symbol identification, the following autocorrelation functions are obtained

$$\begin{aligned} R_{\rho_r}(\tau) &= \langle \rho_r(t) \rangle^2 + R_{\mathrm{b}_1}(\tau)R_{\mathrm{b}_2}(\tau) + R_{\mathrm{b}_1\mathrm{b}_2}(\tau)R_{\mathrm{b}_2\mathrm{b}_1}(\tau) \\ R_{\rho_i}(\tau) &= \left\langle \rho_i(t) \right\rangle^2 + R_{\hat{\mathrm{b}}_1}(\tau)R_{\mathrm{b}_2}(\tau) + R_{\hat{\mathrm{b}}_1\mathrm{b}_2}(\tau)R_{\mathrm{b}_2\hat{\mathrm{b}}_1}(\tau) \end{aligned} \tag{3.104}$$

Carrying out the same procedure, the following cross-correlation functions are easily found

$$\begin{aligned} R_{\rho_r\rho_i}(\tau) &= \langle \rho_r(t) \rangle \left\langle \rho_i(t) \right\rangle + R_{\mathrm{b}_1\hat{\mathrm{b}}_1}(\tau)R_{\mathrm{b}_2}(\tau) + R_{\mathrm{b}_1\mathrm{b}_2}(\tau)R_{\mathrm{b}_2\hat{\mathrm{b}}_1}(\tau) \\ R_{\rho_i\rho_r}(\tau) &= \left\langle \rho_i(t) \right\rangle \langle \rho_r(t) \rangle + R_{\hat{\mathrm{b}}_1\mathrm{b}_1}(\tau)R_{\mathrm{b}_2}(\tau) + R_{\hat{\mathrm{b}}_1\mathrm{b}_2}(\tau)R_{\mathrm{b}_2\mathrm{b}_1}(\tau) \end{aligned} \tag{3.105}$$

The autocorrelation function of $\rho(t) = \rho_r(t) + j\rho_i(t)$ is now obtained by introducing these results into the general expression (1.59). Using furthermore (1.139) along with the properties (1.140) and defining $\rho'(t) = \rho(t) - \langle \rho(t) \rangle$ to utilize (1.49), after some (long) algebra the following compact result is obtained

$$R_{\rho'}(\tau) = R_{b_1}(\tau)R_{\mathrm{b}_2}(\tau) + \frac{1}{4}R_{b_1b_2}(\tau)R_{b_2b_1}(\tau) \tag{3.106}$$

in which the reader should be aware of the italic symbols used in some subscripts to denote analytic signals. In particular, the first term is the product of the autocorrelation function of the analytic signal $b_1(t)$ times that of the real signal $\mathrm{b}_2(t)$. The second term is the product of cross-correlation functions of analytic signals.

To provide an expression for the variance of $\rho(t)$, equal to $R_{\rho'}(0)$, equations (3.82) and (3.84) are first used to make the following substitutions

$$\begin{aligned} \frac{1}{2}R_{b_1b_2}(0) &= \frac{1}{2}R^*_{b_2b_1}(0) = \Gamma_{12} \\ R_{\mathrm{b}_i}(0) &= \frac{1}{2}R_{b_i}(0) = \mathcal{P}_i \quad (i = 1, 2) \end{aligned} \tag{3.107}$$

where the first equation of (1.143) has also been taken into account. Then,

$$\sigma_\rho^2 = 2\mathcal{P}_1\mathcal{P}_2 + |\Gamma_{12}|^2 \tag{3.108}$$

4 Other options provide the same results.

which could also have been obtained directly by particularizing (3.104) for $\tau = 0$. Compared to (3.91), this equation shows an extra additive term that depends on the cross-correlation amplitude and also a factor of 2 multiplying the product of output powers. This factor is the same one encountered in the variance of the instantaneous power when comparing a square law device (2.92) and a quadratic detector (2.103). Clearly, the above result evidences that the instantaneous correlation *before averaging* is noisier when the correlator is implemented with only two multipliers in comparison with a full complex multiplier.

The final correlation measurement $\Gamma(t)$ is the result of averaging $\rho(t)$ during a finite integration time τ_i. Using again the results of section 1.2.6, the standard deviation of $\Gamma(t)$ is approximately equal to the spectral density of $\rho'(t)$ particularized at $f = 0$ and divided by the integration time. Therefore the Fourier transform of (3.106) must be evaluated and its value at the origin selected.

Consider first the second term of (3.106). Using (1.56) to swap subscripts and the properties of sign change, conjugation and multiplication of the Fourier transform given in Table A.1,

$$\begin{aligned}\mathcal{F}\,[R_{b_1b_2}(\tau)R_{b_2b_1}(\tau)] &= S_{b_1b_2}(f) * S^*_{b_1b_2}(f) \\ &= \int_{-\infty}^{\infty} S_{b_1b_2}(f')S^*_{b_1b_2}(f-f')\,df' \end{aligned} \tag{3.109}$$

where $S_{b_1b_2}(f)$ is the cross-power spectral density of the analytic signals $b_1(t)$ and $b_2(t)$. Using (3.51) and the equivalence (1.156),

$$S_{b_1b_2}(f) = 2kH_1(f)H_2^*(f)\mathbb{V}_{12}(f)\,\mathrm{u}(f) \tag{3.110}$$

in which $\mathrm{u}(f)$ is the unit step function (A.31) and $\mathbb{V}_{12}(f)$ the system visibility (3.48).

The contribution of (3.109) to the variance of $\Gamma(t)$ depends only on its value at $f = 0$, at which the product inside the integral becomes $S_{b_1b_2}(f')S^*_{b_1b_2}(-f')$. Taking into account that the spectral density (3.110) is zero for negative frequencies, this product turns out to vanish for all values of f'. As a consequence, *the contribution of the second term of (3.106) to the variance of the filtered correlation is null.* This result is independent of the receivers' filter shape and also of the nature of the source the instrument is pointing to.

To evaluate the contribution of first term of (3.106), the same method is applied but particularized for this case,

$$\mathcal{F}\,[R_{b_1}(\tau)R_{b_2}(\tau)] = S_{b_1}(f) * S_{b_2}(f) = \int_{-\infty}^{\infty} S_{b_1}(f')S^*_{b_2}(f-f')\,df' \tag{3.111}$$

which must be evaluated at $f = 0$. Taking into account that the second term of (3.106) has null contribution at the origin, the variance of the output signal results in

$$\sigma_\Gamma^2 \approx \frac{S_{\rho'}(0)}{\tau_i} = \frac{1}{\tau_i}\int_0^{\infty} S_{b_1}(f)S_{b_2}(f)\,df \tag{3.112}$$

where the limits of integration include only positive frequencies because the integrand is null otherwise. From the above, substituting (2.63) and (2.67) for the corresponding spectral densities,

$$\sigma_{\Gamma}^2 = \frac{P_1 P_2}{B_1 B_2 \tau_i} \int_0^{\infty} |\bar{H}_1(f)|^2 |\bar{H}_2(f)|^2 \, df \tag{3.113}$$

and the standard deviation is

$$\sigma_{\Gamma} = \frac{\sqrt{P_1 P_2}}{\sqrt{B\tau_i}} \alpha_F \tag{3.114}$$

which is *identical to* (3.94) with the same definitions (3.95). Even though the variance of the instantaneous correlation measured with a single signal multiplier (3.108) is larger than the one using a full complex multiplier (3.91), after integration it becomes exactly the same in both cases. This is due to the fact that the cause of the noise increase is the presence of high frequency bands, which are washed out by the integrator. It is exactly the same phenomenon found in total power measurement when comparing square law and quadratic envelope detectors (2.84).

3.4.3.1 Autocorrelation Function of $\rho'(t)$

For completeness, the autocorrelation function of the instantaneous correlation (3.106) is here derived. The first term is immediate after substituting $R_{b_1}(\tau)$ by (2.68) and $R_{b_2}(\tau)$ by (2.64) with the appropriate subscripts. After a simple algebraic manipulation, it can be written as

$$R_{b_1}(\tau) R_{b_2}(\tau) = P_1\, P_2 \tilde{r}_1(\tau) \tilde{r}_2^*(\tau) + P_1\, P_2 \tilde{r}_1(\tau) \tilde{r}_2(\tau)\, e^{j4\pi f_0 \tau} \tag{3.115}$$

which is consistent at $\tau = 0$ with the first term of (3.108) taking into account that $\tilde{r}_i(0) = 1$. This first term is exactly the same as (3.97), the autocorrelation function when all four signal multipliers are used. On the other hand, the second term has a high frequency spectrum, as deduced by the complex exponential term. Since it has been demonstrated that the spectrum of the second term of (3.106) is also located at high frequency, it is inferred that, from the point of view of correlation measurement, *using two multipliers followed by low pass filters is equivalent to using four multipliers.* This conclusion is the same found in relation to total power radiometers when comparing the quadratic envelope and the square law detector (see discussion in section 2.4.4) and assumes only sufficiently high center frequency to avoid bands overlap. It also justifies the increased standard deviation of $\rho(t)$ when comparing (3.108) with (3.91).

For the second term of (3.106), a general closed-form expression cannot be found since it depends on the source extent and position. To facilitate operations and simplify expressions, the following definition is proposed

$$\text{Second term of (3.106):} \quad \mathcal{R}_{12}(\tau) \triangleq \frac{1}{4} R_{b_1 b_2}(\tau) R_{b_2 b_1}(\tau) \tag{3.116}$$

Directly from (3.107) and consistent with (3.108),

$$\mathcal{R}_{12}(0) = |\Gamma_{12}|^2 \tag{3.117}$$

Moreover, the spectrum of $\mathcal{R}_{12}(\tau)$ is one fourth of the convolution integral (3.109). If the product of filter responses in (3.110) is centered at f_0 and has limited bandwidth, then the convolution integral becomes centered at $2f_0$ with a bandwidth about twice that of the filters. This means that $\mathcal{R}_{12}(\tau)$ is itself an analytic signal and can be expressed in terms of a complex envelope $\tilde{\mathcal{R}}_{12}(\tau)$ defined in (1.128),

$$\mathcal{R}_{12}(\tau) = \tilde{\mathcal{R}}_{12}(\tau)e^{j4\pi f_0\tau} = |\Gamma_{12}|^2\bar{\tilde{\mathcal{R}}}_{12}(\tau)e^{j4\pi f_0\tau} \tag{3.118}$$

where the bar over the complex envelope means normalization with respect to its value at $\tau = 0$. Introducing this expression, along with (3.115), into (3.106), the autocorrelation function of $\rho'(t)$ becomes

$$\begin{aligned} R_{\rho'}(\tau) &= \mathcal{P}_1\,\mathcal{P}_2\tilde{r}_1(\tau)\tilde{r}_2^*(\tau) + \mathcal{P}_1\,\mathcal{P}_2\tilde{r}_1(\tau)\tilde{r}_2(\tau)\,e^{j4\pi f_0\tau} \\ &\quad + |\Gamma_{12}|^2\,\bar{\tilde{\mathcal{R}}}_{12}(\tau)e^{j4\pi f_0\tau} \end{aligned} \tag{3.119}$$

where the first term has low frequency spectrum and the two others have high frequency bands centered at $2f_0$. The third summand can only be neglected if $|\Gamma_{12}|^2 \ll \mathcal{P}_1\,\mathcal{P}_2$.

3.4.3.2 Point Source

For illustration purposes, $\bar{\tilde{\mathcal{R}}}_{12}(\tau)$ is now derived for a point source located at an arbitrary spatial direction. Disregarding (or canceling) the noise contribution of the antennas' mismatch and coupling, equation (3.44) states that the system visibility and the antenna visibility coincide, $\mathbb{V}_{12} = \mathcal{V}_{12}$ so using (3.17), provided the antenna pattern phase difference is neglected, the system visibility becomes

$$\mathbb{V}_{12}(f) = \sqrt{T_{a1}\,T_{a2}}\,e^{-j2\pi f\Delta t} \tag{3.120}$$

where Δt is the time difference of the waves arriving to each antenna

$$\Delta t = \frac{r_1 - r_2}{c} \tag{3.121}$$

with c the velocity of light. The inverse Fourier transform of (3.110) is then

$$\begin{aligned} R_{b_1b_2}(\tau) &= 2k\sqrt{T_{a1}\,T_{a2}}\,\mathcal{F}^{-1}[H_1(f)H_2^*(f)e^{-j2\pi f\Delta t}\,\mathrm{u}(f)] \\ &= 2k\sqrt{B_1B_2}\sqrt{G_1G_2}\sqrt{T_{a1}\,T_{a2}}\,\tilde{r}_{12}(\tau - \Delta t)\,e^{j2\pi f_0(\tau-\Delta t)} \end{aligned} \tag{3.122}$$

where the definition of the fringe washing function (3.61) has been used combined with the time delay property of the Fourier transform (see Table A.1). Taking into account that swapping the receiver numbers implies a change of sign in Δt, evaluating (3.116) is straightforward after substituting the above result. The final expression is here written in a more compact form using the correlator gain definition (3.83) and knowing that $G_{21} = G_{12}^*$

$$\mathcal{R}_{12}(\tau) = |G_{12}|^2 T_{a1} T_{a2}\, \bar{\tilde{r}}_{12}(\tau - \Delta t)\bar{\tilde{r}}_{21}(\tau + \Delta t)\, e^{j4\pi f_0 \tau} \tag{3.123}$$

where the bar over the fringe washing function denotes normalization with respect to its value at the origin (3.56). Applying now property 1 of the fringe washing function (section 3.3.2), the normalized complex envelope defined in (3.118) becomes

$$\bar{\tilde{\mathcal{R}}}_{12}(\tau) = \frac{\bar{\tilde{r}}_{12}(\tau - \Delta t)\bar{\tilde{r}}_{12}^*(-\tau - \Delta t)}{|\bar{\tilde{r}}_{12}(-\Delta t)|^2} \quad \text{Point source} \tag{3.124}$$

If the point source is located at boresight ($\Delta t = 0$) then this equation simplifies to

$$\bar{\tilde{\mathcal{R}}}_{12}(\tau) = \bar{\tilde{r}}_{12}(\tau)\bar{\tilde{r}}_{12}^*(-\tau) \quad \text{Point source at boresight} \tag{3.125}$$

For example, if both receivers have identical filters with ideal rectangular response, all fringe washing functions become equal to (3.66)

$$\tilde{r}_1(\tau) = \tilde{r}_2(\tau) = \bar{\tilde{r}}_{12}(\tau) = \mathrm{sinc} B\tau \tag{3.126}$$

with $B = B_1 = B_2$, the common bandwidth in both filters. For a point source at boresight, substitution into (3.125) and (3.119) yields

$$R_{\rho'}(\tau) = \mathcal{P}_1\mathcal{P}_2 \mathrm{sinc}^2 B\tau\,(1 + e^{j4\pi f_0 \tau}) + |\Gamma_{12}|^2 \mathrm{sinc}^2 B\tau\, e^{j4\pi f_0 \tau} \tag{3.127}$$

The corresponding spectral density is the Fourier transform, which is immediate using the tables provided in appendix A.7

$$S_{\rho'}(f) = \frac{\mathcal{P}_1\mathcal{P}_2}{B}\Lambda\left(\frac{f}{B}\right) + \frac{\mathcal{P}_1\mathcal{P}_2}{B}\Lambda\left(\frac{f - 2f_0}{B}\right) + \frac{|\Gamma_{12}|^2}{B}\Lambda\left(\frac{f - 2f_0}{B}\right) \tag{3.128}$$

Here the presence of the two high frequency terms at twice the center frequency is explicit. One depends on the total power and the other on the cross-correlation amplitude, but none of them contributes to the standard deviation of the output signal σ_Γ since this one depends on the spectrum at the origin, so it is given by (3.114) with $\alpha_F = 1$ in this case.

The spectrum (3.128) may now be compared with (3.98), applicable to a full complex correlator. In that case the two high frequency terms are always cancelled out by the combination of signal products either if the source is extended or not.

3.4.4 Analog Multipliers

In analog systems, the signal multipliers of figure 3.8 are in practice implemented by a combination of a signal combiner and quadratic detectors based on the following fundamental equation

$$\left[\mathrm{b}_1(t) \pm \mathrm{b}_2(t)\right]^2 = \mathrm{b}_1^2(t) + \mathrm{b}_2^2(t) \pm 2\mathrm{b}_1(t)\mathrm{b}_2(t) \tag{3.129}$$

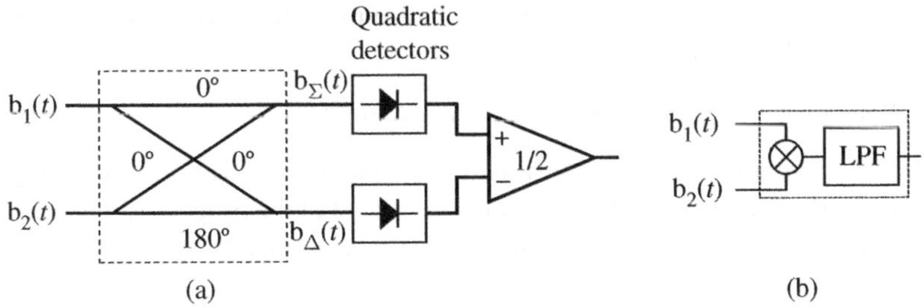

Figure 3.9 Analog multiplier of real signals. (a) Conceptual block diagram and (b) equivalence at system level.

in which the third term contains the multiplication between both signals. It can be retrieved with the block diagram of figure 3.9(a). The subsystem drawn inside the dashed lines box is a 180° hybrid coupler that combines two input signals to produce their sum and difference. The interested reader may consult any basic book on microwave circuits [18] for examples and design methodology of this component. In general, a hybrid coupler is frequency dependent but the schematic of figure 3.9(a) disregards this dependency. Either the frequency response of the hybrid is flat within the receiver bandwidth or it is included in the overall receiver frequency response. Assuming an ideal hybrid, the output signals are, respectively, proportional to the sum and difference

$$b_\Sigma(t) = \frac{1}{\sqrt{2}}[b_1(t) + b_2(t)]; \quad b_\Delta(t) = \frac{1}{\sqrt{2}}[b_1(t) - b_2(t)] \tag{3.130}$$

where the $1/\sqrt{2}$ factor is needed for energy conservation.

In section 2.4.4 a quadratic detector was shown to be equivalent to a square law device followed by a low pass filter. Therefore the output signals of the detectors of figure 3.9 (a) are the filtered versions of (3.129) with the corresponding signs. The differential amplifier (operating at low frequency) subtracts the first two terms in each one but adds the third one, so producing in the end a signal proportional to the product of both input signals followed by a low pass filter. The drawing of figure 3.9(b) shows then the equivalent component at system level.

Replacing the ideal hybrid coupler by an actual design based on the available technology at the operational frequency has an overall impact resulting in a gain, usually denoted as "correlator efficiency," and an offset

$$\text{Output signal} = C_g b_1(t) b_2(t) + C_o \tag{3.131}$$

which are in general estimated through calibration (see chapter 6).

3.4.5 Signal Clipping and Normalized Correlation

Correlation can also be measured through using the arc sine law (section 1.1.4) after clipping the receiver output signals to get only their sign. This seemingly strange approach is very common in digital systems and especially interesting in instruments with a large number of baselines. In this section it is analyzed for analog signals and later in section 5.4 the effect of sampling at discrete epochs is studied.

Given a zero-mean real time domain signal $\mathrm{b}(t)$, its clipped version is

$$\breve{\mathrm{b}}(t) = \begin{cases} +1 & \text{if } \mathrm{b}(t) > 0 \\ -1 & \text{if } \mathrm{b}(t) < 0 \end{cases} \tag{3.132}$$

Since the receiver output signals of figure 3.7 $\mathrm{b}_1(t)$ and $\mathrm{b}_2(t)$ are zero mean jointly normal processes, the arc sine law (1.42) states that

$$\left\langle \breve{\mathrm{b}}_1(t)\breve{\mathrm{b}}_2(t) \right\rangle = \frac{2}{\pi} \arcsin \frac{\langle \mathrm{b}_1(t)\mathrm{b}_2(t) \rangle}{\sqrt{\mathcal{P}_1 \mathcal{P}_2}} \tag{3.133}$$

where $\mathcal{P}_1 = \left\langle \mathrm{b}_1^2(t) \right\rangle$ and $\mathcal{P}_2 = \left\langle \mathrm{b}_2^2(t) \right\rangle$. Therefore, measuring the cross-correlation between clipped signals suffices to get the *normalized correlation* between the original ones by inverting this equation. Using, for example, the first convention of the second lines of (3.99) and (3.100)

$$M_{12} = \sin\left(\frac{\pi}{2}\left\langle \breve{\mathrm{b}}_1(t)\breve{\mathrm{b}}_2(t) \right\rangle\right) + j \sin\left(\frac{\pi}{2}\left\langle \breve{\mathrm{b}}_1(t)\breve{\hat{\mathrm{b}}}_2(t) \right\rangle\right) \tag{3.134}$$

where M_{12} is the normalized *complex* cross-correlation

$$M_{12} = \frac{\Gamma_{12}}{\sqrt{\mathcal{P}_1 \mathcal{P}_2}} \tag{3.135}$$

from which Γ_{12} can be recovered after *denormalization* with respect to the signal powers $\mathcal{P}_1$ and $\mathcal{P}_2$ independently measured [19, 20]. Other conventions of (3.99) and (3.100) translate to corresponding versions of (3.134) with the same M_{12}.

The main advantage of this approach is that the instantaneous cross-correlation between two clipped signals is extremely easy to measure since it is a simple "exclusive NOR": +1 if both inputs are equal and −1 if they are different. For more detail, the reader is referred to the end of section 5.4.1, paragraph titled "one-bit correlation." This technique allows an important hardware simplification, especially when a large number of correlators is required. On the other hand, the measurement uncertainty becomes slightly larger, as it is demonstrated next.

Using (3.82) and (3.84) the visibility is retrieved by an equation similar to (3.86)

$$V_{12} = \frac{M_{12}\sqrt{T_{sys1} T_{sys2}}}{G_{12}} \tag{3.136}$$

where now the complex gain G_{12} is *only the fringe washing function at the origin*

$$G_{12} = \tilde{r}_{12}(0) \quad \text{(for normalized correlation measurement)} \tag{3.137}$$

According to property (7) of the fringe washing function (section 3.3.2), the amplitude of G_{12} is close to unity for filters reasonably similar. On the other hand, the phase can take any value depending on the receivers' design, especially if frequency conversions are included (see section 5.1), and this is an important parameter to calibrate. The reader may wish to consult section 6.2.1 for more details.

The system temperatures needed in (3.136) are measured either by including in each receiver a power detector or, assuming fairly similar hardware, using a single total power radiometer to measure a unique value for it. In any case (2.72) applies, so assuming a different measurement in each receiver,

$$T_{sysi} = \frac{\mathrm{v}_i - \mathrm{v}_{oi}}{G'_i} \quad i = 1, 2 \tag{3.138}$$

where v_i is the detected voltage, v_{oi} the instrumental offset and G'_i the radiometric gain (2.73) in the "power measurement system" chain, which in general differs from the one used for the correlation measurement (hence the prima). These two parameters are obtained by a standard calibration procedure (see section 6.2.6). Note that in this case the radiometric measurement is the system temperature and not the antenna temperature, so the receiver noise temperature needs not be known.

3.4.5.1 Limited Integration Time

If the integration time is finite, the visibility (3.136) becomes a random variable obtained from the product of two distinct measurements, namely the system temperature and the normalized correlation. Since both correlation and power are measured independently, the total uncertainty can be estimated with the aid of equation (1.24), which considers the product of two independent random variables. Denoting, as in (3.87), the standard deviation of the visibility by ΔV_{12} and that of system temperature by ΔT, as in (2.75),

$$(\Delta V_{12})^2 = \frac{T_{sys1} T_{sys2} \sigma_M^2 + |M_{12}|^2 (\Delta T)^2}{|\tilde{r}_{12}(0)|^2} \tag{3.139}$$

where σ_M^2 is the variance of the normalized correlation measurement. Using (2.82) with $T_{sys} = \sqrt{T_{sys1} T_{sys2}}$ for ΔT and (3.136) for M_{12},

$$(\Delta V_{12})^2 = \frac{T_{sys1} T_{sys2}}{|\tilde{r}_{12}(0)|^2} \sigma_M^2 + \frac{|V_{12}|^2}{B' \tau'_i} \alpha_F^2 \tag{3.140}$$

where B' and τ'_i are, respectively, the bandwidth and integration time of the power measurement system, which may differ from those used in the correlation measurement. The second term in this equation is the contribution of the noise associated to the system temperature measurement. It is proportional to the visibility amplitude so it can be negligible in systems imaging large targets, for which all visibilities are much lower than the system temperatures. It is not the case of point (or small) sources, where the visibility amplitude is equal or similar to the antenna temperature [see (3.17)], often with values of the same order of magnitude as the system temperature, provided the receivers' noise do not have a large contribution.

The next step is to find an expression for σ_M^2, which is not straightforward. In agreement with the fundamental definition (1.2), if $M(t) = M_r(t) + jM_i(t)$, then

$$\sigma_M^2 = \sigma_{M_r}^2 + \sigma_{M_i}^2 \tag{3.141}$$

and each term is derived separately from the corresponding measurement of $M_r(t)$ or $M_i(t)$ based on (3.134). The case of using four multipliers is covered later.

The block diagram of figure 3.10 illustrates the setup for the measurement of the real part and a similar one is applicable to the imaginary part only by replacing $\mathrm{b}_1(t)$ by $\hat{\mathrm{b}}_1(t)$ and changing all subscripts r to i. From this figure, the instantaneous correlation of the clipped signals is

$$\breve{\rho}_r(t) = \breve{\mathrm{b}}_1(t)\breve{\mathrm{b}}_2(t) \Rightarrow \begin{cases} \langle \breve{\rho}_r(t) \rangle = \dfrac{2}{\pi} \arcsin \Re e[M_{12}] \\ \langle \breve{\rho}_r^2(t) \rangle = 1 \end{cases} \tag{3.142}$$

Note that $\breve{\rho}_r(t)$ is identical to the clipped version of $\rho_r(t)$ in (3.103), so validating the symbol used. This product is averaged during an integration time τ_i to produce $\varrho_r(t)$ and then the sinus function is applied to get the final measurement $M_r(t)$.

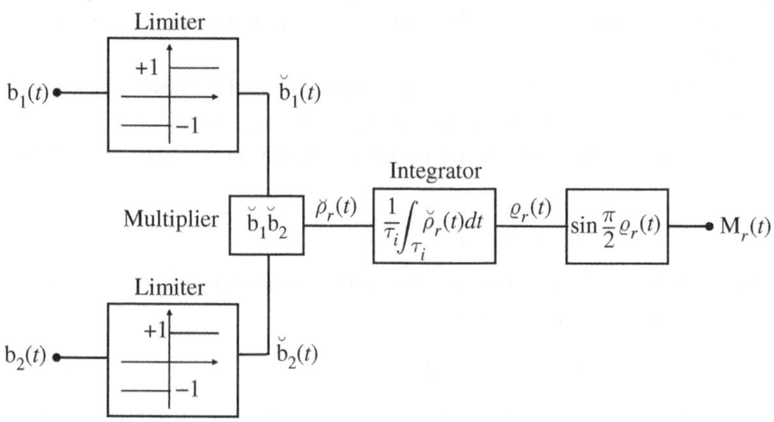

Figure 3.10 Measurement of the real part of the normalized correlation using finite integration time.

The first observation about this setup is that the mean of the output signal is not mathematically identical to the real part of (3.135). Using the approximation (1.45)

$$\langle M_r(t)\rangle \approx \Re e[M_{12}]\left(1-\frac{\pi^2}{8}\sigma^2_{\varrho_r}\right) \tag{3.143}$$

where it has been taken into account that $\langle \varrho_r(t)\rangle = \langle \breve{\rho}_r(t)\rangle$. This technique has thus associated a systematic error in the correlation measurement, which is minimized by increasing the integration time so reducing the variance of $\varrho_r(t)$. In most practical situations it is very small [see (3.149)] so this effect can be neglected. Nevertheless, if needed it can be handled with appropriate calibration. Using directly (1.112), the variance of $\varrho_r(t)$ is

$$\sigma^2_{\varrho_r} = \frac{2}{\tau_i}\int_0^{\infty} R_{\breve{\rho}'_r}(\tau)\,d\tau \tag{3.144}$$

where $R_{\breve{\rho}'_r}(\tau)$ is the autocorrelation function of $\breve{\rho}'_r(t) = \breve{\rho}_r(t) - \langle \breve{\rho}_r(t)\rangle$,

$$R_{\breve{\rho}'_r}(\tau) = R_{\breve{\rho}_r}(\tau) - \langle \breve{\rho}_r(t)\rangle^2 \tag{3.145}$$

where the average of $\breve{\rho}_r(t)$ is provided in (3.142). In general, $R_{\breve{\rho}_r}(\tau)$ cannot be easily evaluated. It can be done, though, in the particular case in which the input signals $\mathrm{b}_1(t)$ and $\mathrm{b}_2(t)$ in figure 3.10 are uncorrelated. This corresponds to having the antennas in figure 3.5 or 3.7 replaced by independent resistors so that the correlation between the output signals is null, $\Gamma_{12} = 0$. It is pertinent to comment now that the standard deviation of $\Gamma(t)$ (3.114), and hence the visibility uncertainty ΔV_{12}, is independent of Γ_{12}, so it does not change when $\Gamma_{12} = 0$. Therefore, the results obtained here assuming uncorrelated signals can be directly compared with those obtained in the general analysis of the sections 3.4.2 and 3.4.3. The impact of clipping in the sensitivity for nonzero visibility is not covered here, although it is expected to be similar, at least for low visibilities, to the one deduced in the ensuing analysis.

Since the input signals in figure 3.10 are jointly normal processes, if they are uncorrelated they are also independent; and hence so they are the clipped signals. In this case (1.62) applies and, taking into account that now $\langle \breve{\rho}_r(t)\rangle = 0$, (3.145) becomes

$$R_{\breve{\rho}'_r}(\tau) = R_{\breve{\rho}_r}(\tau) = R_{\breve{\mathrm{b}}_1}(\tau) R_{\breve{\mathrm{b}}_2}(\tau) \tag{3.146}$$

Using the arcsin law (1.42) the autocorrelation functions of the clipped signals are related to those of the input signals as

$$R_{\breve{\mathrm{b}}_i}(\tau) = \frac{2}{\pi}\arcsin r_{\mathrm{b}_i}(\tau) \quad (i = 1, 2) \tag{3.147}$$

where $r_{\mathrm{b}_i}(\tau)$ is the normalized correlation function defined in (1.55). Using (2.64)

$$R_{\breve{\rho}'_r}(\tau) = \frac{4}{\pi^2}\arcsin \Re e[\tilde{r}_1(\tau)e^{j2\pi f_0\tau}]\,\arcsin \Re e[\tilde{r}_2(\tau)e^{j2\pi f_0\tau}] \tag{3.148}$$

where the fringe washing functions $\tilde{r}_1(\tau)$ and $\tilde{r}_2(\tau)$ follow the definition (2.65). Introducing the above result into (3.144) the variance of ϱ_r can be conveniently written as

$$\sigma_{\varrho_r}^2 = \frac{8}{\pi^2}\frac{\alpha_F^2}{B\tau_l}\mathcal{I} \tag{3.149}$$

where α_F^2 and B are defined in (3.95) and

$$\mathcal{I} = \frac{1}{\alpha_F^2}\int_0^\infty \arcsin \Re e[\tilde{r}_1(\vartheta)e^{j2\pi\bar{f}_0\vartheta}] \arcsin \Re e[\tilde{r}_2(\vartheta)e^{j2\pi\bar{f}_0\vartheta}]\, d\vartheta \tag{3.150}$$

in which $\vartheta = B\tau$ and $\bar{f}_0 = f_0/B$. This integral will be later evaluated numerically. Since $\langle \varrho_r(t)\rangle = 0$ and (for relatively large values of $B\tau_l$) σ_{ϱ_r} is much smaller than one, the output signal in figure 3.10 can be approximated by

$$M_r(t) \approx \frac{\pi}{2}\varrho_r(t) \tag{3.151}$$

and then

$$\sigma_{M_r}^2 \approx \frac{\pi^2}{4}\sigma_{\varrho_r}^2 = 2\frac{\alpha_F^2}{B\tau_l}\mathcal{I} \tag{3.152}$$

The same procedure can now be applied to the imaginary part of (3.134). The instantaneous correlation is now

$$\breve{\rho}_i(t) = \breve{\mathfrak{b}}_1(t)\breve{\mathfrak{b}}_2(t) \Rightarrow \begin{cases} \langle \breve{\rho}_i(t)\rangle = \frac{2}{\pi}\arcsin \Im m[M_{12}] \\ \langle \breve{\rho}_i^2(t)\rangle = 1 \end{cases} \tag{3.153}$$

and identical results are obtained based on property (1.141) of analytic signals. In particular, $R_{\breve{\rho}'_i}(\tau) = R_{\breve{\rho}'_r}(\tau)$ and consequently

$$\sigma_{\varrho_i}^2 = \sigma_{\varrho_r}^2; \quad \sigma_{M_i}^2 = \sigma_{M_r}^2 \tag{3.154}$$

so, from (3.141), $\sigma_M^2 = 2\sigma_{M_r}^2$ and, using (3.152)

$$\sigma_M^2 = 4\frac{\alpha_F^2}{B\tau_l}\mathcal{I} \tag{3.155}$$

which, substituted into (3.140) with $V_{12} = 0$ (uncorrelated input signals) results in

$$\Delta V_{12} = 2\sqrt{\mathcal{I}}\frac{\sqrt{T_{sys1}T_{sys2}}}{\sqrt{B\tau_l}}\frac{\alpha_F}{|\tilde{r}_{12}(0)|} \tag{3.156}$$

In comparison with (3.96) it is apparent that the visibility uncertainty increases by a factor of $2\sqrt{\mathcal{I}}$. Later it will be quantified [see (3.164)], but previously the case of four signal multipliers is analyzed.

3.4.5.2 Four Multipliers

Assuming now full complex multiplication, the real and imaginary parts of the instantaneous correlation of the clipped signals are defined as in (3.101),

$$\begin{aligned} \breve{\rho}_r(t) &= \tfrac{1}{2}\left(\breve{\hat{\mathrm{b}}}_1(t)\breve{\mathrm{b}}_2(t) + \hat{\mathrm{b}}_1(t)\breve{\hat{\mathrm{b}}}_2(t)\right) \\ \breve{\rho}_i(t) &= \tfrac{1}{2}\left(\breve{\hat{\mathrm{b}}}_1(t)\breve{\mathrm{b}}_2(t) - \breve{\mathrm{b}}_1(t)\breve{\hat{\mathrm{b}}}_2(t)\right) \end{aligned} \tag{3.157}$$

and the correlation measurement is $M(t) = M_r(t) + jM_i(t)$ where

$$M_r(t) = \sin\frac{\pi}{2}\varrho_r(t); \quad M_i(t) = \sin\frac{\pi}{2}\varrho_i(t) \tag{3.158}$$

with $\varrho_r(t)$ and $\varrho_i(t)$ the result of averaging of $\breve{\rho}_r(t)$ and $\breve{\rho}_i(t)$, respectively, during an integration time τ_i.

In agreement with the general expression (1.60), the autocorrelation function of either $\breve{\rho}_r(t)$ or $\breve{\rho}_i(t)$ is the sum of four terms. With the assumption of independent $\hat{\mathrm{b}}_1(t)$ and $\hat{\mathrm{b}}_2(t)$, (1.62) allows expressing each one of them as the product of two correlation functions. At this point, the properties of analytic signals (1.141) are applied to write

$$R_{\breve{\hat{\mathrm{b}}}_k}(\tau) = R_{\breve{\mathrm{b}}_k}(\tau) \text{ and } R_{\breve{\mathrm{b}}_k\breve{\hat{\mathrm{b}}}_k}(\tau) = -R_{\breve{\hat{\mathrm{b}}}_k\breve{\mathrm{b}}_k}(\tau) \quad (k = 1, 2) \tag{3.159}$$

These properties hold in principle for the normal input signals $\mathrm{b}_1(t)$ and $\mathrm{b}_2(t)$ but, due to the one to one correspondence of the correlation functions derived from the arcsin law [see (3.147)], they also hold for clipped signals. In the end, after some algebra, the following result is obtained

$$R_{\breve{\rho}_r}(\tau) = R_{\breve{\rho}_i}(\tau) = \frac{1}{2}R_{\breve{\mathrm{b}}_1}(\tau)R_{\breve{\mathrm{b}}_2}(\tau) + \frac{1}{2}R_{\breve{\hat{\mathrm{b}}}_1\breve{\mathrm{b}}_1}(\tau)R_{\breve{\hat{\mathrm{b}}}_2\breve{\mathrm{b}}_2}(\tau) \tag{3.160}$$

The first term is equivalent to (3.146) with a factor of $1/2$. As for the second one, the arc sin law (3.147) must first be used to write it in terms of $R_{\hat{\mathrm{b}}_1\mathrm{b}_1}(\tau)$ and $R_{\hat{\mathrm{b}}_2\mathrm{b}_2}(\tau)$. To get an expression for these correlation functions, the same method applied to derive (2.64) is used together with the second equation (1.148) resulting in

$$R_{\hat{\mathrm{b}}_i\mathrm{b}_i}(\tau) = \Im m[\tilde{r}_i(\tau)e^{j2\pi f_0\tau}] \quad (i = 1, 2) \tag{3.161}$$

and, from this point onwards, the same methodology applied to derive (3.155) from (3.146) is followed from (3.160) to get

$$\sigma_M^2 = 2\frac{\alpha_F^2}{B\tau_i}(\mathcal{I} + \mathcal{J}) \tag{3.162}$$

where

$$\mathcal{J} = \frac{1}{\alpha_F^2}\int_0^\infty \arcsin \Im m\left[\tilde{r}_1(\vartheta)e^{j2\pi\bar{f}_0\vartheta}\right] \arcsin \Im m\left[\tilde{r}_2(\vartheta)e^{j2\pi\bar{f}_0\vartheta}\right] d\vartheta \tag{3.163}$$

Comparing (3.162) and (3.155) leads to the conclusion that the visibility uncertainty is still given by (3.156) but replacing $\mathcal{I}$ by the $(\mathcal{I} + \mathcal{J})/2$. Below, it is assessed numerically that both constants are equivalent (see figure 3.11) demonstrating that, also for clipped signals, the use of two or four signal multipliers does not have impact on the sensitivity.

3.4.5.3 Sensitivity Degradation Due to Clipping

To illustrate the impact of signal clipping in the sensitivity, both integrals (3.150) and (3.163) have been numerically evaluated after substituting the fringe washing functions by their definition (2.65) as a function of the respective frequency responses. For this computation identical filters are assumed in both receivers with the shapes shown in figure 2.10: an ideal rectangular filter and a measured response of an actual filter.

Figure 3.11(a) shows integral $\mathcal{I}$ (3.150) as a function of the center frequency normalized to the bandwidth. In the plot of figure 3.11(b) the average of $\mathcal{I}$ and $\mathcal{J}$ is presented for the same range. The minimum abscise is set to a value slightly larger than $1/2$ since this one corresponds to having the signal spectrum between 0 and B, so avoiding overlapping with the negative part. In both plots there is a rapid saturation to the same constant values for increasing center frequency. This feature is expected since the integral in (3.144) is the spectrum of the instantaneous correlation at the origin ($f = 0$) and this should be quite independent on the location of the signal band.

The decrease on sensitivity due to signal clipping is equal to $2\sqrt{\mathcal{I}}$ [see (3.156)] or $2\sqrt{(\mathcal{I}+\mathcal{J})/2}$ depending on whether two or four signal multipliers are used. Using the asymptotic values of the figure, the following approximate factors can be set, applicable to either two or four signal multipliers:

$$\frac{\Delta V_{12}\text{ [clipped]}}{\Delta V_{12}\text{ [unclipped]}} = \begin{cases} 1.152 & \text{Ideal filter} \\ 1.163 & \text{Real filter} \end{cases} \tag{3.164}$$

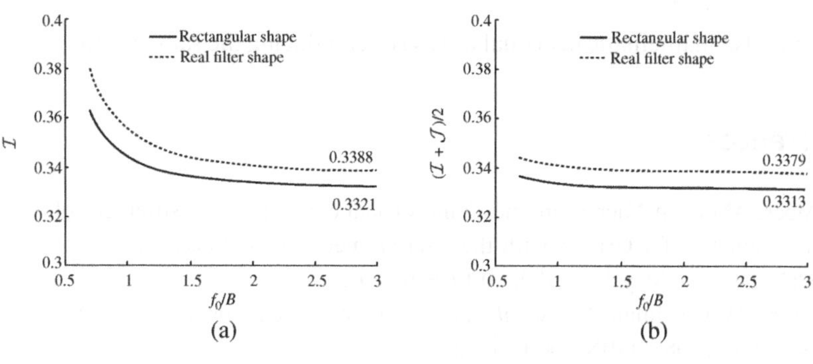

Figure 3.11 Numerical evaluation of (3.150) and (3.163) as a function of center frequency. Plot (a) is for $\mathcal{I}$ and (b) for $(\mathcal{I}+\mathcal{J})/2$.

so, in any case clipping the signal degrades the sensitivity. Be aware that the figure for the real filter is just an example and other filter shapes may result on different values.

3.4.5.4 Unclipped Signals

As an academic exercise it is interesting to apply the same methodology to unclipped signals. Consider first the case of two signal multipliers. Assuming, for compatibility with the above development, uncorrelated signals, equations (3.104) reduce to

$$\begin{aligned} R_{\rho'_r}(\tau) &= R_{\rho_r}(\tau) = R_{\mathfrak{b}_1}(\tau)R_{\mathfrak{b}_2}(\tau) \\ R_{\rho'_i}(\tau) &= R_{\rho_i}(\tau) = R_{\hat{\mathfrak{b}}_1}(\tau)R_{\mathfrak{b}_2}(\tau) \end{aligned} \tag{3.165}$$

which, taking a look to (3.146) evidences that the same approach can be applied by only removing the arcsin operation in the autocorrelation functions. Using then the same steps, the expression for the visibility uncertainty becomes *identical to (3.156)* but now with the integral $\mathcal{I}$ replaced by

$$\mathcal{I}' = \frac{1}{\alpha_F^2}\int_0^\infty \Re e[\tilde{r}_1(\vartheta)e^{j2\pi\bar{f}_0\vartheta}]\,\Re e[\tilde{r}_2(\vartheta)e^{j2\pi\bar{f}_0\vartheta}]\,d\vartheta \equiv \frac{1}{4} \tag{3.166}$$

which, as stated, is identical to $1/4$ independently of the center frequency. This identity is relatively easy to demonstrate using (A.39) and other Fourier transform properties. The reader is invited to check it. Substituting $\mathcal{I} = \mathcal{I}' = 1/4$ into (3.156) provides the correct result for the visibility uncertainty (3.96).

If four signal multipliers are used the result is the same. Now the integral $\mathcal{J}$ (3.163) must be substituted by

$$\mathcal{J}' = \frac{1}{\alpha_F^2}\int_0^\infty \Im m\left[\tilde{r}_1(\vartheta)e^{j2\pi\bar{f}_0\vartheta}\right]\,\Im m\left[\tilde{r}_2(\vartheta)e^{j2\pi\bar{f}_0\vartheta}\right]\,d\vartheta \equiv \frac{1}{4} \tag{3.167}$$

which is also mathematically equal to $1/4$ so reproducing the same result.

References

1 Albert Abraham Michelson and Francis Gladheim Pease. Measurement of the diameter of α-Orionis with the interferometer. *Astrophysical Journal*, 53(5):249–259, May 1921. doi: 10.1086/142603.

2 Joseph W. Goodman. *Statistical Optics*. John Wiley & So ns, Inc., New York, 1st edition, 1985. ISBN 0-471-01502-4.

3 A. Richard Thomson, James M. Moran, and George W. Swenson Jr. *Interferometry and Synthesis in Radio Astronomy*. John Wiley & Sons, Inc., 1986.

4 David M. Le Vine and J. C. Good. Aperture synthesis for microwave radiometers in space. Technical Memorandum TM-85033, NASA Goddard Space Flight Center, Greenbelt, Maryland 20771, August 1983.

5 Christofer S. Ruf, Calvin T. Swift, Alan B. Tanner, and David M. Le Vine. Interferometric synthetic aperture microwave radiometry for the remote sensing of the Earth. *IEEE Transactions on Geoscience and Remote Sensing*, 26(5):597–611, September 1988.

6 Adriano Camps. *Application of Interferometric Radiometry to Earth Observation*. PhD thesis, Universitat Politècnica de Catalunya, November 1996.

7 Manuel Martín-Neira. Introduction to Two-dimensional aperture synthesis microwave radiometry for Earth observation: The Visibility function. Technical Report Internal working paper 1933, ESTEC. European Space Agency, 5 March 1997.

8 Manuel Martín-Neira. Introduction to two-dimensional aperture synthesis microwave radiometry for Earth observation: Polarimetric formulation of the visibility function. Technical Report Internal working paper 2130, ESTE. European Space Agency, 12th October 2001.

9 Kevin McMullan, Michael Brown, Manuel MartNeira, W. Rits, S. Ekholm, J. Marti, and Jerzky Lemanzyk. SMOS: The payload. *IEEE Transactions on Geoscience and Remote Sensing*, 46(3):594–605, March 2008.

10 Manuel Martin-Neira, David M. LeVine, Yann Kerr, Niels Skou, Markus Peichl, Adriano Camps, Ignasi Corbella, Martti Hallikainen, Jordi Font, Ji Wu, Susanne Mecklenburg, and Matthias Drusch. Microwave interferometric radiometry in remote sensing: An invited historical review. *Radio Science*, 49(6):415–449, April 2014. doi: 10.1002/2013RS005230.

11 Manuel Martín-Neira, Martin Suess, and Juha Kainulainen. The flat target transformation. *IEEE Transactions on Geoscience and Remote Sensing*, 46(3):613–620, March 2008.

12 Adriano Camps, Ignasi Corbella, Francesc Torres, Mercè Vall-llossera, and Nuria Duffo. Polarimetric formulation of the visibility function equation including cross-polar antenna patterns. *IEEE Geoscience and Remote Sensing Letters*, 2(3):292–295, July 2005.

13 Sebastilanch Blanch, Jordi Romeu, and Ignasi Corbella. An exact representation of an antenna system diversity performance from its input parameter description. *IEE Electronics Letters*, 39(9):705–706, May 2003.

14 Ignasi Corbella, Nuria Duffo, Mercè Vall-llossera, Adriano Camps, and Francesc Torres. The visibility function in interferometric aperture synthesis radiometry. *IEEE Transactions on Geoscience and Remote Sensing*, 42(8):1677–1682, August 2004.

15 Ignasi Corbella, Francesc Torres, Adriano Camps, Nuria Duffo, Mercè Vall-llossera, Kimmo Rautiainen, Manuel Martín-Neira, and Andreas

Colliander. Analysis of correlation and total power radiometer front-ends using noise waves. *IEEE Transactions on Geoscience and Remote Sensing*, 43(11):2452–2459, November 2005.

16 Scott W. Wedge and David B. Rutledge. Wave techniques for noise modeling and measurement. *IEEE Transactions on Microwave Theory and Techniques*, 40(11):2004–2012, November 1992.

17 Pablo Moreno-Galbis, Juha Kainulainen, and Manuel Martín-Neira. Experimental demonstration of the Corbella equation for aperture synthesis microwave radiometry. *IEEE Transactions on Geoscience and Remote Sensing*, 45(4):945–957, April 2007.

18 David M. Pozar. *Microwave Engineering*. Addison Wesley, 1st edition, 1990. ISBN 0-201-50418-9.

19 Francesc Torres, Ignasi Corbella, Adriano Camps, Nuria Duffo, Mercè Vall-llossera, Santiago Beraza, Carlos Gutierrez, and Manuel Martín-Neira. Denormalization of visibilities for in-orbit calibration of interferometric radiometers. *IEEE Transactions on Geoscience and Remote Sensing*, 44(10):2679–2686, October 2006.

20 Andreas Colliander, Francesc Torres, and Ignasi Corbella. Correlation denormalization in interferometric or polarimetric radiometers: A unified approach. *IEEE Transactions on Geoscience and Remote Sensing*, 47(2):561–568, February 2009.

4

Aperture Synthesis

Antenna temperature and visibility are the primary measurements of interferometric radiometers. The first one characterizes the thermal noise power collected by a single antenna while the second one the complex cross-correlation between the signals collected simultaneously by two antennas. Both of them have units of Kelvin and are related to the brightness temperature of the source through the visibility equation.

Interferometric radiometers measure the visibility of a number of baselines made either of different pairs of fixed antennas or by sequentially moving two single elements. This last option is common in Very Long Baseline Interferometry (VLBI) for Radioastronomy but is rarely used in Earth observation, so it is not contemplated here. Instruments with fixed antennas have operated from both airplane and satellite for sensing the Earth's surface [1]. Some of them use interferometry in one dimension and real aperture in the other while others use two-dimensional interferometry. Only this last case is considered here, which leads to the assumption that the instrument is formed by a certain number of antennas placed at fixed positions on a plane.

The visibility is then a function of the relative antenna spacing and is introduced into the visibility equation to form, after inversion, brightness temperature images. This process is generally referred to as "aperture synthesis." As it will be seen later, it is equivalent to convolving the original (true) image with a two-dimensional function known as "synthetic beam." This one plays a similar role as the antenna pattern in total power radiometers. In particular, the synthetic beam width sets the angular resolution of the interferometer, which depends inversely on the overall instrument size.

The main performances of the instrument, such as spatial resolution, sensitivity, accuracy and others depend not only on the receivers' quality but also on the antenna geometrical layout and the kind of processing performed.

Principles of Interferometric and Polarimetric Radiometry, First Edition. Ignasi Corbella.

4.1 Synthetic Beam

Consider an idealized system with equal antenna patterns, no spatial decorrelation, and no inter-element coupling. According to section 3.3.4, in these conditions a two-dimensional Fourier transform relation between visibility and modified brightness temperature can be assumed. The visibility equation (3.79) is then expressed as

$$V(u,v) = \int_{-\infty}^{\infty}\int_{-\infty}^{\infty} T'_B(\xi,\eta)e^{-j2\pi(\xi u+\eta v)}d\xi d\eta \tag{4.1}$$

where $T'_B(\xi,\eta)$ is the modified brightness temperature (3.81), assumed null outside the unit circle, and $V(u,v)$ is the "visibility function." Variables (ξ,η) and (u,v) are defined, respectively, in (3.72) and (3.74).

Later in this chapter, section 4.3 provides details about visibility spatial sampling. The reader may wish to consult it for better understanding, but for now it is enough to be aware that, due to finite instrument size, the visibility function can only be measured in a limited region of the (u,v) plane denoted as "instrument coverage." This constraint can be mathematically described by assuming that the visibility function has an intrinsically infinite domain but it is multiplied by an "apodization window" $W(u,v)$ that vanishes outside the instrument coverage. In the most simple case, $W(u,v)$ is just one for all measured points and zero otherwise (rectangular window). The *recovered* modified brightness temperature is then the inverse Fourier transform of (4.1) after windowing,

$$\hat{T}'_B(\xi,\eta) = \int_{-\infty}^{\infty}\int_{-\infty}^{\infty} W(u,v)V(u,v)e^{j2\pi(\xi u+\eta v)}dudv \tag{4.2}$$

in which the hat is used to differentiate with respect to the original function. Substituting (4.1) into (4.2) it follows that:

$$\hat{T}'_B(\xi,\eta) = \int_{-\infty}^{\infty}\int_{-\infty}^{\infty} T'_B(\xi',\eta')\mathscr{B}(\xi-\xi',\eta-\eta')d\xi' d\eta' \tag{4.3}$$

where $\mathscr{B}(\xi,\eta) = \mathscr{F}^{-1}[W(u,v)]$ is the "synthetic beam." The above equation is a convolution integral involving the modified brightness temperature and the synthetic beam, so it compares to a similar equation applicable to total power radiometers (2.57) in which the brightness temperature is convolved with the antenna power pattern. It has the effect of smoothing the original image and, depending on the window used, inducing ripples in sharp transitions, as for example coastlines, in what is known as "Gibbs effect."

The synthetic beam is the impulse response of the instrument: if $T'_B(\xi,\eta) = T_0\delta(\xi,\eta)$ where $\delta(\cdot)$ is the two-dimensional Dirac delta function, then (4.3) simplifies to $\hat{T}'_B(\xi,\eta) = T_0\mathscr{B}(\xi,\eta)$. Since the delta function has unit volume, T_0 must be equal to the integral of the modified brightness temperature, which according to (3.78) is the antenna temperature. Using (2.53) for small sources, it follows that $T_0 = \Omega_s/\Omega_a T_s$ where T_s is the source brightness temperature, Ω_s the source solid angle and Ω_a the antenna solid angle. The same result is obtained from (4.2) by assuming that the visibility is a constant equal to (3.17) with zero phase for all (u, v).

The width of the synthetic beam $\mathscr{B}(\xi, \eta)$ sets the angular resolution of the instrument and depends on the particular antenna layout. Consider for example a rectangular instrument with an overall size $D_x \times D_y$ as the one depicted in figure 4.5, section 4.3.1.[1] The corresponding maximum values of (u, v) are $u_M = D_x/\lambda_0$ and $v_M = D_y/\lambda_0$. Assuming rectangular apodization window, the synthetic beam becomes the inverse Fourier transform of a two-dimensional function equal to one in the rectangular domain limited by $(-u_M, +u_M)$ and $(-v_M, +v_M)$ (see figure 4.5(b)) and zero outside. Using basic Fourier theory, the synthetic beam normalized to its maximum value is readily computed as

$$\overline{\mathscr{B}}(\xi, \eta) = \text{sinc } 2u_M\xi \;\; \text{sinc } 2v_M\eta \tag{4.4}$$

where the overbar indicates normalization. A plot of this function is shown in figure 4.1. As expected, there is a maximum at boresight ($\xi = \eta = 0$), but there are also strong ripples, especially along the axes $\xi = 0$ and $\eta = 0$. A cut of this function for $\eta = 0$ is included to show more clearly the shape. The "half-power" beam width is conventionally defined as the range of ξ (or η) for which the beam is larger than 0.5. In this case, it can be approximated by

$$\Delta\xi \approx \frac{0.6}{u_M} = 0.6\frac{\lambda_0}{D_x} \quad \text{Rectangular window} \tag{4.5}$$

which turns out to be inversely proportional to the overall instrument size in this direction D_x. According to (3.72), when $\eta = 0$, the other director cosine is $\xi = \sin\theta$

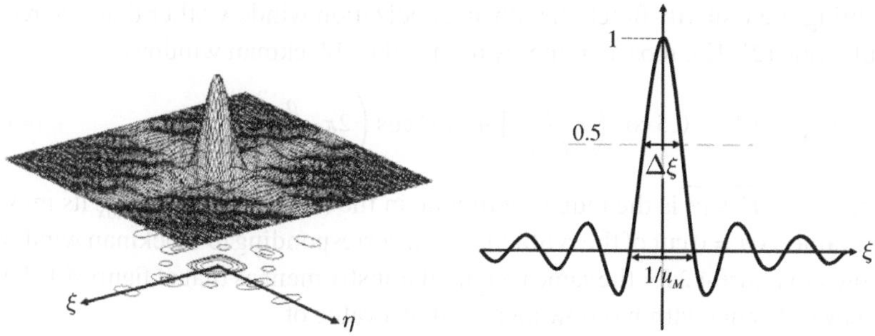

Figure 4.1 Synthetic beam for a rectangular instrument using a rectangular apodization window. The plot at the right is a cut for $\eta = 0$.

1 In the example of the figure, $D_x = 3\Delta x$ and $D_y = 3\Delta y$.

where θ is the elevation angle. In consequence, for small angles, $\Delta\xi \approx \Delta\theta$, so (4.5) can also be expressed in terms of the angle as $\Delta\theta = 34.3^\circ \lambda_0/D_x$. This result should now be compared with those provided in section 2.3.5 in reference to a total power radiometer. An interferometric radiometer has a slightly narrower beam width than a total power radiometer with an antenna having the same size as the whole instrument.

Equation (4.1) is written in terms of the "modified brightness temperature," so (4.3) refers also to this parameter. Taking into account (3.81), the synthetic pattern for brightness temperature would be the same but multiplied by the obliquity factor and divided by the antenna pattern of the individual elements. This correction has a minor effect on beam width and ripples since the obliquity factor is a smooth function and the individual antenna patterns are much wider than the synthetic beam due to the fact that the antennas themselves are much smaller than the whole interferometer.

As clearly seen in figure 4.1, the synthetic beam for rectangular window has a main lobe between $\pm 1/(2u_M)$ and secondary lobes beyond. These ones have alternate signs, so the recovered brightness temperature goes negative at some directions. This is an artifact associated with the technique and has no physical significance. It would never happen in a total power radiometer, where the beam is the antenna pattern, always positive by definition. Main beam efficiency (2.56) is now difficult to interpret.

In any case, the relatively large secondary lobes is an important drawback in this kind of instrument. For example, when imaging the Earth from a satellite, if a strong point source is present, as the Sun or a Radio Frequency Interference (RFI), it will be seen, not only as an intense point, but also with associated ripples across the whole field of view that can spoil all the image.

Ripples can be reduced, at the expense of degrading the spatial resolution, by weighting the visibility function with an apodization window other than the rectangular one [2]. The most commonly used is the "Blackman window,"

$$W_B = 0.42 + 0.5\cos\left(\pi\frac{\rho}{\rho_{max}}\right) + 0.08\cos\left(2\pi\frac{\rho}{\rho_{max}}\right) \tag{4.6}$$

where $\rho = \sqrt{u^2 + v^2}$ is the radial coordinate in the (u, v) plane and ρ_{max} its maximum value. A $\eta = 0$ cut of the synthetic beam corresponding to Blackman window is shown in figure 4.2 for the same rectangular instrument as that of figure 4.1. The half-power beam width has now increased to a value of

$$\Delta\xi \approx \frac{1.17}{u_M} = 1.17\frac{\lambda_0}{D_x} \quad \text{Blackman window} \tag{4.7}$$

which is about twice that of (4.5) for rectangular window. On the other hand there is an important reduction in secondary lobes. While the rectangular window has

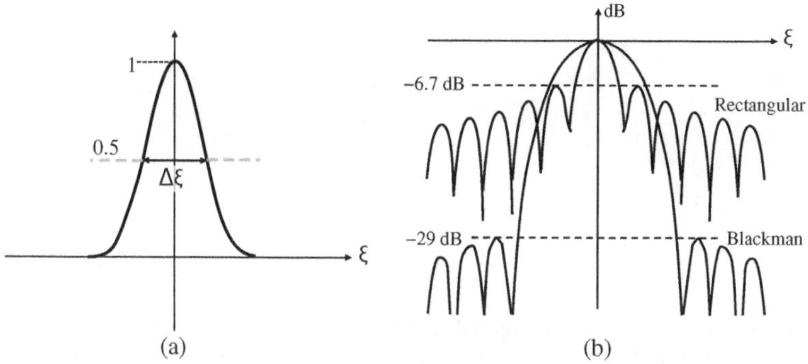

Figure 4.2 (a) Synthetic beam for a rectangular instrument with Blackman apodization window. (b) A comparison between rectangular and Blackman windows in a logarithmic scale.

the maximum secondary lobe at about 21% of its maximum value, the Blackman window is reduced to a tiny 0.14%, impossible to visualize in a linear scale. To be more illustrative, the plot of figure 4.2(b) shows a comparison between both cases on a logarithmic scale using decibels: $10 \log |\overline{\mathscr{B}}|$. The secondary lobes have a maximum level of about −6.7 dB for rectangular window that drops to about −29 dB for Blackman window.

4.1.1 Hexagonal Sampling

The synthetic beams shown so far assume that the visibility function is sampled in a rectangular grid in the (u, v) plane. If the hexagonal grid defined in section 4.3.5 is used instead, the main conclusions hold but ripples become located at six different directions instead of four. Figure 4.3 shows examples for both rectangular and Blackman windows for a Y-shape instrument with 21 antennas per arm separated $0.875\lambda_0$, which are the figures of the Microwave Imaging Radiometer with Aperture Synthesis (MIRAS) instrument [3]. Be aware that concepts developed in section 4.3, which the reader may wish to consult to fully understand the details, have been used to get these results. They have been obtained by direct application of (4.63) by assuming all visibilities equal to 1 for all k_1 and k_2 and normalizing the result with respect to its maximum value. A 512 points inverse DFT algorithm has been used with zero padding in the points outside the instrument coverage.

Again, the ripples are hardly seen in the second case (Blackman) but are perfectly visible in the first one and appear as a star shape with six tails, different than the cross shape seen in figure 4.1 when a rectangular grid is used. On the other hand, the half power beam width and the secondary lobes amplitude remain

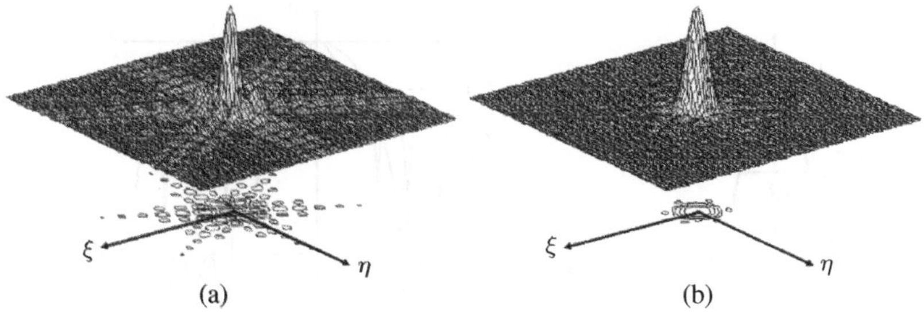

Figure 4.3 Synthetic beams for hexagonal sampling. (a) Rectangular window. (b) Blackman window. Both cases correspond to a Y-shape instrument with 21 elements per arm separated $0.875\lambda_0$.

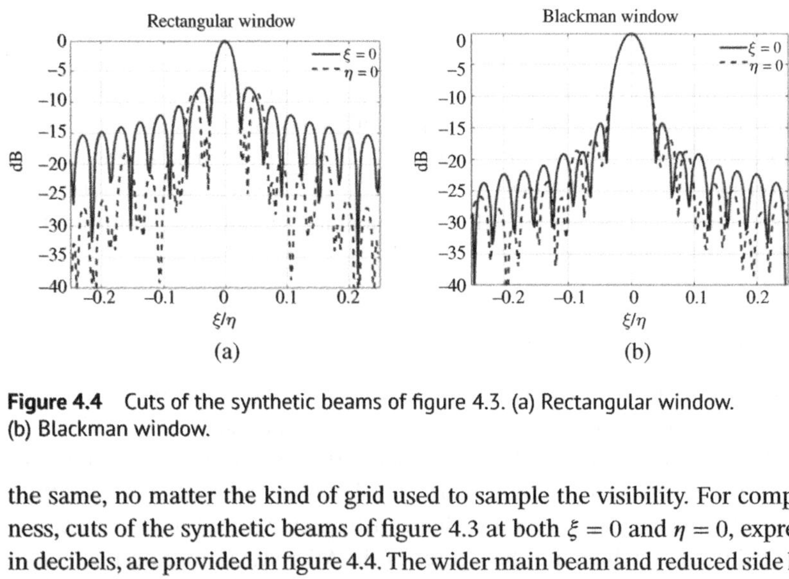

Figure 4.4 Cuts of the synthetic beams of figure 4.3. (a) Rectangular window. (b) Blackman window.

the same, no matter the kind of grid used to sample the visibility. For completeness, cuts of the synthetic beams of figure 4.3 at both $\xi = 0$ and $\eta = 0$, expressed in decibels, are provided in figure 4.4. The wider main beam and reduced side lobes of the Blackman window are in this plots more evident. Examples of a different instrument layout, still with hexagonal sampling, are provided later in figure 4.13.

The main conclusion of this section is that the spatial resolution of an interferometric radiometer is of the same order of magnitude as that of a total power instrument having the antenna as large as the whole interferometer; and that this resolution can be adjusted by signal processing techniques (windowing). In total power radiometers the angular resolution can be equivalently modified by the illumination of the aperture, although with less flexibility to the processing window in aperture synthesis.

4.2 Radiometric Sensitivity

The starting point for estimating the radiometric sensitivity is equation (4.2), which expresses the recovered *modified* brightness temperature as the inverse Fourier transform of the weighted visibility function. Although it neglects important effects that must be taken into account in a quality retrieval, for the purpose of sensitivity computation, it is a reasonable approximation. Antenna pattern differences from one element to the other, the fringe washing function and the "$-T_r$" term are responsible of systematic errors that have only limited impact on the measurement uncertainty due to thermal noise. They are dealt within the calibration and image reconstruction processes.

If the integration time is limited, at a given epoch the visibility measurement of a given baseline is a random variable. Therefore, the recovered brightness temperature (modified or not) is also a random variable. Consistent with the definition (2.76) used in total power radiometers, the radiometric sensitivity ΔT_B of an interferometric radiometer is the standard deviation of the recovered brightness temperature after inversion. Knowing that this one is related to its modified version by (3.81), it follows that

$$\Delta T_B(\xi, \eta) = \frac{\Omega_a \sqrt{1 - \xi^2 - \eta^2}}{t(\xi, \eta)} \sigma_{\hat{T}'_B} \tag{4.8}$$

where $\sigma_{\hat{T}'_B}$ is the standard deviation of the modified brightness temperature, $t(\xi, \eta)$ the single antenna power pattern and Ω_a the equivalent solid angle (1.243). From the above expression, it is apparent that the radiometric sensitivity depends on the spatial direction (ξ, η) with the shape of the inverse of the average antenna pattern and it is optimum at boresight, where the uncertainty becomes equal to $\Omega_a \sigma_{\hat{T}'_B}$. In section 4.2.1 it is demonstrated that $\sigma_{\hat{T}'_B}$ is independent of (ξ, η).

4.2.1 Variance of the Modified Brightness Temperature

As explained below in section 4.3, the visibility function is spatially sampled by the instrument at discrete points, so the inverse Fourier transform (4.2) is discrete

$$\hat{T}'_B(\xi, \eta) = \Delta_{uv} \sum_{n=1}^{N_V} W_n V_n e^{j2\pi(\xi u_n + \eta v_n)} \tag{4.9}$$

where N_V is the total number of visibility points (including the origin) and Δ_{uv} the elementary area in the (u, v) plane, which depends on the sampling grid and is evaluated later in (4.64). The subscript n denotes the discretization points (see for example figure 4.5) and W_n and V_n are, respectively, the apodization window and the visibility function at point (u_n, v_n). This single subscript notation for the visibility is useful in the present context. It is understood that a given value of n corresponds to two receivers with different (or equal) numbers, $n \rightarrow (k,j)$ so $V_n \equiv V_{kj}$ in the terminology of chapter 3. Obviously, a given receiver can be shared by more than one visibility sample.

The variance of $\hat{T}'_B$ is deduced from (1.13) taking also into account (1.3) and (1.7)

$$\sigma^2_{\hat{T}'_B} = \Delta^2_{uv} \sum_{n=1}^{N_V} W_n^2 \sigma^2_{V_n} + 2\,\Delta^2_{uv} \sum_{n=1}^{N_V-1} \sum_{m=n+1}^{N_V} W_n W_m \Re e[C_{V_n V_m}] \tag{4.10}$$

where $\sigma^2_{V_n}$ is the variance of the visibility sample V_n, equal to the square of (3.96) particularized for the corresponding receivers and $C_{V_n V_m}$ the covariance between pairs of visibility samples. The above result implies that, as already pointed out earlier, the variance of the modified brightness temperature is constant, independent on the spatial direction (ξ, η).

In order to simplify the expression for the radiometric sensitivity, all receivers are now assumed identical. This approximation is in line with having all antenna patterns also identical, needed for selecting (4.2) as visibility equation. In this case, (3.96) predicts that all visibility samples have the same uncertainty independently of the particular receiver pair involved: the fringe washing function at the origin is always equal to 1, the system temperature is the same in all receivers, and the constant α_F is the same for all baselines. Putting then $\sigma_{V_n} = \sigma_V$ in (4.10)

$$\sigma^2_{\hat{T}'_B} = \Delta^2_{uv} \sigma^2_V N_V \alpha^2_w + \Delta^2_{uv} \varepsilon \tag{4.11}$$

where ε is the second term excluding the elementary area, and α_w is a constant that depends on the apodization window used,

$$\alpha^2_w = \frac{1}{N_V} \sum_{n=1}^{N_V} W_n^2 \tag{4.12}$$

For a rectangular window, this constant is equal to $\alpha_w = 1$, but otherwise, it is always lower than one and typically of the order of $\alpha_w \sim 0.5$.

The variance of the visibility in (4.11) is equal to the square of (3.96) assuming identical receivers. Replacing the constant α_F by its definition (3.95) it can be expressed as

$$\sigma^2_V = \frac{1}{B^2 \tau_i} \int_0^\infty T^2_{sys} \, |\overline{H}(f)|^4 \, df \tag{4.13}$$

where the system temperature has been placed inside the integral for illustrative purposes.

The second term of (4.10) [termed ε in (4.11)] contains the covariance between two visibility samples. This is duly deduced below in section 4.2.3 and, in the same conditions of identical receivers, it is given by (4.33), rewritten here for convenience

$$C_{V_n V_m} = \frac{1}{B^2 \tau_i} \int_0^{\infty} \mathbb{V}_{n'}(f) \mathbb{V}_{m'}^*(f) \, |\overline{H}(f)|^4 df \tag{4.14}$$

where $\mathbb{V}_{n'}(f)$ is the system visibility (3.48) and the indices n' and m' correspond to rearranging the receiver pairs assigned to n and m in the following way

$$\text{if } n \to (1,2) \ \text{ and } m \to (3,4) \ \text{ then } n' \to (1,3) \ \text{ and } m' \to (2,4) \tag{4.15}$$

which the reader can easily generalize to any receiver numbering.

The two equations (4.13) and (4.14) are equivalent with only replacing system temperature by system visibility. This one depends in general of the geometry of the source and has important differences among baselines both in amplitude and in phase. When imaging an extended source, typical in Earth observation from space, its amplitude is in general small and often much lower than the system temperature. Then it is reasonable to assume the following approximation

$$|\mathbb{V}_n(f)| \ll T_{sys} \qquad (\forall n) \quad \text{(extended source)} \tag{4.16}$$

Furthermore, due to visibility phase differences, the real part of (4.14), which is in fact the parameter appearing in (4.10), becomes positive or negative depending on the pair of visibility samples considered. Therefore, the net contribution to the sum in (4.10) tends to vanish when all summands are included. The conclusion is that the second term of (4.10) is in practice much smaller than the first one and can be neglected, which is equivalent to assuming that the visibility samples are uncorrelated. This important approximation is assumed in section 4.2.2 while the following one (4.2.3) elaborates more in this issue.

4.2.2 Uncorrelated Visibility Samples

Still assuming identical receivers and antennas, if the visibility samples are uncorrelated, the variance of the modified brightness temperature is (4.11) with $\varepsilon = 0$. The standard deviation is the square root of it, which after inserting (3.96), also with identical receivers, becomes

$$\sigma_{\hat{T}'_B} = \Delta_{uv} \frac{T_{sys}}{\sqrt{B\tau_i}} \alpha_F \alpha_w \sqrt{N_V} \tag{4.17}$$

where α_F is defined in (3.95) and depends on the filters' shape. Introducing this result into (4.8)

$$\Delta T_B(\xi, \eta) = \Delta_{uv} \frac{T_{sys}}{\sqrt{B\tau_i}} \alpha_F \alpha_w \sqrt{N_V} \frac{\Omega_a \sqrt{1 - \xi^2 - \eta^2}}{t(\xi, \eta)} \tag{4.18}$$

which is the general expression of the radiometric sensitivity of an interferometric radiometer under the simplifications mentioned in its derivation. It was originally derived in [2]. At boresight $\xi = \eta = 0$ the antenna pattern is one by definition $t(0, 0) = 1$. Therefore,

$$\Delta T_B = \Delta_{uv} \frac{T_{sys}}{\sqrt{B\tau_i}} \alpha_w \alpha_F \sqrt{N_V} \Omega_a \quad \text{(at boresight)} \tag{4.19}$$

Two general remarks are interesting to highlight about this expression,

- Weighting the visibility samples by an apodization window other than rectangular ($\alpha_w < 1$) reduces the measurement noise.
- Reducing the spacing between antennas diminishes the elementary area [see (4.64)] and has a positive impact in noise reduction.

Furthermore, by comparing it with that of a total power radiometer, which is the result of substituting (2.82) into (2.76), it is apparent that

- The radiometric sensitivity is degraded by a factor equal to the product of the window factor times the square root of the number of visibility samples. The more antennas, the less sensitivity.
- The sensitivity is also worsened by a factor approximately equal to the antenna solid angle, which is proportional to the inverse of the antenna directivity. High-directive antennas improve the radiometric sensitivity.
- Even with the same receiver parameters, the system temperature in the numerators of (2.82) and (4.19) may not be comparable, since the pattern of an individual antenna in an interferometric radiometer is usually much wider than that of a total power radiometer, and then the antenna temperature has contributions from a larger source extension.

4.2.2.1 Ratio of Areas Formula

Using (1.279) the antenna solid angle in (4.19) can be substituted by λ_0^2/A_{eff} where A_{eff} is the effective area of an individual antenna. Also, using for example the rectangular setup of figure 4.5, the elementary area in the (u, v) plane becomes $\Delta_{uv} = \Delta x \Delta y / \lambda_0^2$. With these substitutions (4.19) becomes

$$\Delta T_B = \frac{T_{sys}}{\sqrt{B\tau_i}} \alpha_w \alpha_F \frac{N_V \Delta x \Delta y}{\sqrt{N_V} A_{eff}} \tag{4.20}$$

in which the numerator of the second fraction can be interpreted as the total synthesized area A_{synth} and the denominator roughly as the sum of the individual areas of all elements, provided the number of them is approximated by the square root of the number of visibilities. This expression may be useful to get a “rule of thumb” of the radiometric sensitivity, but it will no longer be used in this book. It can be found in some references, as for example [4].

4.2.2.2 Redundant Baselines

The result (4.18) assumes that each visibility sample is used only once. In general there is always some degree of redundancy (see section 4.3) that allows averaging two or more visibility measurements of a given baseline before using the inversion equation. In this case, since samples are uncorrelated, the averaged visibilities have reduced noise and this has a positive impact on the radiometric sensitivity. For a given sample V_n, the redundancy order RO_n is the number of different measurements using different receivers pairs that correspond to the same visibility sample n. Since the variance of this particular sample is reduced by this factor, to take into account redundancy it suffices to multiply, in the first term of (4.10), the window by the inverse of the redundancy factor, or equivalently to compute α_w^2 as

$$\alpha_w^2 = \frac{1}{N_V}\sum_{n=1}^{N_V}\frac{W_n^2}{RO_n} \quad \text{(Redundant baselines)} \tag{4.21}$$

If no redundant baselines are used, all redundancy orders become equal to one and this equation reverts to (4.12). In case of using a rectangular window, then $W_n = 1\,\forall n$ and this is equivalent to replace in (4.18) or (4.19) the number of visibilities N_V by an *effective value*

$$N'_V = \sum_n \frac{1}{RO_n} \tag{4.22}$$

while keeping $\alpha_w = 1$. For another type of window this is not totally correct but still using this effective number of visibilities provides a good estimation of the effect of redundancy on the radiometric sensitivity and is convenient to make quick assessments.

4.2.3 Correlation of Visibility Samples

The general expression (4.18) relies on the assumption that the net contribution of the covariance between visibility samples is negligible, which is equivalent to making $\varepsilon = 0$ in (4.11). The expression for this term is

$$\varepsilon = 2\sum_{n=1}^{N_V-1}\sum_{m=n+1}^{N_V} W_n W_m \Re e[C_{V_nV_m}] \tag{4.23}$$

which has a total of $N_V(N_V - 1)/2$ summands, some of them positive and others negative in general. The covariance between visibility samples $C_{V_nV_m}$ is now computed.

Consider two baselines formed by receiver pairs (k, j) and (l, m), each one using the measurement setup of figure 3.7. Following the one-subscript convention of (4.9), the two baselines are referenced by subscripts 1 and 2, respectively. If the corresponding correlation measurements are $\Gamma_1(t)$ and $\Gamma_2(t)$, equation (3.89) states that

$$\langle \Gamma_1(t) \rangle = G_1 V_1 \; ; \; \langle \Gamma_2(t) \rangle = G_2 V_2 \tag{4.24}$$

and the covariance between visibility samples when the integration time is limited follows from (1.7)

$$C_{V_1 V_2} = \frac{C_{\Gamma_1 \Gamma_2}}{G_1 G_2^*} \tag{4.25}$$

which reverts to the square of (3.87) when a single baseline is considered. In turn, the covariance of the two output signals $\Gamma_1(t)$ and $\Gamma_2(t)$ can be deduced with the aid of equation (1.116)

$$C_{\Gamma_1 \Gamma_2} = \frac{S_{\rho_1' \rho_2'}(0)}{\tau_i} \tag{4.26}$$

where $S_{\rho_1' \rho_2'}(f)$ is the cross-power spectral density of the instantaneous correlations (3.88) after removing their respective means, $\rho_1'(t) = \rho_1(t) - \langle \rho_1(t) \rangle$ and $\rho_2'(t) = \rho_2(t) - \langle \rho_2(t) \rangle$

$$S_{\rho_1' \rho_2'}(f) = \mathcal{F}\left[R_{\rho_1' \rho_2'}(\tau)\right] = \mathcal{F}\left[R_{\rho_1 \rho_2}(\tau) - \langle \rho_1(t) \rangle \langle \rho_2(t) \rangle^*\right] \tag{4.27}$$

where equation (1.54) has been used.

On the other hand, directly from (3.88) and taking into account that baseline 1 corresponds to receivers (k, j) and baseline 2 to receivers (l, m),

$$R_{\rho_1 \rho_2}(\tau) = \frac{1}{4} \left\langle b_k(t) b_j^*(t) b_l^*(t - \tau) b_m(t - \tau) \right\rangle \tag{4.28}$$

where $b_k(t)$ is the analytic signal coming out of the receiver identified by the corresponding subscript. Since these signals are circular jointly normal processes [see remark at the end of section 1.3], equation (1.35) applies. Identifying in this equation $z_1 = b_k(t)$, $z_2 = b_m(t - \tau)$, $z_3 = b_j(t)$ and $z_4 = b_l(t - \tau)$ the following result is readily obtained:

$$R_{\rho_1' \rho_2'}(\tau) = \frac{1}{4} R_{b_k b_l}(\tau) R_{b_j b_m}^*(\tau) \tag{4.29}$$

which, as expected, reduces to (3.90) when $k = l = 1$ and $j = m = 2$. Using the conjugation and multiplication properties of the Fourier transform (see table A.1),

$$S_{\rho_1' \rho_2'}(f) = \frac{1}{4} S_{b_k b_l}(f) * S_{b_j b_m}^*(-f) = \frac{1}{4} \int_{-\infty}^{\infty} S_{b_k b_l}(f') S_{b_j b_m}^*(-f + f')\, df' \tag{4.30}$$

where the convolution definition (A.42) has been used. Particularizing at $f = 0$, substituting the result in (4.26) and then in (4.25) and replacing the dummy variable of integration f' by f yields

$$C_{V_1 V_2} = \frac{1}{G_1 G_2^*} \frac{1}{4\tau_i} \int_0^{\infty} S_{b_k b_l}(f) S_{b_j b_m}^*(f) df \tag{4.31}$$

where the limits of the integral include only positive frequencies since the spectral densities of analytic signals are zero otherwise. Again, it is interesting to observe that this equation is the generalization of (3.93) [substituted into (3.87)] for two baselines.

Equation (3.110) is now used to express the spectral densities of (4.31) as functions of the system visibilities $\mathbb{V}_{kj}(f)$ defined in (3.48) [or their polarimetric versions]. Taking into account also the definition of the complex gain (3.83) and restoring the two-subscript convention of the visibility,

$$C_{V_{kj}V_{lm}} = \frac{1}{\tau_l\sqrt{B_kB_jB_lB_m}\tilde{r}_{kj}(0)\tilde{r}^*_{lm}(0)} \times \int_0^\infty \overline{H}_k(f)\overline{H}^*_l(f)\overline{H}_j(f)\overline{H}^*_m(f)\mathbb{V}_{kl}(f)\mathbb{V}^*_{jm}(f)df \tag{4.32}$$

where $\overline{H}_k(f)$ is the normalized frequency response of receiver k, $H_k(f)/\sqrt{G_k}$, B_k its noise equivalent bandwidth (3.55) and $\tilde{r}_{kj}(\tau)$ the fringe washing function (3.54) corresponding to receivers k and j. As expected, if $k = l = 1$ and $j = m = 2$, this equation reduces exactly to the square of (3.96) since, in view of (3.45), the system visibilities become system temperatures, assumed frequency independent.

Equation (4.32) must be evaluated for all pairs of visibility samples and introduced into (4.23) to determine the additive error term in the variance of the modified brightness temperature (4.10); used later to infer the radiometric sensitivity. This procedure is not straightforward in general since (4.32) depends on the system visibility, which in turn depends on the source geometry and on the receiver parameters. An approximation and a special case are given below for illustrative purposes.

If, consistent with the assumption made in (4.17), all four frequency responses are identical and equal to $H(f)$ with bandwidth B, then

$$C_{V_{kj}V_{lm}} = \frac{1}{\tau_l B^2}\int_0^\infty |\overline{H}(f)|^4\mathbb{V}_{kl}(f)\mathbb{V}^*_{jm}(f)df \tag{4.33}$$

If furthermore the visibility is assumed frequency independent, the small source approximation (3.59) is used, and the filters have rectangular shape, then $C_{V_{kj}V_{lm}} \approx V_{kl}V^*_{jm}/(B\tau_l)$, which is the expression provided in [5]. The validity of this result is limited to the mentioned approximations, and it is not further used here.

4.2.3.1 Point Source

As an example, equation (4.33) is now evaluated for a point source located at a given spatial direction. In this case, the system visibility is (3.120), valid under the assumptions stated in reference to this equation. Assuming that all antenna temperatures are the same,

$$\mathbb{V}_{kl}(f) = T_a e^{-j2\pi\Delta t_{kl}f} \tag{4.34}$$

where Δt_{kl} is the arrival time delay difference (3.121) at antennas k and l. Introducing this equation into (4.33)

$$C_{V_{kj}V_{lm}} = \frac{T_a^2}{\tau_i B^2} \int_0^{\infty} |\overline{H}(f)|^4 e^{-j2\pi\Delta t f} df \tag{4.35}$$

where Δt is

$$\Delta t = \Delta t_{kl} - \Delta t_{jm} = \frac{r_k - r_l}{c} - \frac{r_j - r_m}{c} \tag{4.36}$$

Only the real part of this covariance goes into the error term (4.23). Assuming for simplicity, an ideal rectangular filter shape and noting that (4.35) is an inverse Fourier transform evaluated at $t = \Delta t$,

$$\Re e[C_{V_{kj}V_{lm}}] = \frac{T_a^2}{B\tau_i} \operatorname{sinc}(B\Delta t) \cos 2\pi f_0 \Delta t \tag{4.37}$$

where the symmetry of the sinc and cosine functions has been taken into account. For a different filter shape, the function sinc($B\Delta t$) should be replaced by one similar to the fringe washing function (3.62) but involving the fourth power of the frequency response amplitude instead of the square.

In order to properly sample the visibility (see section 4.3), a typical imaging instrument must have baselines at different antenna spacing from zero to several wavelengths. Therefore, the phase $2\pi f_0 \Delta t$ in (4.37) spans several cycles and then the cosinus gets both positive and negative values depending on the baseline pair considered. In consequence, after summing up all contributions in (4.23), the result tends to vanish.

For example, consider an interferometer for which the maximum distance between antennas is $\kappa\lambda_0$ where κ is much larger than one. The associated time difference for the largest baseline is $\kappa\lambda_0 \cos\alpha/c$ where c is the velocity of light and α the angle of arrival of the radiation (a glimpse to figure 3.1 may help the reader). The maximum time difference Δt in (4.36) is then twice this value, corresponding to subtracting the delay of the largest baseline from the one resulting of swapping the same antennas. Then $2\pi f_0 \Delta t(\max) = 4\pi\kappa \cos\alpha$ while the minimum phase can go negative depending on the baselines chosen. Unless the source is located at boresight ($\alpha = \pi/2$), if κ is much larger than one, then the associated phase difference spans more than one complete cycle of 2π radians. Additionally, for relatively large time differences, the sinc function in (4.37) is always lower than one so making this term even lower.

In extended sources, the main conclusion does not change since they can be considered as superposition of many elementary point sources so the oscillatory behavior remains. Besides, as already pointed out before, in this case, the amplitude of the visibility is much lower than the antenna temperature in all baselines, resulting in all covariances (4.33) having low amplitude. In a practical situation then, the error term in (4.11) can be neglected.

Nevertheless, there is only one *very particular case* in which the visibilities are correlated. It is a boresight point source, for which $\Delta t = 0$ in all baselines and (4.37) reduces to $T_a^2/(B\tau_i)$. In this case, assuming rectangular window, substituting (4.23) into (4.11) and using also (3.96), the variance of the modified brightness temperature becomes

$$\sigma^2_{\hat{T}'_B} = \Delta^2_{uv} \frac{T^2_{sys}N_V + T_a^2 N_V(N_V - 1)}{B\tau_i} \quad \text{Point source at boresight} \tag{4.38}$$

This result is not surprising since now all signals arriving at different antennas are replicas of exactly the same one, since they are originated in a single source and traveled exactly the same distance. Note that, if the receiver noise is disregarded, then $T_{sys} = T_a$ and the above equation reduces to $\Delta^2_{uv} N_V^2 T_a^2/(B\tau_i)$, which is equivalent to multiplying the variance of the visibility by N_V^2, consistent with the second equation (1.15) knowing that the visibility samples are now correlated. However, this is a very impractical example. It makes no sense using aperture synthesis to image a pure point source located at boresight.

4.3 Spatial Sampling

Samples of the visibility function (4.1) at discrete points on the (u, v) plane are obtained by measuring the cross-correlation between the output signals of receiver pairs with corresponding antennas spread along a mechanical structure. The way this spatial sampling is performed has important implications for the methods and algorithms used to produce brightness temperature images in this kind of instruments. Consequently, the specific antenna layout is a key design parameter of interferometric radiometers.

4.3.1 Visibility Coverage

Figure 4.5(a) shows an example of an instrument with twelve antennas on the $z = 0$ plane arranged as a rectangular shape with, in general, different spacing between elements in each direction. Each antenna pair (baseline) has associated a sample of the visibility function at coordinates (u, v), duly defined in (3.74) as the relative distance between both antennas normalized to the wavelength. Due to the hermiticity property (3.23), each baseline actually corresponds to *two* visibility samples, one conjugate of the other. Additionally, for each single element, the antenna temperature is a measurement of the visibility at the origin, $u = v = 0$, sometimes termed "zero-spacing baseline" (see section 3.3.4).

Different baselines may correspond to a single point in the (u, v) plane. For example, in the instrument of figure 4.5 all six baselines formed by adjacent elements in the horizontal arms have the same coordinates $u = \Delta x/\lambda_0$, $v = 0$. The

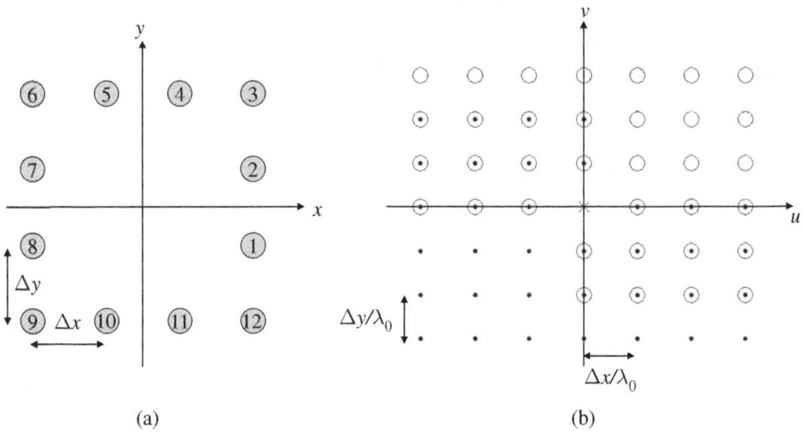

Figure 4.5 Interferometric radiometer with rectangular layout. (a) Antenna positions. (b) Visibility sampling in the u, v domain.

most extreme case is the visibility at the origin, which has a redundancy order equal to the number of antennas. In practice all redundant measurements can be averaged to reduce thermal noise and so improving radiometric sensitivity (see sections 4.2 and 4.4.2).

The drawing of figure 4.5(b) shows the sampling points in the (u, v) plane for all antenna pairs (k, j) with the numbering shown in figure 4.5(a). The dots correspond to pairs with $k < j$ and the circles to their conjugate $(k > j)$. The point at the origin is marked with a star. In this example, the total number of (u, v) points is 49 while the number of antenna pairs is 66, or 132 when the conjugated points are considered. For a given instrument layout, the (u, v) grid is unique but the distinction between "nominal" and "conjugated" points depends on the antenna numbering criterion. This distinction is irrelevant and will be omitted from now on.

4.3.2 Reciprocal Grids

Still assuming the rectangular instrument of figure 4.5, but now with $\Delta x = \Delta y = \Delta$ to simplify the formulation, all points in the (u, v) plane can be generated by

$$u = k_1 d \; ; \; v = k_2 d \tag{4.39}$$

with $d = \Delta/\lambda_0$ and k_1 and k_2 sets of sequential integers within the following bounds

$$[N_m \cdots N_M] = \begin{cases} \left[-\frac{N_T}{2} \cdots \frac{N_T}{2} - 1\right] & N_T \text{ even} \\ \left[-\frac{N_T-1}{2} \cdots \frac{N_T-1}{2}\right] & N_T \text{ odd} \end{cases} \tag{4.40}$$

where N_T is the total number of points in one direction. In the example of figure 4.5, $N_T = 7$ so both k_1 and k_2 are equal to $[-3, -2, -1, 0, +1, +2, +3]$. In any case, either for N_T even or odd, it follows that $N_M = N_m + N_T - 1$.

The value of N_T can be arbitrarily increased so that the indices of (4.39) become just a subset of the complete sequence (4.40). In order that all (u, v) points of measured visibilities have corresponding indices (k_1, k_2) within the bounds (4.40), the following condition must be met

$$N_T \leq 2N_{EL} + 1 \tag{4.41}$$

where N_{EL} is the number of elements per arm, each one including one vertex. In the example of figure 4.5 $N_{EL} = 3$.

The brightness temperature is recovered by inverting (3.77) in a limited set of director cosines (ξ, η) (see section 4.4). In principle, the choice of the (ξ,η) grid is irrelevant, but efficient inversion methods recommend using a "reciprocal grid" defined as

$$\xi = \frac{n_2}{N_T d} \; ; \; \eta = \frac{n_1}{N_T d} \tag{4.42}$$

where n_2 and n_1 are integers with the same bounds (4.40) so that both (k_1, k_2) and (n_2, n_1) are *identical* regular grids[2] of $N_T \times N_T$ integers and N_T^2 is the total number of grid points in each of the two domains.

As an example, figure 4.6 shows the (u, v) and (ξ, η) grids of the instrument of figure 4.5 with equal spacing in both directions. In each domain, all grid points generated by (4.40) form the so-called "principal period." As shown in the figure,

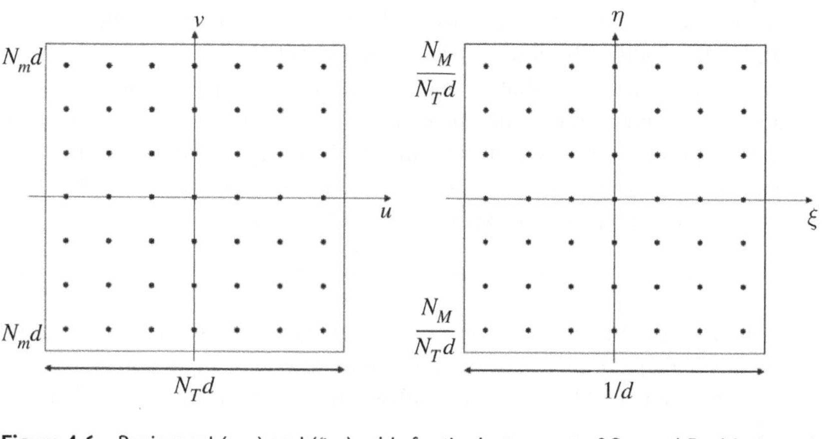

Figure 4.6 Reciprocal (u, v) and (ξ, η) grids for the instrument of figure 4.5 with $\Delta x = \Delta y$. The rectangles limit the principal period.

2 Swapping of subscripts 1 and 2 is due to historical reasons.

all of them lay inside a square of size $N_T d$ in (u, v) and $1/d$ in (ξ, η). Adjacent periods in any direction are obtained by adding or subtracting N_T to the indices of the principal period.

Although not strictly needed, increasing the number of points N_T can be useful. It can be done by extending the (u, v) principal period with new points and assigning zero visibility to them, a procedure commonly referred to as "zero padding." With this operation the (u, v) spacing d remains unchanged and the principal period in this domain enlarges to include the extra points. On the other hand, the (ξ, η) grid becomes thicker to allocate the new points in the principal period, which keeps the same size $1/d$ independent of N_T. Be aware, however, that a denser grid does not mean increased spatial resolution, always determined by the width of the synthetic beam (section 4.1). Zero padding is a way of interpolating the brightness temperature to a new grid. The (ξ, η) principal period can only be enlarged by reducing the distance between antennas.

4.3.3 Aliasing

The periodicity of the complex exponential function is responsible of the known aliasing effect associated with discrete sampling. Applying to (4.1) the change of variables (4.39) and (4.42),

$$V(k_1, k_2) = \Delta_{\xi\eta} \sum_{(n_2,n_1)\subset UC} T'_B(n_2, n_1) e^{-j\frac{2\pi}{N_T}(k_1 n_2 + k_2 n_1)} \tag{4.43}$$

where $\Delta_{\xi\eta} = 1/(N_T^2 d^2)$ is the Jacobian determinant of (4.42), interpreted as elementary area, and UC stands for "unit circle," that is the region of (n_2,n_1) for which the corresponding (ξ, η) values lay inside the region $(\xi^2 + \eta^2 < 1)$. Outside it, by definition, the modified brightness temperature is null.

All values of n_2 and/or n_1 that do not belong to the principal period are beyond the range (4.40) and therefore they are equal to corresponding indices of the principal period plus or minus N_T. In consequence the complex exponential function is the same. For example, if $n'_2 = n_2 + N_T$ where n_2 belongs to the set (4.40), then the corresponding phase in the exponent of (4.43) becomes

$$\frac{2\pi}{N_T}[k_1(n_2 + N_T)] = \frac{2\pi}{N_T} k_1 n_2 + 2\pi k_1 = \frac{2\pi}{N_T} k_1 n_2 \tag{4.44}$$

where the last equality holds because k_1 is an integer. Therefore, each grid point outside the principal period has the same phase as one or more points inside it. The contribution to (4.43) of two "equal phase" points (n'_2, n'_1) outside and (n_2, n_1) inside is then

$$\begin{aligned} &T'_B(n_2, n_1) e^{-j\frac{2\pi}{N_T}(k_1 n_2 + k_2 n_1)} + T'_B(n'_2, n'_1) e^{-j\frac{2\pi}{N_T}(k_1 n'_2 + k_2 n'_1)} \\ &= \left[T'_B(n_2, n_1) + T'_B(n'_2, n'_1)\right] e^{-j\frac{2\pi}{N_T}(k_1 n_2 + k_2 n_1)} \end{aligned} \tag{4.45}$$

which means that (4.43) can be written using *only the points inside the principal period* as long as the modified brightness temperature outside it is added to the one inside at positions having the same phase.

$$V(k_1, k_2) = \Delta_{\xi\eta} \sum_{n_2=N_m}^{N_M} \sum_{n_1=N_m}^{N_M} \hat{T}'_B(n_2, n_1) e^{-j\frac{2\pi}{N_T}(k_1 n_2 + k_2 n_1)} \tag{4.46}$$

where $\hat{T}'_B(n_2, n_1)$ is the sum of the different contributions of equal phase points. The ones not belonging to the "true" point (n_2, n_1) are known as "aliases."

To have the whole (ξ, η) space inside the principal period, the unit circle must be allocated inside the square of side $1/d$ (see figure 4.6). Then all points outside the principal period are also outside the unit circle and $T'_B(n'_2, n'_1) = 0$ in (4.45), so there are no aliases. Otherwise, if the antenna spacing is larger, some director cosines lay in zones outside the principal period and nonzero aliases are present in (4.46).

Since $d = \Delta/\lambda_0$, this condition translates into $\Delta < \lambda_0/2$, meaning that the antenna spacing must be half a wavelength or less to avoid aliasing. This result is recognized as the Nyquist sampling criterion.

4.3.4 Field of View

Figure 4.7 may help understanding the aliasing effect just explained. Figure 4.7(a) shows a theoretical example of a modified brightness temperature image in the whole unit circle. The elliptical shape corresponds roughly to the Earth disc as seen from a Low Earth Orbit (LEO) satellite at 770 km altitude and a tilt angle of 32.5°,

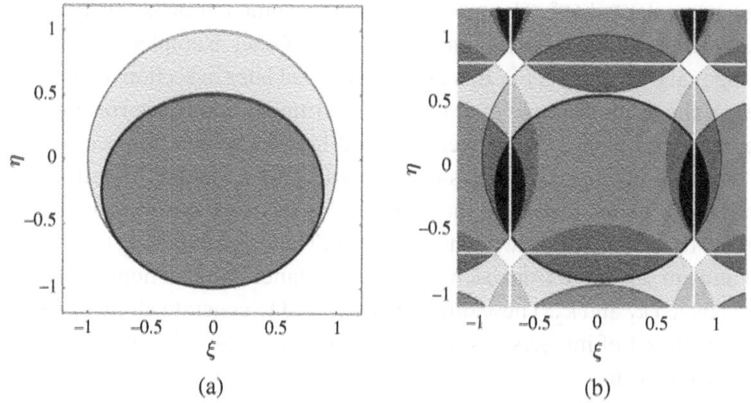

Figure 4.7 (a) An example of modified brightness temperature image in the whole unit circle to illustrate aliasing. (b) Recovered image under aliasing conditions.

which are the parameters of the Soil Moisture and Ocean Salinity (SMOS) satellite. To simplify the example for illustration purposes, only two levels of modified brightness temperature are considered, a low value in the "sky" zone and a larger value on the "Earth." Figure 4.7(b) shows the way it would be seen by an interferometric radiometer with spacing $d = 0.65$, clearly in an undersampling condition. The principal period is marked with a white line. As expected, the image repeats itself in adjacent periods. The original image is entirely inside the principal period, but the parts that were originally outside it are "folded" over the principal period. Only in the small central area between the unit circle aliases the retrieval is fully correct. This zone is usually denoted as "alias-free field of view." It has a width at boresight of $2(1/d - 1)$ which, in this case, is equal to about 1.

When imaging the Earth from a LEO satellite, as in this example, the brightness temperature of the sky induces artifacts in the zone where its aliases overlap with the Earth points in the principal period. Since the brightness temperature of the sky is much lower than that of the Earth, the impact in the final image is small, although non-negligible. Furthermore, it can be corrected by using a priori information (see section 4.4.8) coming, for example, from ground-based radiometers. For this reason, these areas of the image, where only sky aliases are added to the Earth, are still usable and enlarge de-facto the field of view. The zone is called "extended alias-free field of view" and is easily recognized in the example of figure 4.7.

4.3.5 Hexagonal Grids: Y-shape Instrument

The alias-free field of view can be enlarged by using a hexagonal grid instead of a rectangular one. Hexagonal grids are obtained with instruments having a layout with some arms slanted 60°. The most common implementation is the so-called Y-shape, which has three arms separated 120° apart. Other examples are triangular or hexagonal instruments [6], this last one detailed later in section 4.3.6.

Figure 4.8 shows an example of a Y-shape instrument and the corresponding coverage in the u, v plane. The convention used here for orientation is a rotated Y-shape with one side horizontal (instead of vertical for an actual "Y" letter). All points in the u, v plane, including the zero-spacing baseline and the conjugated ones, lay into the star shape drawn in figure 4.8(b). These points belong to a hexagonal grid and are separated from each other by a distance d in two nonorthogonal directions labeled k_1 and k_2. These directions are used to generate all (u, v) points from a complete set of integers. Using directly the proposal of [7], in hexagonal grids (4.39) converts to

$$u = \frac{d}{2}(2k_1 - k_2)\ ;\ v = \frac{d\sqrt{3}}{2}k_2 \tag{4.47}$$

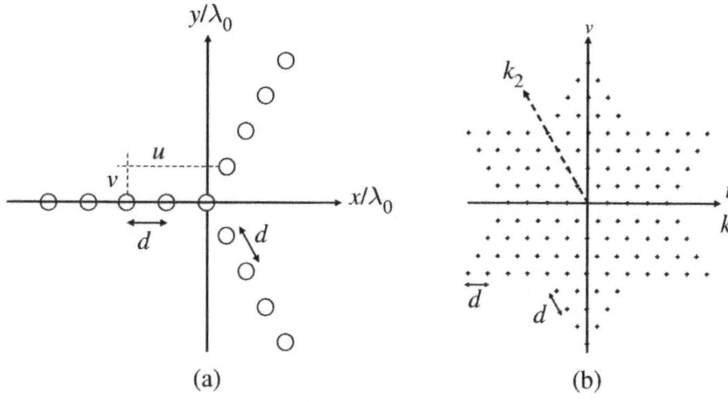

Figure 4.8 Example of Y-shape instrument. (a) Antenna layout. (b) $u - v$ coverage.

When k_1 and k_2 in these equations are substituted by the complete set of integers (4.40), a grid with a rhomboidal shape is obtained. However, by adding N_T to specific points, the principal period can be converted to a hexagon. This is further explained with the aid of figure 4.9. On figure 4.9(a), all grid points generated by introducing (4.40) into (4.47) are drawn. All of them lay inside a rhomboidal principal period that is eventually replicated in the same way as the rectangular case considered before. However, this rhomboid can be converted to a hexagon by using points of adjacent periods. In reference to the figure, all points in the triangles labeled A' to D' have the same values of k_1 and k_2 as triangles A through D but adding or subtracting N_T to one of them or both. In this way, a new principal period is defined, this time with a hexagonal shape. It is important to point out that the criterion for selecting which points lay just in the hexagon border is not unique and can differ from the one shown in the figure. Here, each hexagon side has one half with points and the other half without any, but one can consider, for example, three sides with points and the other three without. In any case, the total number of points inside the hexagon must be the same as those of the rhomboid and the hexagonal period is replicated in a form similar to a honeycomb. The six centers of the alias hexagons are located in a circumference of radius $N_T d$ at angles from 0° to 300° spaced 60°.

The points of the star defined by the instrument sampling (figure 4.8) always belong to the hexagonal grid. They are marked as circles in figure 4.9(b). To have the whole star fit into the principal period, the following restrictions apply

$$\begin{aligned} &N_T \geq 3N_{EL} + 1 \quad \text{To have the star inside the hexagon} \\ &N_T \geq 4N_{EL} + 1 \quad \text{To have the star inside the rhombus} \end{aligned} \tag{4.48}$$

where N_{EL} is the number of elements per arm (not including the center antenna, if present). In the example of figure 4.8, $N_{EL} = 4$, so to fit the whole star into the

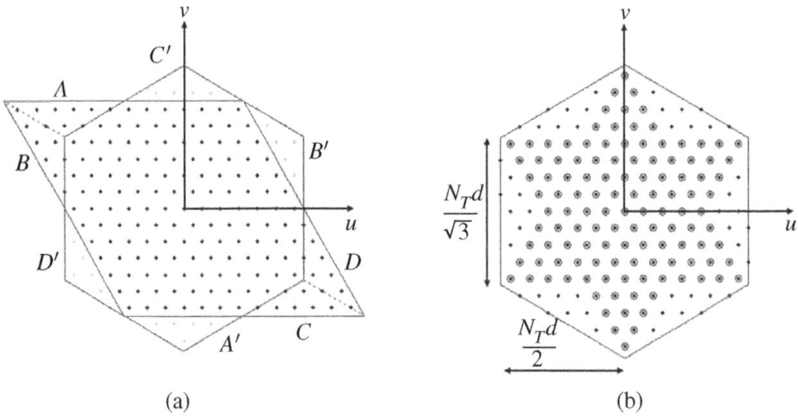

Figure 4.9 Grid points in the (u, v) plane. (a) Rhomboidal principal period and conversion to hexagon. (b) Fundamental hexagon and instrument coverage.

principal hexagon, the total number of grid points should be N_T^2 where $N_T = 13$. This is the value used in figure 4.9.

In a hexagonal sampling there are always some grid points that do not correspond to any baseline, so no visibility measurement can be assigned to them. They are identified in figure 4.9 by having a dot with no circle and must be ignored or zero-padded in the image reconstruction process. If the value of N_T is increased beyond (4.48), then the (u, v) hexagon becomes larger to allocate all points, but the star remains the same as it corresponds to measured samples. All new points must then be ignored or zero-padded if used in image reconstruction.

Using again [7], the reciprocal grid in hexagonal sampling is defined with equations similar to (4.42) for the director cosines:

$$\xi = \frac{n_2}{N_T d} \; ; \; \eta = \frac{2n_1 + n_2}{\sqrt{3} N_T d} \tag{4.49}$$

Substituting in these equations n_2 and n_1 by the whole set of integers in the principal period (4.40) results also in a grid with rhomboidal shape but, with the same method used in the (u, v) plane, it can be converted to a hexagon. The result is shown in figure 4.10, where a slightly different criterion for including points in the hexagon borders is used. Note that the fundamental hexagon is rotated 30° with respect to that of the visibility grid. The principal hexagon is also replicated as a honeycomb, and the centers of the aliases are located in a circumference of radius $2/(\sqrt{3}d)$ at angles from 30° to 330° spaced 60°.

In order to have the whole unit circle inside the principal hexagon, the distance to the alias centers must be equal to or larger than 2 and this translates into the condition that $d \leq 1/\sqrt{3} = 0.577$, which is the Nyquist criterion for hexagonal grids.

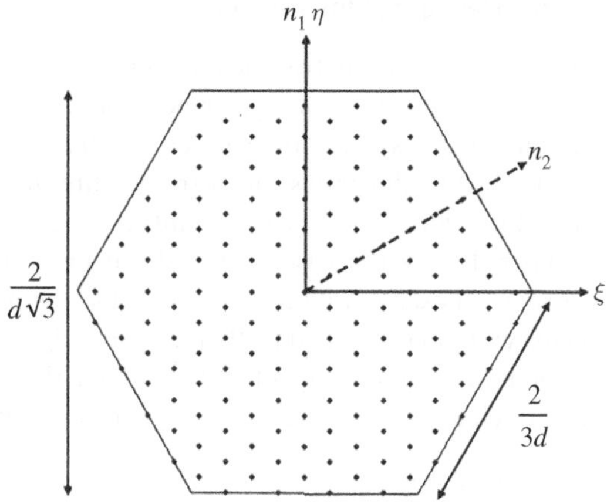

Figure 4.10 Grid points in the principal hexagon of ξ, η for hexagonal sampling.

Compared to the case of a rectangular instrument ($d \leq 0.5$) an instrument with hexagonal sampling allows lager spacing, so in the end, reduced number of elements for the same overall size, and hence, spatial resolution.

For a given antenna spacing with $d > 1/\sqrt{3}$, the alias-free field of view is larger in hexagonal sampling when compared to rectangular sampling. The drawing of figure 4.11 shows the aliasing corresponding to the same image of figure 4.7 for a spacing of also $d = 0.65$. The larger alias-free field of view with respect to figure 4.7 is apparent. As a matter of fact, this is the main advantage of using hexagonal sampling.

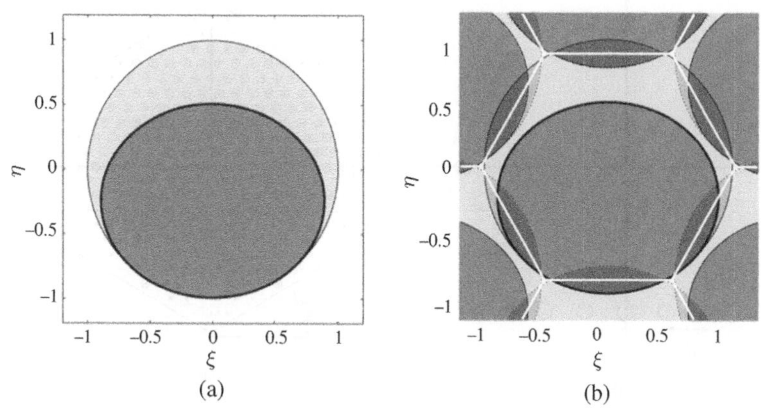

Figure 4.11 Aliasing in hexagonal grids. (a) Original image. (b) Recovered image in alias conditions.

4.3.6 Hexagonal Instrument

An interesting alternative for implementing a hexagonal grid is an instrument with six arms forming a regular hexagon. As an example, figure 4.12(a) shows, the antenna positions for one such instrument with four antennas per arm. Figure 4.12(b) illustrates the corresponding hexagonal grid in the (u, v) plane indicating with circles the grid points corresponding to the instrument coverage. Compared with the case of a Y-shape instrument (figure 4.9), now the star becomes a hexagon rotated with respect to the principal period. Again, there are points with no visibility data that have to be ignored or zero-padded.

The minimum number of points to have the instrument coverage fully inside the principal period is the same as the second condition of (4.48)

$$N_T \geq 4N_{EL} + 1 \tag{4.50}$$

and this is what has been used in the figure. In this case, $N_{EL} = 4$ and therefore $N_T = 17$. Increasing this value enlarges the principal period while keeping the same measured points (inner hexagon). Zero padding of added points would be needed in case of using them in image reconstruction.

The (ξ, η) grid, in turn, is the same as that of figure 4.10; in this example, slightly thicker due to the increased number of points.

One of the main advantages of the hexagonal setup with respect to the Y-shape is that the synthetic beam has reduced side lobes due to having a smoother contour of the instrument coverage. The sharp vertices of the star of figure 4.9 are responsible of increasing secondary lobes. This fact is illustrated in figure 4.13, which shows

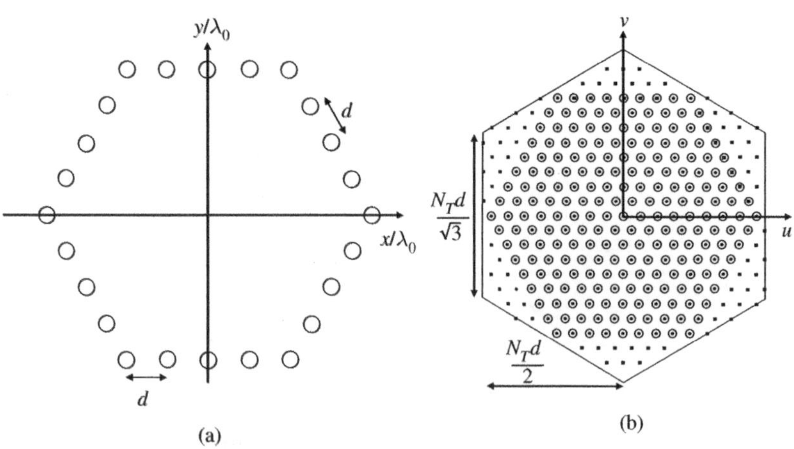

Figure 4.12 Example of hexagonal instrument. (a) Antenna layout. (b) u-v coverage including additional points to complete the principal period.

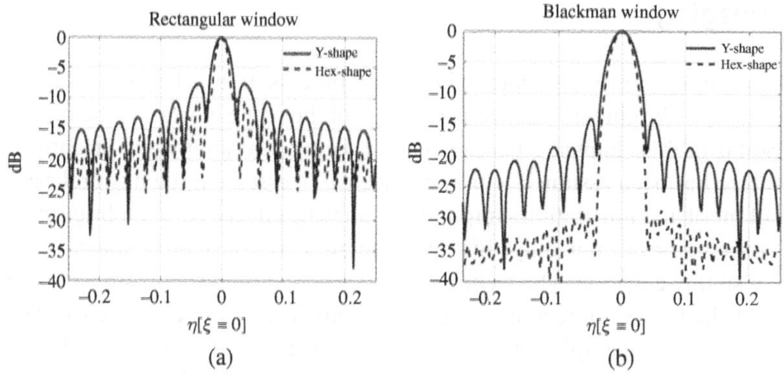

Figure 4.13 Cuts for $\xi = 0$ of synthetic beams for hexagonal sampling comparing Y-shape and hexagonal instruments. (a) Rectangular window. (b) Blackman window.

a comparison of synthetic beam cuts for both instruments assuming 21 antennas per arm and a distance between elements of $0.875\lambda_0$. It is the same example used in figure 4.4.

As seen, the improvement is dramatic when Blackman window is used but not so important for rectangular window. The slight main lobe narrowing in the hexagonal case is expected since the maximum antenna separation is slightly larger for the same number of elements per arm. In this example, the most distant antennas in the hexagon are separated $36.75\lambda_0$ while in the Y-shape instrument they are at $31.82\lambda_0$.

A further benefit of using a hexagon is that the baseline redundancy is increased with respect to the Y-shape. In the latter, redundancy is only present between elements along the arms while the hexagon has redundant baselines between different arms. A useful metric to quantify the radiometric resolution improvement is the square root of the ratio between the effective number of visibilities (4.22) and its actual number. For the same instrument considered in the previous example, the following figures are obtained

$$\sqrt{\frac{N'_V}{N_V}} = \begin{cases} 0.98 & \text{(Y-shape instrument)} \\ 0.69 & \text{(Hexagonal instrument)} \end{cases} \tag{4.51}$$

meaning that redundancy is almost irrelevant in a Y-shape instrument while it reduces the brightness temperature uncertainty ΔT_B by a factor of about 0.7 in a hexagonal one (in this example).

On the other hand, a hexagonal instrument has twice as many receivers and antennas as one having a Y-shape with similar overall size, increasing thus cost, power consumption and weight among others.

4.4 Imaging

Imaging, or image reconstruction, refers to the process of inverting the visibility equation to get a brightness temperature image of the scene in front of the instrument. So far in this chapter, a Fourier relation between visibility and brightness temperature has been assumed with the objective of assessing the most relevant figures of merit and illustrating sampling and aliasing effects. Now, for brightness temperature retrieval, not only this approximation is considered but also the general visibility equation. The techniques described in this section are based on those published in [8], in turn heritage of previous work reported in [9, 10]. The procedure is currently implemented in a dedicated software that regularly obtains both dual-polarization and full polarimetric images from the SMOS instrument for ocean, land or ice [11]. Other methods based on [12], used in the nominal SMOS processing, are not covered here.

4.4.1 System of Equations

The starting point is equation (3.77) for single polarization or its extension to the full polarimetric case obtained by introducing the director cosines change of variables (3.72) and the fringe washing function (3.54) into (3.49), properly expanded. If the "$-T_r$" term [see section 3.2.6] cannot be neglected, the first step is to remove it by measuring a flat target and applying (6.47), as duly explained in section 6.2.4.[3] Once this is done, the single polarization equation is still (3.77) but without the $-T_r$ term. It is here repeated for convenience with a slightly different notation.

$$V_{kj} = \iint_{\xi^2+\eta^2<1} T'_{kj}(\xi,\eta) e^{-j2\pi(u_{kj}\xi+v_{kj}\eta)} d\xi d\eta \tag{4.52}$$

where $T'_{kj}(\xi,\eta)$ is the modified brightness temperature, now defined as

$$T'_{kj}(\xi,\eta) = T_B(\xi,\eta) \frac{F_k(\xi,\eta)F_j^*(\xi,\eta)}{\sqrt{1-\xi^2-\eta^2}\sqrt{\Omega_{ak}\Omega_{aj}}} \bar{\bar{r}}_{kj}\left(-\frac{u_{kj}\xi+v_{kj}\eta}{f_0}\right) \tag{4.53}$$

in which $T_B(\xi,\eta)$ is the scene brightness temperature, $F_{k,j}(\xi,\eta)$ the field antenna patterns (1.238) for the two elements k and j, $\Omega_{ak,j}$ their corresponding antenna solid angles (1.243) and $\bar{\bar{r}}_{kj}()$ the fringe washing function normalized to its value at the origin (3.56). The coordinates u_{kj} and v_{kj} are those defined in (3.74) but they include subscripts to stress on the fact that they depend on the specific antenna pair. Finally, ξ and η are the director cosines (3.72). In regard of property 1 of

3 The prima used there to denote the visibility after removing this term is not kept in this section to simplify notation.

the fringe washing function (see section 3.3.2), swapping the indices of (4.53) is equivalent to complex conjugating, $T'_{jk}(\xi,\eta) = T'^{*}_{kj}(\xi,\eta)$. In consequence, $V_{jk} = V^{*}_{kj}$ in agreement with the conjugation property of the visibility (3.23) particularized for the single polarization case.

The *zero-spacing baselines* ($u = v = 0$), applicable for single antennas ($k = j$), must also be included in the system of equations. The "visibility" of any such baseline is the antenna temperature (3.78), repeated also here for convenience,

$$T_{a_k} = \frac{1}{\Omega_{ak}} \iint_{\xi^2+\eta^2<1} \frac{T_B(\xi,\eta)|F_k(\xi,\eta)|^2}{\sqrt{1-\xi^2-\eta^2}} d\xi d\eta \tag{4.54}$$

The above equations assume implicitly that all antenna patterns have the *same polarization vector* (1.235) and therefore the retrieved brightness temperature corresponds also to this polarization. The same procedure applies to the *ideal* polarimetric interferometric radiometer based on equations (3.18) and (3.19) and, in this case, the retrieval corresponds to the polarimetric brightness temperature. Subtleties arising from the fact that each antenna has two patterns or that the polarimetric antenna temperature is complex are not considered here for simplicity but they are highlighted in section 4.4.5. The general full polarimetric case is discussed in section 4.4.6.

4.4.1.1 Optional Subtraction of Antenna Temperature

The average antenna temperature of both receivers involved in a given baseline $T_{a_{kj}} = (T_{a_k} + T_{a_j})/2$ can be subtracted from $T_B(\xi,\eta)$ in (4.53) before image reconstruction. Substituting this difference into (4.52), the following *modified* visibility function is obtained.

$$\check{V}_{kj} = V_{kj} - T_{a_{kj}} \text{FTR}_{kj} \tag{4.55}$$

where FTR_{kj} is the Flat Target response (3.13), provided the back lobes are neglected. It can be estimated from sky measurements using (6.46).

Similarly, the modified visibility for the zero-spacing baseline k is

$$\check{T}_{a_k} = T_{a_k} - T_{a_k} \text{FTR}_k = 0 \tag{4.56}$$

so vanishing for all antennas due to (3.14).

Image reconstruction can be carried out from this modified visibility using exactly the same methods explained in sections 4.4.2–4.4.8, including the discretization. The final brightness temperature image is then approximated by the sum of the reconstructed one plus a constant antenna temperature that can be taken as the average of those of all antennas. This approach, which can be combined with the removal of the $-T_r$ term (6.45), aims at reducing the impact of possible antenna pattern errors. It was proposed in [10] and, slightly different and with the name of "Flat Target Transformation," also in [13]. In what follows, it is not considered.

4.4.1.2 Antenna Positions

Before entering in the methods and techniques to invert (4.52), it is important to understand the inter-relation between antenna patterns and their positions in the mechanical structure.

For the above equations to be consistent, the assumed location of a given antenna must coincide with the center of the local axis its pattern is referenced to. As long as this consistency is fulfilled, the equations are correct. But, beyond that, there is margin to decide where the antennas are actually located. One can decide, for example, that the antennas are nominally placed at their respective phase centers [see section 1.6.6] or, in another example, at a given geometrical location properly defined for the particular antenna shape (for example the center of the radiating area). But these options, or similar ones, are unpractical since they imply having an irregular grid of (u, v) points, even though this irregularity may be minor if the mechanics and the antennas are well designed.

The best option is then to force all antenna patterns to be referenced to local coordinates centered in the *theoretical antenna positions*. Depending on the instrument shape, these locations are visualized in figures 4.5(a), 4.8(a), or 4.12(a). Obviously, other instrument layouts may have different, but known in advance, antenna distributions. With this option, the (u_{kj}, v_{kj}) coordinates in (4.52) and (4.53) are *exactly* the regular grids in the (u, v) plane defined in sections 4.3.2 or 4.3.5. Depending on the specific antenna design, this definition may result in phase ramps in some patterns, but as long as they are known and properly introduced in the equations, they do not represent any inaccuracy by themselves.

4.4.1.3 Discretization

Equation (4.52) is thus discretized at precisely the reciprocal grids (4.39) and (4.42) (rectangular case) or (4.47) and (4.49) (hexagonal case). This is carried out by applying to (4.52) the change of variables defined by these equations, which leads to an equation similar to (4.43)

$$V_{kj} = \Delta_{\xi\eta} \sum_{(n_2,n_1)\subset UC} T'_{kj}(n_2, n_1) e^{-j\frac{2\pi}{N_T}(k_1 n_2 + k_2 n_1)} \tag{4.57}$$

where $\Delta_{\xi\eta}$ is the Jacobian determinant of the transformation from (ξ, η) to (n_2, n_1) [(4.42) or (4.49)],

$$\Delta_{\xi\eta} = \begin{cases} \dfrac{1}{N_T^2 d^2} & \text{(rectangular grid)} \\ \dfrac{2}{\sqrt{3} N_T^2 d^2} & \text{(hexagonal grid)} \end{cases} \tag{4.58}$$

with N_T^2 the total number of grid points in the principal period and d the antenna spacing. As expected, assuming that all antenna patterns are identical and neglecting the fringe washing function effect, T'_{kj} becomes a function *only* of n_2 and n_1 and (4.57) reduces to a digitized Fourier integral.

4.4.1.4 The G-matrix

In general, when (4.53) is substituted into (4.57), the visibility equation becomes a linear system of equations $\boldsymbol{V} = \boldsymbol{G}\,\boldsymbol{T}_B$ where $\boldsymbol{G}$ is the so-called "G-matrix," $\boldsymbol{V}$ the column vector of measured visibilities including the ones for zero spacing and $\boldsymbol{T}_B$ that of brightness temperatures. The G-matrix has as many rows as antenna pairs plus the ones corresponding to the zero spacing. For an instrument having N antennas, there are thus $N(N-1)/2$ *complex* rows plus N *real* rows. The number of columns is equal to the total number of (ξ, η) grid points that fall inside the unit circle $\xi^2 + \eta^2 < 1$.

Using (4.52) and (4.53), the row of the G-matrix corresponding to the receiver pair (k, j) has elements given by

$$\Delta_{\xi\eta}\frac{F_k(\xi,\eta)F_j^*(\xi,\eta)}{\sqrt{1-\xi^2-\eta^2}\sqrt{\Omega_k\Omega_j}}\tilde{r}_{kj}\left(-\frac{k_1 n_2 + k_2 n_1}{N_T f_0}\right)e^{-j\frac{2\pi}{N_T}(k_1 n_2 + k_2 n_1)} \tag{4.59}$$

where the values of (ξ, η) in the antenna patterns and obliquity factor are those corresponding to the discrete grid with indices (n_2, n_1) derived from (4.42) or (4.49). The G-matrix rows for the zero spacing baselines ($k_1 = k_2 = 0$ and $k = j$) come directly from (4.54)

$$\Delta_{\xi\eta}\frac{|F_k(\xi,\eta)|^2}{\sqrt{1-\xi^2-\eta^2}\Omega_k} \tag{4.60}$$

4.4.2 Conjugate Extension and Redundant Baselines Averaging

Since $V_{jk} = V_{kj}^*$, in order to have the complete system, for each complex row of the system of equations (4.57), an additional one must be added by conjugating both sides and swapping the receivers indices. This operation must be performed before dealing with the redundant baselines explained here below.

As already mentioned in section 4.3.1, baselines having identical (u, v) values but formed by different receiver pairs are redundant. Even though belonging to the same visibility sample, they provide slightly different measurements with respect to each other because of the unequal antenna patterns of the involved elements, the fringe washing function and other effects such as noise and calibration errors. Considering two redundant baselines, the corresponding two distinct rows of the discretized visibility equation, written in terms of the G-matrix, are:

$$V_l = \sum_m G_{lm} T_{B_m} \quad ; \quad V_n = \sum_m G_{nm} T_{B_m} \tag{4.61}$$

where the row subscripts l and n refer to two (k,j) pairs of redundant baselines and the column subscript m ranges all (ξ,η) pairs in the unit circle. Both V_l and V_n are different visibility measurements assigned to the *same* point in the (u,v) grid.

The above two equations can be averaged to form a third one relating to the visibility of the very same (u,v) point to the same scene brightness temperature

$$\overline{V}_l = \sum_m \overline{G}_{lm} T_{B_m} \tag{4.62}$$

where $\overline{V}_l = (V_l + V_n)/2$ and $\overline{G}_{lm} = (G_{lm} + G_{nm})/2$. Apparently, this last equation can be used in brightness temperature retrieval instead of any of the other two without any loss of information. Different complete sets of visibility samples are obtained by randomly choosing unique sets of non-redundant baselines. For each one, the corresponding visibility function becomes related to *the same* brightness temperature image, so the image reconstruction algorithm should yield the same result in the absence of noise and errors. In other words, only one set of non-redundant visibilities is enough to fully recover the brightness temperature. Averaging all measurements of the same (u,v) points is not needed in the ideal case but has the effect of thermal noise and antenna error reduction in practice.

The averaging operation must also be performed for the zero-spacing visibility, which has a redundancy order equal to the number of antennas. Instrumental constraints may reduce the number of antennas that are actually used to measure the visibility at the origin, but at least one of them must be dedicated to this purpose.

Image reconstruction is carried out by inverting the system of equations that results after conjugate extension and averaging of redundant baselines, either assuming a Fourier integral approximation or with the G-matrix. After performing these two operations, the number of rows of the resultant G-matrix becomes equal to the total number of unique points in the (u,v) domain. For a square instrument they are those within the square depicted in figure 4.6(a). In a Y-shape instrument they correspond to the star drawn in figure 4.8(b) [or 4.9(b)], and for a hexagonal setup, they are formed by the set of points in the inner hexagon of figure 4.12(b). Columns are not affected by these operations, as they still correspond to the points (ξ,η) that fall inside the unit circle.

4.4.3 Fourier Image Reconstruction

If all antennas can be considered reasonably alike, a common mean pattern can be used for all of them in (4.53). If, furthermore, the field of view and/or the signal bandwidth are small enough so that the normalized fringe washing function can be approximated to unity for all values of its argument, then (4.52) becomes a Fourier integral with $T'_{kj}(\xi,\eta)$ equal to $T'_B(\xi,\eta)$ defined in (3.81), independent of the specific antenna pair. This approximation has a big impact on retrieval since it implies an important reduction in computer resources requirements. On the other hand, in some systems, it may lead to inaccuracies beyond what it is acceptable.

Assuming in (4.57) $T'_{kj} = T'_B$ independent of the specific antenna pair, the aliasing analysis of section 4.3.3 applies. Therefore, (4.46) holds for all visibility samples and their conjugates after averaging redundant visibilities. Assuming zero padding of those outside the instrument coverage, the Fourier inversion formula provides then the *reconstructed* modified brightness temperature,

$$\hat{T}'_B(n_2, n_1) = \Delta_{uv} \sum_{k_1=N_m}^{N_M} \sum_{k_2=N_m}^{N_M} W(k_1, k_2) V(k_1, k_2) e^{j\frac{2\pi}{N_T}(k_1 n_2 + k_2 n_1)} \tag{4.63}$$

where $W(k_1, k_2)$ is the apodization window [see section 4.1] and Δ_{uv} the Jacobian determinant of (4.39) or (4.47), interpreted as elementary area

$$\Delta_{uv} = \begin{cases} d^2 & \text{(rectangular grid)} \\ \dfrac{d^2\sqrt{3}}{2} & \text{(hexagonal grid)} \end{cases} \tag{4.64}$$

In practice, (4.63) is evaluated by a two-dimensional discrete Fourier transform (A.60). The apodization window needs not to be necessarily applied at this step. It can be done later in frequency components, as explained in section 4.4.7. Since the single polarization visibility equation is being assumed, $\hat{T}'_B(n_2, n_1)$ must be real by definition. If correctly implemented, practical computation of (4.63) results in complex values with negligible imaginary part, only due to computational errors. The final result is then the real part of (4.63).

The reconstructed *modified* brightness temperature (4.63) is defined in the (ξ, η) principal period and has periodic repetition in all others. If the antenna spacing does not meet the Nyquist criterion, it is also affected by strong aliasing like the one illustrated in figures 4.7 or 4.11. In the alias-free field of view the retrieval has no error except for ripples and smoothing due to limited instrument coverage, which is function of the window applied.

The final result of the reconstructed brightness temperature is obtained by solving (3.81) using the above retrieval $\hat{T}'_B(n_2, n_1)$. Expressing it back as a function of director cosines through (4.42) or (4.49), it becomes

$$\hat{T}_B(\xi, \eta) = \hat{T}'_B(\xi, \eta) \frac{\Omega_a \sqrt{1 - \xi^2 - \eta^2}}{t(\xi, \eta)} \tag{4.65}$$

where $t(\xi, \eta) = |F(\xi, \eta)|^2$ can be taken as the mean antenna power pattern along all the elements. Note that, even though $\hat{T}'_B(\xi, \eta)$ is periodic, $\hat{T}_B(\xi, \eta)$ is not by virtue of the multiplication by the obliquity factor and the power pattern. Errors in alias regions are notably increased due to the inverse of the antenna pattern evaluated at these spatial directions. A method to mitigate this error is given in section 4.4.8.

4.4.4 G-matrix Image Reconstruction

Consider now the complete equation (4.57), expressed in terms of the G-matrix,

$$\boldsymbol{V} = \boldsymbol{G}\,\boldsymbol{T}_B \tag{4.66}$$

Assuming that both conjugate extension and redundant baselines averaging (section 4.4.2) have been performed, the G-matrix used in this equation has as many rows as number of *unique* (u, v) points including the origin. The number of columns is still equal to that of (ξ, η) points in the unit circle, so in general it is not square. However, thanks to having applied the two operations just mentioned, it is fairly well conditioned and therefore can be efficiently inverted using the Moore–Penrose algorithm, resulting in a matrix $\boldsymbol{G}^{-1}$ with a size swapped with respect to that of $\boldsymbol{G}$. The reader may also consider other inversion methods if he/she deems to be needed, but further discussion about them is out of the scope of this book. As a matter of fact, following the procedures proposed here, G-matrix inversion should not be an issue in image reconstruction.

If the antenna spacing meets the Nyquist criterion (oversampling case), all (ξ, η) grid points inside the unit circle are a subset of those in the principal period. The reconstructed brightness temperature, expressed now as a column vector for all points in the unit circle, is then directly

$$\hat{\boldsymbol{T}}_B = \boldsymbol{G}^{-1}\boldsymbol{V} \quad \text{(oversampling case)} \tag{4.67}$$

in which $\boldsymbol{V}$ is the column vector of visibility samples. It includes all measured ones and their complex conjugates plus the ones at the origin, all of them after having averaged those of redundant baselines.

Of course, even with perfectly calibrated visibilities, if the G-matrix fails to accurately model the instrument, for example, with poor knowledge of antenna patterns, the reconstructed brightness temperature has systematic errors with respect to that of the actual scene. But, even with a perfect antenna characterization, $\hat{\boldsymbol{T}}_B$ differs from $\boldsymbol{T}_B$ due to the limited instrument coverage pointed out in section 4.1. It can be assessed by substituting (4.66) into (4.67)

$$\hat{\boldsymbol{T}}_B = \boldsymbol{G}^{-1}\boldsymbol{G}\,\boldsymbol{T}_B \tag{4.68}$$

which is obviously a linear relation. The corresponding impulse response can be deduced by setting all elements of $\boldsymbol{T}_B$ equal to zero except the ones corresponding to $\xi = \eta = 0$, fixed to one. The result is the synthetic beam

$$\mathscr{B} = \boldsymbol{G}^{-1}\boldsymbol{G}_0 \tag{4.69}$$

where $\boldsymbol{G}_0$ is the column of $\boldsymbol{G}$ corresponding to $\xi = \eta = 0$. Note that, according to (4.59), it is fairly constant and equal to the inverse of the geometric mean of antenna solid angles. The reconstructed brightness temperature is then the convolution of the synthetic beam and the true image

$$\hat{\boldsymbol{T}}_B = \mathscr{B} * \boldsymbol{T}_B \tag{4.70}$$

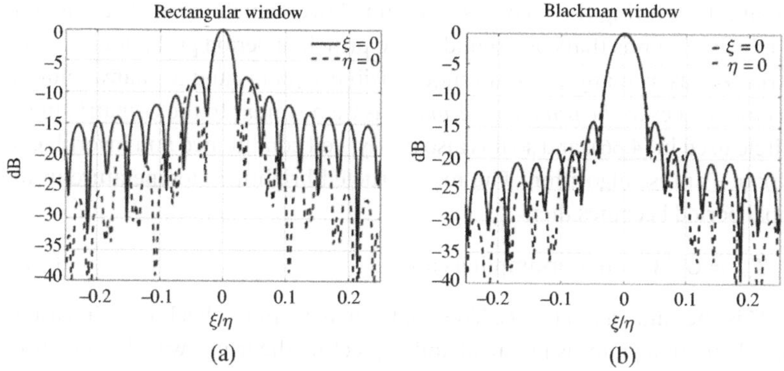

Figure 4.14 Cuts of the synthetic beams computed from the G-matrix of the MIRAS instrument. (a) Rectangular window. (b) Blackman window.

Since the visibility samples have not been weighted, this synthetic beam corresponds to a rectangular apodization window. Other ones, for example Blackman (4.6), can be efficiently applied using the method of spatial components described in section 4.4.7, consisting of changing to the (u, v) domain, multiply the result by the window and going back to the (ξ, η) domain.

As an example, figure 4.14 shows cuts of the normalized synthetic beams for both rectangular and Blackman windows computed by (4.69) using the G-matrix of the MIRAS instrument [3], which was derived from measured antenna patterns. Comparing this figure with (4.4), obtained by Fourier transform and only based on antenna layout, shows an amazing consistency, as differences are hardly detected.

4.4.4.1 Undersampling Case

If the antenna spacing does not meet the Nyquist criterion, there exist some points of the unit circle that lay outside the principal period. The G-matrix can then be split into two: one with the columns of the principal period $\boldsymbol{G}_P$ and other with those external to it $\boldsymbol{G}_X$. The discretized visibility equation (4.66) can then be written as

$$\boldsymbol{V} = \begin{bmatrix} \boldsymbol{G}_P & \boldsymbol{G}_X \end{bmatrix} \begin{bmatrix} \boldsymbol{T}_{B_P} \\ \boldsymbol{T}_{B_X} \end{bmatrix} = \boldsymbol{G}_P \boldsymbol{T}_{B_P} + \boldsymbol{G}_X \boldsymbol{T}_{B_X} \tag{4.71}$$

where the same subscript nomenclature applies to $\boldsymbol{T}_{B_P}$ and $\boldsymbol{T}_{B_X}$.

As indicated in relation to (4.45), the complex exponential phases of all matrix elements of the second term above are identical to some others belonging to the first one, and this feature tends to make the whole system ill-conditioned. On the other hand, Fourier image reconstruction (section 4.4.3) demonstrates that a retrieval restricted to the principal period results in perfect imaging within the alias-free field of view at the expense of having large errors in some zones due to overlapping aliases. The same concept applied to G-matrix reconstruction means simply ignoring the second summand above, which means keeping only the G-matrix columns that correspond to the principal period grid points.

Matrix G_P has as many rows as unique visibility points and as many columns as director cosines *in the principal period*. It is then similar to the G-matrix in the unit circle used in (4.66), and so it is also pretty well conditioned and invertible by the Moore–Penrose algorithm. The reconstructed brightness temperature in the principal period becomes then

$$\hat{\boldsymbol{T}}_{B_P} = \boldsymbol{G}_P^{-1}\boldsymbol{V} \quad \text{(undersampling case)} \tag{4.72}$$

where $\boldsymbol{V}$ is the same as in (4.67). The reader may want to check that, in case of having all antenna patterns identical and neglecting the fringe washing function, the above equation can be rewritten using inverse Fourier transform with zero padding of visibilities and rectangular apodization window. The final expression becomes then *identical* to (4.65) and therefore (4.72) may be interpreted as a sort of generalization of the Fourier reconstruction to the general case.

Equation (4.72) can be interpreted as the convolution of a synthetic pattern with the original image in the principal period plus an added error term. Substituting (4.71)

$$\hat{\boldsymbol{T}}_{B_P} = \mathcal{B} * \boldsymbol{T}_{B_P} + \boldsymbol{E}\,\boldsymbol{T}_{B_X} \tag{4.73}$$

where $\boldsymbol{E} = \boldsymbol{G}_P^{-1}\boldsymbol{G}_X$ is the "reconstruction error matrix" and the synthetic beam is now

$$\mathcal{B} = \boldsymbol{G}_P^{-1}\boldsymbol{G}_0 \tag{4.74}$$

which is the same as (4.69) but keeping the principal period columns of the G-matrix prior to inverting. The examples in figure 4.14 correspond actually to this case since MIRAS layout does not comply with the Nyquist criterium. The error term in (4.73) depends only on the brightness temperature outside the principal period so it vanishes in the oversampling case. In general it is small in the alias-free field of view and takes large values in the aliasing zones [8]. For the approximations leading to Fourier reconstruction, this term becomes just the aliasing. A method to correct this error is covered in section 4.4.8.

4.4.4.2 System Reduction

As already pointed out, the brightness temperature, either reconstructed by (4.67) or (4.72), is real by definition. Then, denoting in general by $\mathrm{I}\boldsymbol{G}$, the corresponding inverse G-matrix ($\mathrm{I}\boldsymbol{G} = \boldsymbol{G}^{-1}$ or $\mathrm{I}\boldsymbol{G} = \boldsymbol{G}_P^{-1}$), the solution can be written as

$$\mathrm{I}\boldsymbol{G}\,\boldsymbol{V} = \left[\mathrm{I}\boldsymbol{G}_0\ \mathrm{I}\boldsymbol{G}_A\ \mathrm{I}\boldsymbol{G}_B\right]\begin{bmatrix} V_0 \\ \boldsymbol{V}_A \\ \boldsymbol{V}_B \end{bmatrix} = \mathrm{I}\boldsymbol{G}_0 V_0 + 2\Re e\left[\mathrm{I}\boldsymbol{G}_A\boldsymbol{V}_A\right] \tag{4.75}$$

where the subscript 0 refers to the column for $u = v = 0$ and A and B to two subsets of grid points (matrix columns) that have opposite (u, v) signs. The subscripts in the visibilities are consistently defined. The last equality holds because of the conjugation property of the visibility (3.23). Taking into account that both $\mathbf{IG}_0$ and V_0 (a single scalar) are real, this equation can be written in a more compact form as

$$\mathbf{IG}\,\mathbf{V} = \Re e\left\{ \left[\mathbf{IG}_0 \; 2\,\mathbf{IG}_A\right] \begin{bmatrix} V_0 \\ \mathbf{V}_A \end{bmatrix} \right\} \tag{4.76}$$

which reduces the system of equations in about one half.

4.4.5 Polarimetric Retrieval: Ideal Case

Consider now that all baselines are formed by two dual-polarization antennas as the one depicted in figure 3.4 and assume *initially* that they all have infinite cross-coupling isolation. Selecting all p-ports or all q-ports reverts to a system of equations essentially the same as that of section 4.4.1. Replacing the antenna patterns by $F^p_{k,j}$ or $F^q_{k,j}$, respectively, leads to reconstructing $T_{B_p}(\xi, \eta)$ or $T_{B_q}(\xi, \eta)$.

Polarimetric retrieval is carried out by forming all baselines with one p-antenna and one q-antenna. In this case, equation (3.18) applies, so (4.53) becomes

$$T'_{kj}(\xi, \eta) = T_{B_{pq}}(\xi, \eta) \frac{F^p_k(\xi, \eta) F^{q*}_j(\xi, \eta)}{\sqrt{1 - \xi^2 - \eta^2}\sqrt{\Omega_k \Omega_j}} \tilde{\bar{r}}_{kj}\left(-\frac{u_{kj}\xi + v_{kj}\eta}{f_0}\right) \tag{4.77}$$

while (4.54) is now substituted by the polarimetric antenna temperature (3.19),

$$T^{pq}_{ak} = \frac{1}{\sqrt{\Omega^p_{ak} \Omega^q_{ak}}} \iint_{\xi^2+\eta^2<1} \frac{T_{B_{pq}}(\xi, \eta) F^p_k(\xi, \eta) F^{q*}_k(\xi, \eta)}{\sqrt{1 - \xi^2 - \eta^2}} d\xi d\eta \tag{4.78}$$

Therefore, the polarimetric brightness temperature $T_{B_{pq}}(\xi, \eta)$ is retrieved in this case. The procedure to do it mimics the one described for single polarization but replacing the antenna patterns in (4.59) and (4.60) consistently with the ones used in the above two expressions. Also, the conjugate extension of section 4.4.2 must now use the complete conjugation property (3.23) so $V^{pq}_{kj} = V^{qp*}_{jk}$. Finally, the system reduction is no longer applicable since the polarimetric brightness temperature is now complex.

4.4.6 Full Polarimetric Case

If antenna patterns have finite cross-coupling isolation, the starting point is the set of equations that result from expanding (3.49) without the $-T_r$ term, either neglected or canceled with a flat target, including the fringe washing function and applying the change of variables (3.72). Discretization is carried out using also the reciprocal grids in (u, v) and (ξ, η), either in rectangular or hexagonal sampling, depending on the instrument layout, and the resultant system of equations is expressed as a set of polarimetric G-matrices as

$$\begin{aligned} \boldsymbol{V}^{p} &= \boldsymbol{G}_{pp}^{pp}\boldsymbol{T}_{B_p} + \boldsymbol{G}_{qq}^{pp}\boldsymbol{T}_{B_q} + \boldsymbol{G}_{pq}^{pp}\boldsymbol{T}_{B_{pq}} + \boldsymbol{G}_{qp}^{pp}\boldsymbol{T}_{B_{qp}} \\ \boldsymbol{V}^{q} &= \boldsymbol{G}_{pp}^{qq}\boldsymbol{T}_{B_p} + \boldsymbol{G}_{qq}^{qq}\boldsymbol{T}_{B_q} + \boldsymbol{G}_{pq}^{qq}\boldsymbol{T}_{B_{pq}} + \boldsymbol{G}_{qp}^{qq}\boldsymbol{T}_{B_{qp}} \\ \boldsymbol{V}^{pq} &= \boldsymbol{G}_{pp}^{pq}\boldsymbol{T}_{B_p} + \boldsymbol{G}_{qq}^{pq}\boldsymbol{T}_{B_q} + \boldsymbol{G}_{pq}^{pq}\boldsymbol{T}_{B_{pq}} + \boldsymbol{G}_{qp}^{pq}\boldsymbol{T}_{B_{qp}} \\ \boldsymbol{V}^{qp} &= \boldsymbol{G}_{pp}^{qp}\boldsymbol{T}_{B_p} + \boldsymbol{G}_{qq}^{qp}\boldsymbol{T}_{B_q} + \boldsymbol{G}_{pq}^{qp}\boldsymbol{T}_{B_{pq}} + \boldsymbol{G}_{qp}^{qp}\boldsymbol{T}_{B_{qp}} \end{aligned} \tag{4.79}$$

where, for example, the row of matrix $\boldsymbol{G}_{pp}^{pq}$ corresponding to the receiver pair (k, j) is

$$\Delta_{\xi\eta} \frac{F_{k_p}^{p}(\xi,\eta) F_{j_p}^{q*}(\xi,\eta)}{\sqrt{1-\xi^2-\eta^2}\sqrt{\Omega_k \Omega_j}} \bar{\bar{r}}_{kj}\left(-\frac{k_1 n_2 + k_2 n_1}{N_T f_0}\right) e^{-j\frac{2\pi}{N_T}(k_1 n_2 + k_2 n_1)} \tag{4.80}$$

which the reader may want to compare to (4.59), also to consult, if needed, the meaning of most symbols. In this expression, the patterns are exactly the ones defined in (3.24). Other matrices are formed similarly.

Adding conjugated rows and averaging those of redundant baselines is here carried out for each of the submatrices as explained in section 4.4.2 taking into account the conjugation property of the polarimetric visibility (3.23).

If the antenna spacing does not meet the Nyquist criterion, only the columns of the submatrices corresponding to the principal period must be kept. Otherwise, all of them are used. Once all this is done, the combination of the four equations (4.79) can be again written as $\boldsymbol{V} = \boldsymbol{G}\,\boldsymbol{T}_B$ where now $\boldsymbol{G}$ is a matrix four times larger than the one used for single polarization in both rows and columns. It requires more computer resources but it can still be inverted using the Moore–Penrose algorithm, so the reconstructed polarimetric brightness temperature is directly $\hat{\boldsymbol{T}}_B = \boldsymbol{G}^{-1}\,\boldsymbol{V}$.

4.4.6.1 System Reduction

The system of equations that results after inversion can also be reduced by making use of the conjugation property of the visibility. To this end, the inverse G-matrix is first split into submatrices corresponding to the four polarization combinations,

$$\boldsymbol{G}^{-1} = \begin{bmatrix} \mathbf{IG}^{11} & \mathbf{IG}^{12} & \mathbf{IG}^{13} & \mathbf{IG}^{14} \\ \mathbf{IG}^{21} & \mathbf{IG}^{22} & \mathbf{IG}^{23} & \mathbf{IG}^{24} \\ \mathbf{IG}^{31} & \mathbf{IG}^{32} & \mathbf{IG}^{33} & \mathbf{IG}^{34} \\ \mathbf{IG}^{41} & \mathbf{IG}^{42} & \mathbf{IG}^{43} & \mathbf{IG}^{44} \end{bmatrix} \tag{4.81}$$

Then, applying the conjugation property to the sub-matrices of the first two rows using exactly the procedures of section 4.4.4, the following expressions for the reconstructed brightness temperatures at polarizations p and q are obtained:

$$\hat{\boldsymbol{T}}_{B_p} = \Re e\left\{ \left[\mathbf{IG}_0^{11}\ 2\mathbf{IG}_A^{11}\right] \begin{bmatrix} V_0^p \\ \boldsymbol{V}_A^p \end{bmatrix} + \left[\mathbf{IG}_0^{12}\ 2\mathbf{IG}_A^{12}\right] \begin{bmatrix} V_0^q \\ \boldsymbol{V}_A^q \end{bmatrix} \right.$$
$$\left. + \left[\mathbf{IG}_0^{13}\ 2\mathbf{IG}_A^{13}\right] \begin{bmatrix} V_0^{pq} \\ \boldsymbol{V}_A^{pq} \end{bmatrix} + \left[\mathbf{IG}_0^{14}\ 2\mathbf{IG}_A^{14}\right] \begin{bmatrix} V_0^{qp} \\ \boldsymbol{V}_A^{qp} \end{bmatrix} \right\} \tag{4.82}$$

$$\hat{\boldsymbol{T}}_{B_q} = \Re e\left\{ \left[\mathbf{IG}_0^{21}\ 2\mathbf{IG}_A^{21}\right] \begin{bmatrix} V_0^p \\ \boldsymbol{V}_A^p \end{bmatrix} + \left[\mathbf{IG}_0^{22}\ 2\mathbf{IG}_A^{22}\right] \begin{bmatrix} V_0^q \\ \boldsymbol{V}_A^q \end{bmatrix} \right.$$
$$\left. + \left[\mathbf{IG}_0^{23}\ 2\mathbf{IG}_A^{23}\right] \begin{bmatrix} V_0^{pq} \\ \boldsymbol{V}_A^{pq} \end{bmatrix} + \left[\mathbf{IG}_0^{24}\ 2\mathbf{IG}_A^{24}\right] \begin{bmatrix} V_0^{qp} \\ \boldsymbol{V}_A^{qp} \end{bmatrix} \right\} \tag{4.83}$$

where, again, subscripts A and B refer to columns with opposite (u, v) signs and 0 to the origin. The reconstructed polarimetric brightness temperature is retrieved from the third row of (4.81),

$$\hat{\boldsymbol{T}}_{B_{pq}} = \mathbf{IG}^{31}\boldsymbol{V}^p + \mathbf{IG}^{32}\boldsymbol{V}^q + \mathbf{IG}^{33}\boldsymbol{V}^{pq} + \mathbf{IG}^{34}\boldsymbol{V}^{qp} \tag{4.84}$$

while the fourth row needs not be considered since, by definition, $T_{B_{qp}} = (T_{B_{pq}})^*$ [see (2.18)].

4.4.7 Spatial Frequency Components

The reconstructed brightness temperature, either using Fourier transform or G-matrix and for any polarization combination, is an image defined in the (ξ, η) principal period (with zero padding if the unit circle is contained in it) sampled at a regular grid. The discrete Fourier transform applied to this image is formed by the so-called spatial frequency components

$$\hat{V}(k_1, k_2) = \Delta_{\xi\eta} \sum_{n_2=N_m}^{N_M} \sum_{n_1=N_m}^{N_M} \hat{T}_B(n_2, n_1) e^{-j\frac{2\pi}{N_T}(k_1 n_2 + k_2 n_1)} \tag{4.85}$$

which are defined in the principal period of the (u, v) grid, including the points not belonging to the instrument coverage. The above expression uses exactly the same reciprocal grids applied in image reconstruction. For illustration, the reader may want to take a look at the drawing of figure 4.9(b) or 4.12(b).

Spatial frequency components are useful to interpolate the reconstructed brightness temperature to different grids and/or to apply an apodization window if not done during inversion. Consider, for example, a new principal period defined by (4.40) but with a larger number of points $N_T' > N_T$. As it is explained at the end of section 4.3.2, this operation implies enlarging the (u, v) grid and make thinner the (ξ, η) grid. Assuming zero padding of spatial frequencies in the new points, the reconstructed brightness temperature defined in the new grid is the discrete inverse Fourier transform of $\hat{V}(k_1, k_2)$,

$$\hat{T}_B(n_2, n_1) = \Delta_{uv} \sum_{k_1=N'_m}^{N'_M} \sum_{k_2=N'_m}^{N'_M} W(k_1, k_2)\hat{V}(k_1, k_2)e^{j\frac{2\pi}{N'_T}(k_1 n_2 + k_2 n_1)} \tag{4.86}$$

where $W(k_1, k_2)$ is the apodization window. It is important to point out that windowing must be applied *only once*, either during inversion or afterwards with the above mechanism. Otherwise the results are difficult to interpret if not directly wrong. The advantage of doing it in spatial frequencies is that it increases flexibility if different windows are to be applied. If using the above equation to just windowing, the number of points may be kept to N_T unchanged. If larger, the result is an image sampled in a thinner grid that may be useful for better visualization. However, the spatial resolution remains the same.

The discrete Fourier transform (4.85) assumes implicitly that the reconstructed brightness temperature $\hat{T}_B$ is a periodic function in (ξ, η) replicating itself in periods adjacent to the principal one. Based on the scene brightness temperature at the border of the principal period, abrupt changes may be present at the transitions to other ones. Consequently, depending on the window used, they may induce ripples that extend to other parts of the image due to the well known "Gibbs effect." Consider for example the image illustrated in figure 4.7(a), now representing brightness temperature. The portion of this image that belongs to the principal period (the square in white drawn in the image of figure 4.7(b)) would be the reconstructed brightness temperature in the absence of errors (see section 4.4.8) and thus it is the function $\hat{T}_B(n_2, n_1)$ in (4.85). Clearly, there are abrupt changes at the border of the principal period that may be responsible of ripples. They can be mitigated to some extent by subtracting suitable constants from the reconstructed image in "sky" and "Earth" zones so as to have a zero mean image. The constant temperature subtracted to the sky grid points can be just the median of the recovered image in them, while the value subtracted to the "Earth" points can be computed so as to cancel the spatial frequency at the origin. The consequent reduction of the contrasts within the image causes minimization of the associated ripples. The constant temperatures are added back after Fourier inversion (4.86). Of course, this technique must be customized to the expected image shape.

Another common application of the spatial frequency components is to interpolate the reconstructed brightness temperature to an arbitrary point (ξ_0, η_0),

$$\hat{T}_B(\xi_0, \eta_0) = \Delta_{uv} \sum_n W_n \hat{V}_n e^{j2\pi(\xi_0 u_n + \eta_0 v_n)} \tag{4.87}$$

where the subscript n spans all the (u, v) plane. As indicated previously, a window other than rectangular should only be included if it is not applied during inversion.

4.4.7.1 Point Source Removal

Equation (4.85) provides also an effective way for removing unwanted point sources that may be present in the image, as for example RFI or the Sun. The idea behind this cancellation is to guess the frequency components of the point source and subtract them from those of the image.

A point source is characterized by a Dirac delta function in brightness temperature, so in the spatial frequency domain, it is a complex exponential,

$$\hat{V}_\delta(u,v) = A\,e^{-j2\pi(u\xi_p + v\eta_p)} \tag{4.88}$$

where A is the amplitude and (ξ_p, η_p) the position. The correction consists of subtracting the estimated point source frequency components from the original ones (4.85),

$$\hat{V}(\text{corrected}) = \hat{V} - \hat{V}_\delta \tag{4.89}$$

and reconstruct the brightness temperature using (4.86) or (4.87) applying, if needed, the apodization window, which should *not* have been applied in image reconstruction.

Both the position and the amplitude of the source are estimated from the same frequency components to which the correction is to be applied. To this end, (4.86) is used with rectangular window and a large number of points N'_T in order to obtain a brightness temperature image interpolated to a thin grid. A rough estimation of the point source location is then readily obtained by searching for the maximum value within this image. A priori information can be of help at this stage to reduce the search area but, in the end, the first estimation of the point source position is the grid point having maximum brightness temperature. To refine this position, a small circle containing several grid points around it is considered and the "center of mass" is determined. The radius of this circle depends on the instrument's impulse response and a good choice is twice the one of the synthetic beam between zeroes (see for example figure 4.1). The center of mass can be determined through a simple formula by expressing the grid coordinates as a complex number $\zeta = \xi + j\eta$,

$$\xi_p + j\eta_p = \frac{1}{\sum_{i=1}^{n} \hat{T}_{Bi}} \sum_{i=1}^{n} \zeta_i \hat{T}_{Bi} \tag{4.90}$$

where $\hat{T}_{Bi}$ is the reconstructed brightness temperature at grid point i, $\zeta_i = \xi_i + j\eta_i$ is the corresponding location, and n is the number of grid points inside the circle. The coordinates (ξ_p, η_p) correspond thus to the refined position of the source, which in general does not belong to the nominal grid. The above equation may be iterated once or twice to improve accuracy, if needed.

The spatial frequency of a *unit* point source in this location is then (4.88) with $A = 1$. Introducing it into (4.86) with the same number of points N'_T and no window, the expected contribution to the reconstructed brightness temperature of a unit point source, $\hat{T}_1(n_2, n_1)$, is obtained at all grid points.

The brightness temperature in the grid point nearest to the actual source location (ξ_{pg}, η_{pg}), which is known, can then be modelled by the sum of two contributions, the source point and the scene beneath it

$$\hat{T}_B(\xi_{pg}, \eta_{pg}) = A\,\hat{T}_1(\xi_{pg}, \eta_{pg}) + \hat{T}_{sc} \tag{4.91}$$

from which the source amplitude A is immediately retrieved provided $\hat{T}_{sc}$ is estimated as the median of $\hat{T}_B(\xi, \eta)$ in a small area around the source location. A circular corona may be considered to exclude the main lobe of the synthetic beam.

Once ξ_p, η_p and A are computed using (4.90) and (4.91), they are substituted in (4.88) and then (4.89) is used to obtain the corrected frequency components. This algorithm is based on the one published in [14], although in this reference it is applied to visibilities instead of spatial frequency components.

4.4.8 Reconstruction Error and Alias Mitigation

If the antenna spacing does not meet the Nyquist criterion, the reconstructed brightness temperature is affected by errors. In the Fourier formulation, they are limited to the aliasing illustrated in figures 4.7 or 4.11. For the G-matrix approach they spread also into the alias-free field of view, although with much less amplitude in general as most of the artifacts are concentrated in alias overlapping zones.

This reconstruction error, sometimes denoted as "floor error", is thus *always* present if the antennas are separated more than $\lambda_0/2$ in the case of rectangular grids or $\lambda_0/\sqrt{3}$ in hexagonal grids. However, it can be negligible if the source extent is small, and its brightness temperature vanishes outside the principal period. This is the case of imaging the Earth from a geostationary orbit where antenna spacing can be conveniently designed to have the whole Earth disc within the principal period, so having negligible brightness temperature outside. Another example is a Radioastronomy interferometer imaging a small portion of the sky and having highly directive antennas that block the radiation coming from other directions. For applications of Earth observation from low-orbit satellites or airplanes this error cannot be ignored except when the antennas are closer than the above-mentioned bounds. Note, however, that, even in the absence of reconstruction/aliasing errors, due to the limited coverage of visibilities in the (u, v) plane, images are affected by ripples, which can only be mitigated by appropriate windowing.

The reconstruction error originates in neglecting the second summand of (4.71) $\boldsymbol{G}_X \boldsymbol{T}_{B_X}$, so retrieving the brightness temperature using (4.72). For full polarimetric imaging, the same operation is performed in each of the sub-matrices of (4.79). The total G-matrix is then also split into columns belonging to the principal period and the rest, which are discarded to form the inverse (4.81) so reconstruction error is also present and equally defined.

In general, it can be mitigated by subtracting from the measured visibilities an estimation of the second term of (4.71) [or the corresponding one for the full polarimetric equation (4.79)] based on a brightness temperature model supposedly known. The equation to invert becomes then

$$\boldsymbol{V} - \boldsymbol{G}_X \boldsymbol{T}_{M_X} \approx \boldsymbol{G}_P \boldsymbol{T}_{B_P} \tag{4.92}$$

where $\boldsymbol{T}_{M_X}$ is the vector of model brightness temperatures particularized for the points outside the principal period. Therefore,

$$\hat{\boldsymbol{T}}_{B_P} \approx \boldsymbol{G}_P^{-1}(\boldsymbol{V} - \boldsymbol{G}_X \boldsymbol{T}_{M_X}) = \boldsymbol{G}_P^{-1}\boldsymbol{V} - \boldsymbol{E}\,\boldsymbol{T}_{M_X} \tag{4.93}$$

where $\boldsymbol{E} = \boldsymbol{G}_P^{-1}\boldsymbol{G}_X$ is the reconstruction error matrix already defined in (4.73). Needless to say, the closer the model to the actual image, the higher error reduction. Even though it may be computationally expensive depending on the instrument size, the error matrix $\boldsymbol{E}$ is specific of the sensor and thus needs to be computed only once.

System reduction due to conjugation property can also be applied to this matrix. For example, for the single polarization case (4.76),

$$\boldsymbol{E} = \Re e\left\{ \left[\mathrm{I}\boldsymbol{G}_0 \; 2\,\mathrm{I}\boldsymbol{G}_A\right] \begin{bmatrix} \boldsymbol{G}_{X_0} \\ \boldsymbol{G}_{X_A} \end{bmatrix} \right\} \tag{4.94}$$

where again $\mathrm{I}\boldsymbol{G} = \boldsymbol{G}_P^{-1}$ and the meaning of subscripts A and 0 remains the same. Similar equations based on (4.82) and (4.83) hold for the full polarimetric case.

This technique can only be applied to specific zones of the whole (ξ, η) range where the knowledge of the model is thought to be particularly accurate. Consider, for example the scene of figure 4.7, in which there exist two distinct zones, corresponding, respectively, to Earth and sky. For example, the model outside the principal period $\boldsymbol{T}_{M_X}$ may include only the sky brightness temperature (known from externally available data) in those directions pointing to it and zero otherwise. In this case, the aliases of the sky seen in figure 4.7(b) become canceled, or largely reduced, so that the alias-free is effectively enlarged. This technique was successfully applied to the SMOS processing, already at the start of the mission, to define the extended alias-free field of view, which is used in all data provided. Later, a more sophisticated modeling of the Earth was introduced, which resulted in improved imaging [8].

4.4.8.1 Subtracting the Visibility of a Model

A conceptually different approach but practically equivalent consists of reconstructing the difference between the measured visibility and that estimated from the brightness temperature model $\boldsymbol{T}_M$ in the whole unit circle; and then adding back the model. In this case, (4.72) is replaced by

$$\hat{\boldsymbol{T}}_{B_P} \approx \boldsymbol{G}_P^{-1}(\boldsymbol{V} - \boldsymbol{G}\boldsymbol{T}_M) + \left[\mathscr{B} * \boldsymbol{T}_M\right]_P \tag{4.95}$$

where $\mathscr{B}$ is the synthetic beam and the subscript P refers to the principal period. Convolution is done in the spatial frequency domain as explained in section 4.4.7 using the model in the whole unit circle and with proper sampling to avoid aliasing. The result is then interpolated to the points of the principal period. This method was proposed by [15] and it is implemented in versions 7 and higher of the SMOS operational processor.

Expanding $\boldsymbol{G}$ into submatrices in the two regions as in (4.71), the above equation is rewritten equivalently as

$$\hat{\boldsymbol{T}}_{B_P} \approx \boldsymbol{G}_P^{-1}\boldsymbol{V} - \boldsymbol{G}_P^{-1}\boldsymbol{G}_X\,\boldsymbol{T}_{M_X} - \boldsymbol{G}_P^{-1}\boldsymbol{G}_p\boldsymbol{T}_{M_P} + \left[\mathscr{B} * \boldsymbol{T}_M\right]_P \tag{4.96}$$

where the subscript X denotes the region external to the principal period. The first two terms are immediately recognized as (4.93) so the only difference between this approach and the previous one is the sum of the other two. Using the synthetic pattern of (4.74) it can be expressed as

$$\left[\mathscr{B} * \boldsymbol{T}_M\right]_P - \mathscr{B} * \boldsymbol{T}_{M_P} \tag{4.97}$$

which is not mathematically equal to zero. In the first term filtering is applied to the model in the unit circle and then it is interpolated to the points of the principal period. In the second one, filtering is applied to the model already restricted to the principal period. This is a subtle difference that has an impact only in the points near the edge of the principal period.

References

1 Manuel Martin-Neira, David M. LeVine, Yann Kerr, Niels Skou, Markus Peichl, Adriano Camps, Ignasi Corbella, Martti Hallikainen, Jordi Font, Ji Wu, Susanne Mecklenburg, and Matthias Drusch. Microwave interferometric radiometry in remote sensing: An invited historical review. *Radio Science*, 49(6):415–449, April 2014. doi: 10.1002/2013RS005230.

2 Adriano Camps, Ignasi Corbella, Javier Bará, and Francesc Torres. Radiometric sensitivity computation in aperture synthesis interferometric radiometry. *IEEE Transactions on Geoscience and Remote Sensing*, GRS-36(2):680–685, March 1998.

3 Kevin McMullan, Michael Brown, Manuel Martin-Neira, W. Rits, S. Ekholm, J. Marti, and Jerzky Lemanzyk. SMOS: The payload. *IEEE Transactions on Geoscience and Remote Sensing*, 46(3):594–605, March 2008.

4 David M. Le Vine, Michael Kao, Alan Tanner, Calvin T. Swift, and Andrew Griffis. Initial results in the development of a synthetic aperture microwave radiometer. *IEEE Transactions on Geoscience and Remote Sensing*, 28(4):614–619, July 1990.

5 Javier Bará, Adriano Camps, Francesc Torres, and Ignasi Corbella. The correlation of visibility noise and its impact on the radiometric resolution of an aperture synthesis radiometer. *IEEE Transactions on Geoscience and Remote Sensing*, GRS-38(5 Part 2):2423–2426, September 2000.

6 Alberto M. Zurita, Ignasi Corbella, Manuel Martin-Neira, Miguel A. Plaza, Francesc Torres, and F. Javier Benito. Towards a SMOS operational mission: SMOSOps-Hexagonal. *IEEE Journal of Selected Topics in Applied Earth Observations and Remote Sensing*, 6(3):1769–1780, June 2013.

7 Adriano Camps, Javier Bará, Ignasi Corbella, and Francesc Torres. The processing of hexagonally sampled signals with standard rectangular techniques: Application to 2D large aperture synthesis interferometric radiometers. *IEEE Transactions on Geoscience and Remote Sensing*, GRS-35(1):183–190, January 1997.

8 Ignasi Corbella, Francesc Torres, Nuria Duffo, Israel Duran, Vera González-Gambau, and Manuel Martin-Neira. Wide field of view microwave interferometric radiometer imaging. *Remote Sensing*, 11(6):682, March 2019. ISSN 2072-4292. doi: 10.3390/rs11060682.

9 Adriano Camps, Mercè Vall-llossera, Ignasi Corbella, Nuria Duffo, and Francesc Torres. Improved image reconstruction algorithms for aperture synthesis radiometers. *IEEE Transactions on Geoscience and Remote Sensing*, 46(1):146–158, January 2008.

10 Ignasi Corbella, Francesc Torres, Adriano Camps, Nuria Duffo, and Mercè Vall-llossera. Brightness temperature retrieval methods in synthetic aperture radiometers. *IEEE Transactions on Geoscience and Remote Sensing*, 47(1):285–294, January 2009.

11 Verónica González-Gambau, Estrella Olmedo, Antonio Turiel, Cristina González-Haro, Aina García-Espriu, Justino Martínez, Pekka Alenius, Laura Tuomi, Rafael Catany, Manuel Arias, Carolina Gabarró, Nina Hoareau, Marta Umbert, Roberto Sabia, and Diego Fernández. First SMOS Sea surface salinity dedicated products over the Baltic Sea. *Earth System Science Data*, 14:2343–2368, 2022. doi: 10.5194/essd-14-2343-2022.

12 Eric Anterrieu. A resolving matrix approach for synthetic aperture imaging radiometers. *IEEE Transactions on Geoscience and Remote Sensing*, 42(8):1649–1659, August 2004. doi: 10.1109/TGRS.2004.830940.

13 Manuel Martín-Neira, Martin Suess, and Juha Kainulainen. The flat target transformation. *IEEE Transactions on Geoscience and Remote Sensing*, 46(3):613–620, March 2008.

14 Adriano Camps, Mercè Vall-llossera, Nuria Duffo, Miguel Zapata, Ignasi Corbella, Francesc Torres, and Valentín Barrena. Sun effects in 2D aperture synthesis radiometry imaging and their cancellation. *IEEE Geoscience and Remote Sensing Letters*, 42(6):1161–1167, June 2004.

15 Ali Khaz, Philippe Richaume, Frans Cabot, Eric Anterrieu, Arnaud Mialon, and Yann H. Kerr. Improving the spatial bias correction algorithm in SMOS image reconstruction processor: Validation of soil moisture retrievals with in situ data. *IEEE Transactions on Geoscience and Remote Sensing*, 57(1):277–290, January 2019. doi: 10.1109/TGRS.2018.2853619.

5

Instrument Techniques

The concepts presented in the four preceding chapters have not mentioned the particular techniques used at instrument level to provide the measured quantities involved. This chapter deals with some of the most common techniques to measure correlation and power, which are the two fundamental magnitudes needed for processing both interferometric and polarimetric radiometer data.

5.1 Frequency Conversion

Frequency conversion is commonly used in all kinds of radiometers, especially those designed to operate at high microwave or millimeter wave frequencies. The purpose of this technique is to shift the radio frequency (RF) band to an intermediate frequency (IF) by mixing the signal with a sinusoidal local oscillator (LO). Figure 5.1(a) shows the block diagram of a typical frequency converter. It includes an RF filter to discard out-of-band input signals, a mixer and an IF filter. The function of the mixer is to multiply the two time-domain signals applied to its input ports (RF and LO). The IF filter is used to reject all undesired harmonics produced by the mixing so keeping only the intermediate frequency.

5.1.1 Frequency Bands

For a given radiometric band of interest at RF, there are two options for locating the local oscillator: either at a higher frequency ($f_{LO} > f_{RF}$) or at a lower one ($f_{LO} < f_{RF}$). In the first case, the RF would be the "lower band" depicted in figure 5.1(b) ($f_{RF} = f_L$) while the second case corresponds to the RF defined as the "upper band" ($f_{RF} = f_U$). Setting the LO frequency in the middle of the RF band (conversion to DC) needs different analysis tools and is not covered in this book. Therefore, the *nominal* RF band, either upper or lower, is always assumed to vanish (or become residual) at the local oscillator frequency (as depicted in the figure).

Principles of Interferometric and Polarimetric Radiometry, First Edition. Ignasi Corbella.

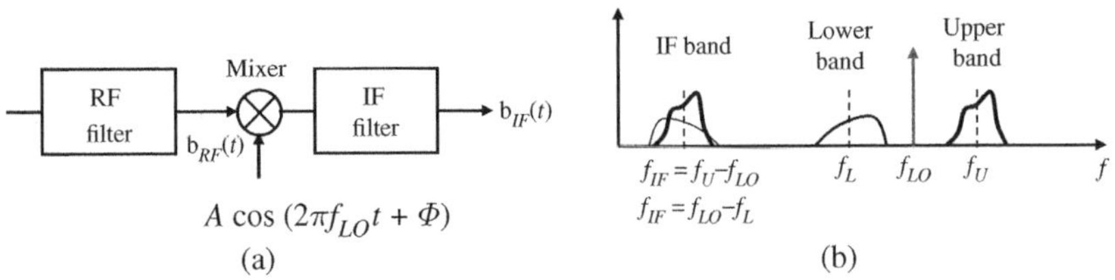

Figure 5.1 (a) Block diagram of a frequency converter. (b) Sketch of the signals spectra. The nominal RF band can be either the upper or the lower band. The image band is the other.

This restriction does *not* necessarily apply to the RF filter frequency response, which in principle does not need to accurately tailor the band of the radiometric signal. When frequency conversion is used, the overall frequency response includes the combined contributions of both RF and IF sections. Directly from the block diagram of figure 5.1,

$$\overline{H}(f) = \overline{H}_{RF}(f)\overline{H}_{IF}(f - f_{LO})e^{\pm j\Phi \,\mathrm{sgn}\, f} \tag{5.1}$$

where $H_{RF}(f)$ and $H_{IF}(f)$ are the frequency responses of the RF and IF filters, respectively, and the overbar indicates normalization with respect to the nominal center frequency, either at RF or IF. The signum function sgn is defined in (A.30). The constant phase Φ is directly that of the local oscillator defined in the block diagram. It sets a time origin and, if a single receiver is considered, it can be set to zero without loss of generality. As will be shown later, the positive sign in the exponent holds when the RF is the lower band and the negative otherwise. It is important to point out that all the band sketches drawn in figure 5.1 are based on the above definition. Different and asymmetrical shapes are used only for illustration purposes.

Both upper and lower bands are converted to the same intermediate frequency, so the RF filter must be effective in suppressing the unwanted one, usually denoted "image band." A technique to suppress it in the mixer itself, so reducing the requirements of the RF filter, is explained later.

Using (5.1), the time delay property of the fringe washing function (3.64) becomes

$$\tilde{r}'_{12}(t) = \tilde{r}_{12}(t - \Delta t_{RF} - \Delta t_{IF})e^{-j2\pi f_0 \Delta t_{RF}}e^{-j2\pi f_{IF}\Delta t_{IF}}e^{\pm j(\Phi_1 - \Phi_2)\,\mathrm{sgn}\, f} \tag{5.2}$$

where Δt_{RF} and Δt_{IF} are the delay differences introduced at RF and IF, respectively, always defined as the delay in receiver 1 minus the delay in receiver 2 $(t_1 - t_2)$. Recall that f_0 is the center frequency of the band and f_{IF} is $f_U - f_{LO}$ (or $f_{LO} - f_L$).

5.1.2 Mixer Operation

The mixer usually includes a signal combiner and a nonlinear element, such as a transistor or a diode, to multiply the input signals by means of the fundamental equation (3.129), also used in analog correlators. Other approaches are possible depending on the frequency of operation, but in general, due to the intrinsic nonlinearity of the operation, the output signal contains a large number of intermodulation products that have to be properly filtered to keep only the intermediate frequency. The interested reader may want to consult the classical book by S. A. Maas [1] for more details. For the sake of simplicity, here the mixer is modeled as a device that multiplies the two input time domain waveforms and rejects from its output all signals other than the intermediate frequency.

The mixer operation is thus well described in the time domain with the aid of the analytic signal (section 1.3). Specifically, all signals involved are represented by their corresponding complex envelopes (1.129), each one with its own reference frequency

$$\begin{aligned}
\text{Lower band:}\quad & \mathrm{b}_L(t) = \Re e[\tilde{b}_L(t)e^{j2\pi f_L t}] \\
\text{Upper band:}\quad & \mathrm{b}_U(t) = \Re e[\tilde{b}_U(t)e^{j2\pi f_U t}] \\
\text{Local oscillator:}\quad & \mathrm{b}_{LO}(t) = \Re e[\tilde{b}_{LO}e^{j2\pi f_{LO} t}] \\
\text{Intermediate frequency:}\quad & \mathrm{b}_{IF}(t) = \Re e[\tilde{b}_{IF}(t)e^{j2\pi f_{IF} t}]
\end{aligned} \tag{5.3}$$

in which, since the local oscillator is a pure sinusoidal signal, $\tilde{b}_{LO}$ is the phasor (1.119), a complex constant with amplitude A and phase Φ consistent with the block diagram of figure 5.1. Taking into account that the mixer input signal includes in general contributions from both the lower and the upper bands, the mixer output becomes

$$\mathrm{b}_{IF}(t) = G_c\,\text{Filtered}\left\{\mathrm{b}_L(t)\mathrm{b}_{LO}(t) + \mathrm{b}_U(t)\mathrm{b}_{LO}(t)\right\} \tag{5.4}$$

where G_c is a constant known as "conversion gain" and the word "Filtered" means discarding all bands except the one at intermediate frequency. The conversion gain is intrinsic to the specific mixer design and is usually lower than one, so it is a "loss." It depends on the local oscillator amplitude and other parameters such as DC bias and it may be different for the upper and lower bands. However, for the sake of simplicity and without loss of generality, in what follows this conversion gain is set to unity. Substituting now (5.3) and using the identity (A.11) but keeping only the first term due to the filtering,

$$\mathrm{b}_{IF}(t) = \frac{1}{2}\Re e\left[[\tilde{b}_L^*(t)\tilde{b}_{LO} + \tilde{b}_U(t)\tilde{b}_{LO}^*]e^{j2\pi f_{IF} t}\right] \tag{5.5}$$

where $f_{IF} = f_U - f_{LO} = -(f_L - f_{LO})$. The first term above is conjugated in order to convert the negative exponential into positive based on the fact that the real part of a complex number is identical to that of its conjugate. Consistent with the conjugation property of the Fourier transform (see table A.1), the corresponding spectrum is mirrored with respect to the origin, and this is indicated also in figure 5.1.

For example, assume that RF is the upper band and that the lower one is eliminated either via filtering or with an image rejection mixer. Then

$$b_{IF}(t) = K\Re e\left[\tilde{b}_{RF}e^{-j\Phi}e^{j2\pi f_{IF}t}\right] \tag{5.6}$$

where K is a constant that takes into account all the multiplicative factors, including the LO amplitude, conversion gain, and filter loss. The minus sign in the oscillator phase justifies the assertion made in relation to (5.1). According to (5.5), for the same local oscillator, it becomes positive when the RF is set to the lower band.

Assuming $K = 1$ to simplify interpretation, (5.6) indicates that the complex envelope of the output signal with respect to f_{IF} is equal to the complex envelope of the input signal with respect to f_{RF} with an additional constant phase shift that depends on the oscillator. Or, equivalently, *the analytic signal at the output is equal to the analytic signal at the input, but frequency shifted to IF and phase shifted by a constant phase.* At this point it should be recalled that, according to (1.127), a quadrature filter is equivalent to multiplying the analytic signal by $-j$. Therefore, if $\Phi = 90°$, frequency conversion, besides a frequency shift, implements also a quadrature filter. This is compatible with the overall frequency response (5.1) in which the last term becomes identical to (1.123) when $\Phi = 90°$.

5.1.3 Image Rejection Mixer

By properly combining two mixers, it is possible to cancel either the upper or the lower bands while enhancing the other one. This is a very interesting option that allows relaxing the RF filter specifications in terms of selectivity. The procedure is schematized in figure 5.2, where the block diagram labeled (a) keeps only the upper band while (b) keeps the lower band. The components labeled $H_q(f)$ are ideal quadrature filters, which in practice are implemented with 90° hybrid

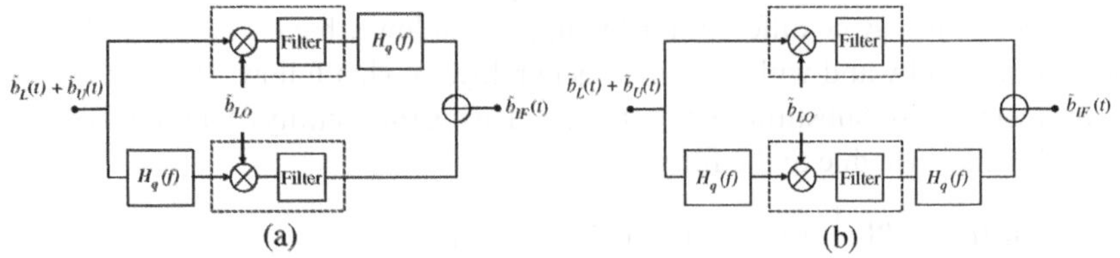

Figure 5.2 Image rejection mixer. Setup (a) rejects the lower band while (b) rejects the upper band.

couplers. One operates at high frequency and must cover both the lower and the upper bands while the other is designed to work at intermediate frequency. Needless to say, departure from ideality of any of these filters, particularly in the phase response, affects the rejection performance of the mixer. Maybe the most critical design is the RF quadrature filter, which should provide 90° phase shift in quite a large bandwidth. Detailed analysis of actual implementations is out of the scope of this book and, therefore, in the analysis that follows all components are assumed ideal.

As indicated by the tilde in the symbols, all signals in this figure represent complex envelopes. In the setup of figure 5.2(a), taking into account that a quadrature filter multiplies the analytic signal, and hence the complex envelope, by $-j$, the complex envelope of the output signal is immediate after applying (5.5) to both branches,

$$\begin{aligned}\tilde{b}_{IF}(t) &= \frac{1}{2}\left[\tilde{b}_L^*(t)\tilde{b}_{LO} + \tilde{b}_U(t)\tilde{b}_{LO}^*\right](-j) \\ &+ \frac{1}{2}\left[[-j\tilde{b}_L(t)]^*\tilde{b}_{LO} + [-j\tilde{b}_U(t)]\tilde{b}_{LO}^*\right] \\ &= -j\tilde{b}_U(t)\tilde{b}_{LO}^*\end{aligned} \tag{5.7}$$

so effectively cancels the lower band and enhances the upper band. If the RF band is this one, the overall response of this mixer is equivalent to having a perfect image rejection filter. On the other hand, if the quadrature filter at IF is placed in the lower branch, as shown in the block diagram on figure 5.2(b), the same procedure yields

$$\tilde{b}_{IF}(t) = \tilde{b}_L^*(t)\tilde{b}_{LO} \tag{5.8}$$

which cancels the upper band in this case, so this is applicable when the RF band is located in the lower band.

5.2 In-phase and Quadrature (IQ) Mixer

5.2.1 Concept

An IQ mixer is a common method of separating the real and imaginary parts of the analytic signal in order to get individually the signal itself and its Hilbert transform. The method, shown schematically in figure 5.3, uses a combination of two mixers driven by corresponding local oscillators at the same frequency but in quadrature respect to each other. An additional phase in one of the branches produces the quadrature error covered in section (5.2.3). In the ideal case, if the two

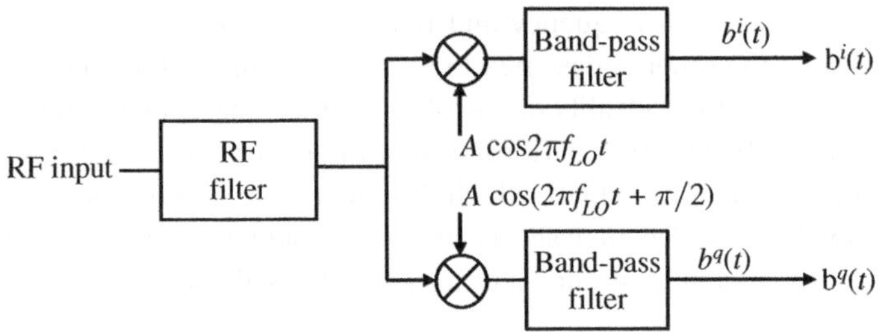

Figure 5.3 In-phase and quadrature mixer to separate the real and imaginary parts of the analytic signal. Italic letters denote analytic signals and upright letters real signals.

mixers and IF filters are identical, using (5.6), the analytic signals at each output port, labeled here "i" and "q" for in-phase and quadrature components,[1] are

$$b^i(t) = K\tilde{b}_{RF}e^{j2\pi f_{IF}t} \ ; \ b^q(t) = -jK\tilde{b}_{RF}e^{j2\pi f_{IF}t} \tag{5.9}$$

which are identical except for the $-j$ multiplication in the lower branch. According to (1.127), this means that the *real* output signals $\mathrm{b}^i(t)$ and $\mathrm{b}^q(t)$ are, respectively, the real and imaginary parts of the analytic signal $b^i(t)$. In consequence, the lower branch implements a quadrature filter. Its output is the Hilbert transform of the output signal in the upper branch.

If two receivers are now used as a baseline of an interferometer, equations (3.99) and (3.100) apply. Using, for example, the first one of the simplified options in both,

$$\begin{aligned} \Gamma_{12}^{ii} &= \left\langle \mathrm{b}_1^i(t)\mathrm{b}_2^i(t) \right\rangle = \Re e\left[\Gamma_{12}\right] \\ \Gamma_{12}^{qi} &= \left\langle \mathrm{b}_1^q(t)\mathrm{b}_2^i(t) \right\rangle = \Im m\left[\Gamma_{12}\right] \end{aligned} \tag{5.10}$$

where the symbols Γ_{12}^{ii} and Γ_{12}^{qi} are here used for the cross-correlations between the corresponding real signals. Of course, other combinations of output signals are possible using different options of (3.99) and (3.100).

5.2.2 General Analysis

According to (3.82), the cross-correlation between any two real signals coming from different receivers can be written in general as

$$\Gamma_{12}^{\alpha\beta} = \left\langle \mathrm{b}_1^{\alpha}(t)\mathrm{b}_2^{\beta}(t) \right\rangle = \Re e[G_{12}^{\alpha\beta}V_{12}^{\alpha\beta}e^{j\Delta\Phi_{LO}}] \tag{5.11}$$

1 In some textbooks, these names are used for the real and imaginary parts of the complex envelope. It is not the case here.

where α and β should be substituted by i or q, $\Delta\Phi_{LO}$ is

$$\Delta\Phi_{LO} = \begin{cases} 0 & \text{if } \alpha\beta = ii \text{ or } \alpha\beta = qq \\ -\pi/2 & \text{if } \alpha\beta = qi \\ \pi/2 & \text{if } \alpha\beta = iq \end{cases} \tag{5.12}$$

the visibility $V_{12}^{\alpha\beta}$ is (3.58) with the fringe washing function (3.54) corresponding to the overall filter in each branch,

$$\tilde{r}_{12}^{\alpha\beta}(t) = \frac{e^{-j2\pi f_0 t}}{\sqrt{B_1^\alpha B_2^\beta}} \int_0^\infty \overline{H}_1^\alpha(f)\overline{H}_2^{\beta *}(f) e^{j2\pi f t} df \tag{5.13}$$

and the correlator gains $G_{12}^{\alpha\beta}$ are equal to (3.83) after replacing the corresponding fringe washing functions at the origin

$$G_{12}^{\alpha\beta} = k\sqrt{B_1^\alpha B_2^\beta}\sqrt{G_{T1}^\alpha G_{T2}^\beta}\,\tilde{r}_{12}^{\alpha\beta}(0) \tag{5.14}$$

It is important to understand that $\tilde{r}_{12}^{\alpha\beta}(0)$ is the primed version of (5.2) so it includes any difference of local oscillator phase between both receivers. Equation (5.11) makes use of (1.143) to write the cross-correlation between any two real signals as one-half the real part of the correlation between the corresponding analytic signals. Even though it is expressed as the real part, when $\Delta\Phi_{LO}$ is $\pm\pi/2$, the equation actually provides the imaginary part. In particular, for $\alpha\beta = ii$ and $\alpha\beta = qi$, if the IF filters are identical in both branches, equations (5.10) are reproduced. Differences in the frequency response of each branch produce measurement errors.

The frequency responses in (5.13) are the combination of the RF and IF blocks (5.1) without including the 90° LO phase shift that has already been taken into account:

$$H_k^\alpha(f) = H_{RF_k}(f)H_{IF_k}^\alpha(f - f_{LO})e^{\pm j\Phi_k \operatorname{sgn} f} \quad \alpha = i, q \quad k = 1, 2 \tag{5.15}$$

The correlation between the i and q outputs of *the same receiver* can be derived using the methodology of section 3.3.1 for this particular case, resulting in an equation similar to (3.57). Taking the real part in order to obtain the correlation between the real signals and including the local oscillator phase difference (5.12), the result is

$$\langle \mathrm{b}_k^q(t)\mathrm{b}_k^i(t)\rangle = k\sqrt{B_k^q B_k^i}\sqrt{G_k^q G_k^i}\,T_{sysk}\,\Im m[\tilde{r}_{kk}^{qi}(0)] \quad k = 1, 2 \tag{5.16}$$

where T_{sysk} is the system temperature of this receiver, assumed the same for both branches. The other symbols are duly defined in (3.57), adding the superscripts q and i to identify the IQ mixer output.

In the ideal case of identical IF filters, the imaginary part of the fringe washing function is odd (property 2 in section 3.3.2) so the above correlation vanishes.

This is a totally expected result since, as pointed out in section 1.3 with reference to equation (1.142), a signal is always uncorrelated with its Hilbert transform. The reader is invited to check that interchanging the order of the signals in the left-hand side of (5.16) produces exactly the same result.

Taking into account that the total power of any output signal is $kG_T T_{sys}$, the normalized correlation (3.135) in this case reduces to

$$\mathrm{M}_{kk}^{qi} = \frac{\left\langle \mathrm{b}_k^q(t)\mathrm{b}_k^i(t)\right\rangle}{\sqrt{\left\langle \mathrm{b}_k^q(t)^2\right\rangle \left\langle \mathrm{b}_k^i(t)^2\right\rangle}} = \Im m[\tilde{r}_{kk}^{qi}(0)] \quad k = 1, 2 \tag{5.17}$$

which is used in section 5.2.3 to device a method to measure the quadrature error.

5.2.3 Quadrature Error

If the IF filters in the i and q branches have important differences with respect to each other, the visibility retrieved from each one $V_{12}^{\alpha\beta}$ is not unique and the interferometric processing is very hard, if even possible. These filters must then be designed with a high degree of similarity. However, a constant phase difference between branches produces a quadrature error that can be corrected. This phase difference can come from the filters themselves and/or from an error in the phase of the local oscillator departing quadrature.

Assume that for a given receiver, the overall frequency response (5.15) can be expressed as

$$H_k^\alpha(f) = H_k(f)e^{j\varphi_k^\alpha} \quad \alpha = i, q \quad k = 1, 2 \tag{5.18}$$

where φ_k^α is a *constant* phase. This assumption implies that the transducer power gain and noise bandwidth for each receiver are the same in both branches. The fringe washing function (5.13) becomes

$$\tilde{r}_{12}^{\alpha\beta}(t) = \tilde{r}_{12}(t)e^{j(\varphi_1^\alpha - \varphi_2^\beta)} \tag{5.19}$$

where $\tilde{r}_{12}(t)$ is now a unique fringe washing function expressed as (3.54) with only the common frequency responses $H_1(f)$ and $H_2(f)$. Since normalizing (5.19) with respect to its value at the origin is independent of the branch, the visibility (3.58) is unique

$$V_{12}^{\alpha\beta} = V_{12} \quad \forall \alpha, \beta \tag{5.20}$$

For a given receiver, the quadrature error is defined as the difference between the phases in both i and q branches defined in (5.18)

$$\text{Quadrature error:} \, \theta_{qk} = \varphi_k^i - \varphi_k^q \quad k = 1, 2 \tag{5.21}$$

and it can be measured from the normalized correlation between the corresponding output signals (5.17). Using (5.19) and taking into account that $\tilde{r}_{kk}(0) = 1$ (see property 7 of the fringe washing function in section 3.3.2), it follows that

$$M_{kk}^{qi} = \Im m[\tilde{r}_{kk}^{qi}(0)] = \Im m[e^{j(\varphi_k^q - \varphi_k^i)}] = -\sin\,\theta_{qk} \tag{5.22}$$

meaning that the quadrature error θ_{qk} can be retrieved as the arc sinus of the normalized correlation between the I and Q outputs. If there is no quadrature error, then this correlation vanishes, as expected.

5.2.4 Correction of Phase Errors

The phase difference appearing in the exponential term of (5.19) can be expressed as sum of a *common* term α_{12}, sometimes referred to as phase error, and another one depending in the quadrature errors of both receivers:

$$\begin{aligned} \varphi_1^i - \varphi_2^i &= \alpha_{12} - Q_{12} \\ \varphi_1^q - \varphi_2^i &= \alpha_{12} - Q_{12}' \\ \varphi_1^i - \varphi_2^q &= \alpha_{12} + Q_{12}' \\ \varphi_1^q - \varphi_2^q &= \alpha_{12} + Q_{12} \end{aligned} \tag{5.23}$$

with

$$\alpha_{12} = \frac{1}{2}(\varphi_1^i + \varphi_1^q) - \frac{1}{2}(\varphi_2^i + \varphi_2^q) \tag{5.24}$$

and Q_{12} and Q_{12}' the semi-sum and semi-difference of quadrature errors:

$$Q_{12} = -\frac{1}{2}(\theta_{q1} - \theta_{q2}) \qquad Q_{12}' = \frac{1}{2}(\theta_{q1} + \theta_{q2}) \tag{5.25}$$

The phase α_{12} (5.24) is the difference between the mean phases of each receiver, not included in the common frequency responses $H_k(f)$ in (5.18). Therefore it adds to the common fringe washing function phase to properly define the common complex gain from (3.83) in this case:

$$G_{12} = k\sqrt{B_1 B_2}\sqrt{G_{T1}G_{T2}}\tilde{r}_{12}(0)e^{j\alpha_{12}} \tag{5.26}$$

As pointed out in relation to (5.14), $\tilde{r}_{12}(0)$ is the primed version of (5.2) so it includes any difference of local oscillator phase between both receivers. Introducing now (5.19) into (5.14), using (5.23) and the above definition of G_{12}, equation (5.11) is expanded to

$$\begin{aligned} \Gamma_{12}^{ii} &= \Re e\left[e^{-jQ_{12}}\Gamma_{12}\right] \\ \Gamma_{12}^{qi} &= \Im m\left[e^{-jQ_{12}'}\Gamma_{12}\right] \\ \Gamma_{12}^{iq} &= -\Im m\left[e^{jQ_{12}'}\Gamma_{12}\right] \\ \Gamma_{12}^{qq} &= \Re e\left[e^{jQ_{12}}\Gamma_{12}\right] \end{aligned} \tag{5.27}$$

where $\Gamma_{12} = G_{12}V_{12}$ with V_{12} the unique visibility (5.20). Consistent with (3.82), Γ_{12} can be interpreted as *quadrature corrected correlation*. If now the *measured* complex correlation is build from two real correlations as

$$\Gamma'_{12} = \Gamma^{ii}_{12} + j\Gamma^{qi}_{12} \tag{5.28}$$

then, after substituting the first two equations (5.27), the quadrature corrected correlation can be solved using the method described in appendix A.5. The result is

$$\Gamma_{12} = \frac{1}{\cos\theta_{q1}} \left(\Re e[M_1\Gamma'_{12}] + j\Im m[M_2\Gamma'_{12}]\right) \tag{5.29}$$

where the constants M_1 and M_2 are

$$M_1 = \cos Q'_{12} + j\ \sin\ Q_{12}\ ;\ M_2 = \cos Q_{12} + j\ \sin\ Q'_{12} \tag{5.30}$$

Other combinations of i and q outputs can be used instead. The corresponding expressions for Γ_{12} are provided below for reference

$$\begin{aligned}
\Gamma'_{12} &= \Gamma^{qq}_{12} - j\Gamma^{iq}_{12} \Rightarrow \Gamma_{12} = \frac{1}{\cos\theta_{q1}} \left(\Re e[M_1^*\Gamma'_{12}] + j\Im m[M_2^*\Gamma'_{12}]\right) \\
\Gamma'_{12} &= \Gamma^{ii}_{12} - j\Gamma^{iq}_{12} \Rightarrow \Gamma_{12} = \frac{1}{\cos\theta_{q2}} \left(\Re e[M_1\Gamma'_{12}] + j\Im m[M_2^*\Gamma'_{12}]\right) \\
\Gamma'_{12} &= \Gamma^{qq}_{12} + j\Gamma^{qi}_{12} \Rightarrow \Gamma_{12} = \frac{1}{\cos\theta_{q2}} \left(\Re e[M_1^*\Gamma'_{12}] + j\Im m[M_2\Gamma'_{12}]\right)
\end{aligned} \tag{5.31}$$

with the same expressions (5.30) for M_1 and M_2. Once Γ_{12} is obtained from these equations, the visibility is retrieved using (3.86) with (5.26) for the correlator gain, to be determined from calibration as explained in section 6.2.1.

5.2.5 Normalized Correlations

The same procedure can also be applied when normalized correlations are considered (section 3.4.5). In this case, all measured correlations $\Gamma^{\alpha\beta}_{12}$ in (5.27) are normalized:

$$\mathrm{M}^{\alpha\beta}_{12} = \frac{\left\langle \mathrm{b}^{\alpha}_1(t)\mathrm{b}^{\beta}_2(t)\right\rangle}{\sqrt{\left\langle \mathrm{b}^{\alpha}_1(t)^2\right\rangle \left\langle \mathrm{b}^{\beta}_2(t)^2\right\rangle}} \tag{5.32}$$

and, consistent with (3.136) the quadrature corrected normalized correlation is:

$$M_{12} = \frac{G_{12}V_{12}}{\sqrt{T_{sys1}T_{sys2}}} \tag{5.33}$$

where, according to (3.137) now G_{12} is just the fringe washing function at the origin but with the added phase α_{12} as in (5.26)

$$G_{12} = \tilde{r}_{12}(0)e^{j\alpha_{12}} \tag{5.34}$$

The measured complex correlation M'_{12} is now constructed from the normalized values in the same way. For example (5.28) and (5.29) become

$$M'_{12} = \mathrm{M}^{ii}_{12} + j\mathrm{M}^{qi}_{12} \Rightarrow M_{12} = \frac{1}{\cos\theta_{q1}}\left(\Re e[M_1 M'_{12}] + j\Im m[M_2 M'_{12}]\right) \tag{5.35}$$

and similar substitutions for all other cases (5.31). The visibility is finally retrieved by solving (5.33), that is, using (3.136), with the corresponding calibrated gain and system temperatures. This is the approach followed in SMOS data processing [2].

5.3 Quarter Period Delay

5.3.1 Concept

The frequency response of an ideal quadrature filter (1.123) has unit amplitude and constant 90° phase at all (positive) frequencies. This response can be approximated by a constant time delay, implemented, for example, by a transmission line section or digitally (see section 5.4.5). The frequency response of a time delay t_d is

$$H_d(f) = e^{-j2\pi f t_d} \tag{5.36}$$

so, at $f = 1/(4t_d)$ it is the same as that of a quadrature filter. Therefore, *for narrow band signals*, a time delay equal to a quarter of a period at the center frequency can be used as a quadrature filter to generate the Hilbert transform of the input signal.

Including a time delay in one of the two receivers of an interferometer is straightforward. Based on equation (3.57), taking the real part to get the cross-correlation between real signals [see (1.143)] and introducing a delay in the second one taking into account property 4 of the fringe washing function (section 3.3.2),

$$\langle \mathrm{b}_1(t)\mathrm{b}_2(t - t_d)\rangle = k\sqrt{B_1 B_2}\sqrt{G_1 G_2}\,\Re e\left[\tilde{r}_{12}(t_d)e^{j2\pi f_0 t_d} V'_{12}\right] \tag{5.37}$$

where t_d is the time delay applied to receiver 2, $\tilde{r}_{12}(t)$ the fringe washing function referenced to f_0 corresponding to the two filters *without adding the delay* and V'_{12} the visibility (3.58) but using in the integral the normalized fringe washing function that includes the delay:

$$\bar{\tilde{r}}'_{12}\left(^{-(r_1-r_2)}/_c\right) = \frac{\tilde{r}_{12}\left(^{-(r_1-r_2)}/_c + t_d\right)}{\tilde{r}_{12}(t_d)} \tag{5.38}$$

It is important to point out that this technique implies that V'_{12} in (5.37) is implicitly approximated by V_{12}, corresponding to the correlation without delay. Otherwise the complex correlation is not related to a single visibility function and the equation is inconsistent. The impact of this approximation in the brightness temperature retrieval must be studied by simulation in each particular case. In general, it gets worse with increasing relative bandwidth and/or field of view. This important limitation refrains from using this technique in practical applications.

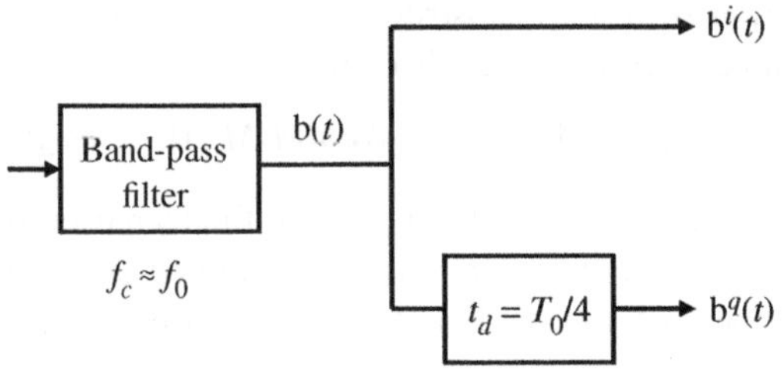

Figure 5.4 A quarter period delay used as quadrature filter. The period corresponds to the nominal center frequency $T_0 = 1/f_0$.

Nevertheless, even if the visibility has been assumed unique, some corrections must still be applied to compensate for the term $\tilde{r}_{12}(t_d)$ in (5.37). The methodology described here below is based on the one published in [3].

Figure 5.4 shows a simplified block diagram for separating the i and q signals, both named as in section 5.2. The i output (in-phase component) is directly the input signal and the q output (quadrature component) is the delayed signal. The time delay is a quarter of a period of the *nominal* center frequency f_0 at which the system is designed. Of course, the filter is expected to be reasonably centered on f_0 but a different center frequency f_c is assumed so that any deviation from the nominal is an error that will be quantified in section 5.3.2.

A complex correlator based on this concept measures all correlation products between the i and q signals of figure 5.4. Particularizing (5.37) with $f_0 t_d = 1/4$,

$$\begin{aligned}
\Gamma_{12}^{ii} &= \left\langle b_1^i(t) b_2^i(t) \right\rangle = \Re e\left[\Gamma_{12}\right] \\
\Gamma_{12}^{qi} &= \left\langle b_1^q(t) b_2^i(t) \right\rangle = \Im m\left[\bar{\tilde{r}}_{12}(-t_d)\Gamma_{12}\right] \\
\Gamma_{12}^{iq} &= \left\langle b_1^i(t) b_2^q(t) \right\rangle = -\Im m\left[\bar{\tilde{r}}_{12}(t_d)\Gamma_{12}\right] \\
\Gamma_{12}^{qq} &= \left\langle b_1^q(t) b_2^q(t) \right\rangle = \Re e\left[\Gamma_{12}\right]
\end{aligned} \tag{5.39}$$

where $\Gamma_{12} = G_{12} V_{12}$ with G_{12} the correlator gain (3.83). The fringe washing function used in this gain is $\tilde{r}_{12}(0)$, which does not include the delay.

These quadrature equations substitute (5.27) for this approach. As expected, only the imaginary part is affected by an error. They can also be solved using the method of appendix A.5 to yield the quadrature-corrected complex correlation Γ_{12} out of real correlation measurements. The results for two signal selection options are

$$\begin{aligned}
\Gamma'_{12} &= \Gamma_{12}^{ii} + j\Gamma_{12}^{qi} \Rightarrow \Gamma_{12} = \Re e[\Gamma'_{12}] + j\Im m[M_1 \Gamma'_{12}] \\
\Gamma'_{12} &= \Gamma_{12}^{qq} - j\Gamma_{12}^{iq} \Rightarrow \Gamma_{12} = \Re e[\Gamma'_{12}] + j\Im m[M_2 \Gamma'_{12}]
\end{aligned} \tag{5.40}$$

where

$$M_1 = \frac{1 - j\Im m[\bar{\tilde{r}}_{12}(-t_s)]}{\Re e[\bar{\tilde{r}}_{12}(-t_s)]} \; ; \; M_2 = \frac{1 - j\Im m[\bar{\tilde{r}}_{12}(t_s)]}{\Re e[\bar{\tilde{r}}_{12}(t_s)]} \tag{5.41}$$

Equations for the other two signal choices, corresponding to the last two equations (5.31) are straightforward. Provided that the normalized fringe washing function at $\pm t_d$ is known, the above equations allow retrieving the corrected correlation Γ_{12} out of the raw measurements.

In summary, generating the quadrature signal by a delay induces an error in the imaginary part that can be corrected if the fringe washing function is known. If it is approximated by a sinc [see (3.66)], then (5.41) simplifies to $M_1 = M_2 = 1/\text{sinc } Bt_d$ with B the signal bandwidth. Then (5.40) approximates to

$$\Gamma_{12} = \Re e[\Gamma'_{12}] + j\frac{\Im m[\Gamma'_{12}]}{\text{sinc } Bt_d} \tag{5.42}$$

so the imaginary part needs a correction quantified by the inverse of sinc Bt_d, close to unity if the product $Bt_d = B/(4f_0) \ll 1$. For example, assuming a nominal center frequency $f_0 = 30$ MHz and a bandwidth of $B = 20$ MHz the correction is a factor of 1.047.

5.3.2 Center Frequency Error

Using (5.16), the cross-correlation between the i and q outputs of a single receiver is

$$\left\langle b_k^q(t) b_k^i(t) \right\rangle = kB_k G_k T_{sysk} \Im m\left[\tilde{r}_k(-t_d)\right] \quad k = 1, 2 \tag{5.43}$$

where $\tilde{r}_k$ is the fringe washing function of the single receiver k (3.62). Ideally, if the time delay is designed to implement a quadrature filter, this correlation vanishes. The error is then quantified by the imaginary part of the fringe washing function evaluated at $-t_d$.

Consider that the frequency response of the receiver is centered around $f = f_c$ and define a center frequency error as

$$\Delta f = f_0 - f_c \tag{5.44}$$

where f_0 is the nominal value. Then, still using the sinc approximation, the fringe washing function of the receiver with respect to its center frequency f_c is just equal to sinc Bt. Applying property 5 of the fringe washing function (section 3.3.2) to change the reference frequency to f_0,

$$\tilde{r}_k(t) \approx \text{ sinc } Bt\, e^{-j2\pi\, \Delta f\, t} \quad k = 1, 2 \tag{5.45}$$

and introducing this result into (5.43),

$$\left\langle b_k^q(t) b_k^i(t) \right\rangle = kB_k G_k T_{sysk} \text{ sinc } Bt_d \ \sin\ 2\pi\, \Delta f\, t_d \tag{5.46}$$

If the center frequency coincides with the nominal value ($\Delta f = 0$), then this correlation vanishes, as expected. Otherwise, it can be used to estimate the center frequency,

$$f_c = f_0 - \frac{2f_0}{\pi}\arcsin\left(\frac{\mathrm{M}_{kk}^{qi}}{\mathrm{sinc}\ Bt_d}\right) \tag{5.47}$$

where M_{kk}^{qi} is the normalized correlation defined as in (5.17).

This result can be used to write the correction constants M_1 and M_2 (5.41) as a function of the center frequency error estimated for each receiver. Using still the sinc approximation, the *normalized* fringe washing function corresponding to both receivers becomes identical to that of a single receiver (5.45). It can thus be written as:

$$\bar{\tilde{r}}_{12}(t) \approx \ \mathrm{sinc}\ Bt(\cos 2\pi\overline{\Delta f}\,t - j\ \sin\ 2\pi\overline{\Delta f}\,t) \tag{5.48}$$

where $\overline{\Delta f}$ is the center frequency error for the pair, which can be set equal to the average of those estimated from (5.47) using the definition (5.44). The above equation can be directly introduced into (5.41) to get the correction factors M_1 and M_2 for the imaginary part of the correlation

$$M_{1,2} = \frac{1 \pm j\ \mathrm{sinc}\ Bt_d\ \sin\ 2\pi\frac{\overline{\Delta f}}{f_s}}{\mathrm{sinc}\ Bt_d\ \cos 2\pi\frac{\overline{\Delta f}}{f_s}} \quad \textit{(Plus sign for } M_2 \textit{ and minus for } M_1\textit{)} \tag{5.49}$$

which in turn must be substituted into (5.40) to obtain the corrected correlation Γ_{12}. Without additional information, this is the best correction that can be applied to compensate the decorrelation of the imaginary part.

5.3.3 Normalized Correlations

This technique can obviously be applied to normalized correlations. In this case, the correction equations (5.40) become

$$\begin{aligned} M'_{12} &= \mathrm{M}_{12}^{ii} + j\mathrm{M}_{12}^{qi} \Rightarrow M_{12} = \Re e[M'_{12}] + j\Im m[M_1 M'_{12}] \\ M_{12} &= \mathrm{M}_{12}^{qq} - j\mathrm{M}_{12}^{iq} \Rightarrow M_{12} = \Re e[M'_{12}] + j\Im m[M_2 M'_{12}] \end{aligned} \tag{5.50}$$

where M_1 and M_2 are the same (5.41) and M_{12} is the quadrature corrected normalized correlation (5.33).

5.4 Digital Techniques

In present days, radiometric measurements are always converted to numerical format and saved in computer data files for further processing. This is achieved through the use of analog-to-digital converters which perform two functions at the same time: sampling a signal at regular time intervals and providing the corresponding values in digital words. There exist then two associated errors, one due to keeping the signal at only specific epochs and other relative to the necessary quantization of the samples to represent their value with a finite number of bits. The first one is the *sampling error* and depends on the rate at which samples are taken, the so-called sampling frequency or sampling rate, here denoted by f_s. The second one, known as *quantization error*, is not covered in the present book except for the specific case of one bit correlation. For more information, the interested reader may consult more specific treatises on digital signal processing for example the classical [4] or others.

The outcome of a digital system consists then of a set of discrete samples representing the measured quantity, either power or cross-correlation in the present context. They define a random variable with mean equal to the equivalent measurement of an analog system with the same design parameters (gain, bandwidth, frequency band, ...). This mean is independent of the sampling rate, but the standard deviation is highly affected by the way digitation is implemented. This section deals with the impact of sampling rate on the radiometric performance, either for power or cross-correlation measurements, including for this last one two quantization strategies: multibit and one bit.

5.4.1 Sampling

Sampling analog signals to produce digital data streams can be done at three different planes in the chains of figure 2.9 (power measurement) or 3.7 (cross-correlation measurement),

1. Integrated output signals $\mathrm{v}(t)$ or $\Gamma(t)$
2. Instantaneous power $\mathrm{p}(t)$ or correlation $\rho(t)$
3. Receiver output signals $\mathrm{b}(t)$ or $\mathrm{b}_1(t)$ and $\mathrm{b}_2(t)$

Only the third case is conventionally known as "digital radiometer" since the other two (especially the first one) are commonly regarded as just a way of digitizing the output signal of an analog instrument. Nonetheless, all three of them are studied.

5.4.1.1 First Case

The first case converts to digital format a slowly varying function of time. Although the raw measurements have already been integrated with an analog filter, it is frequently convenient to further reduce the standard deviation by averaging N samples,

$$\mathrm{v}_a = \frac{1}{N}\sum_{i=1}^{N}\mathrm{v}(t_i)\ ;\ \Gamma_a = \frac{1}{N}\sum_{i=1}^{N}\Gamma(t_i) \tag{5.51}$$

where t_i are the sampling epochs and the subscript a states for "averaged." As indicated in (1.18), averaging preserves the mean, so the mean of v_a and Γ_a are, respectively, equal to $\langle \mathrm{v}(t)\rangle$ and $\langle \Gamma(t)\rangle$, in turn proportional to the total integrated power and cross-correlation.

Given a total time t_a allocated to digital averaging, at a rate f_s the total number of samples is $N = f_s t_a$. Based on (1.110) and the text that follows, it is apparent that two consecutive samples are uncorrelated if they are separated a time interval at least equal to the analog integration time τ_i, which corresponds to $f_s = 1/\tau_i$. At this rate, averaging N consecutive samples reduces the standard deviation by a factor of $1/\sqrt{N}$. On the other hand, within a total time t_a available for averaging, the number of samples at the above rate is $N = t_a/\tau_i$ so the radiometric sensitivity (2.82) or visibility uncertainty (3.96) become, respectively

$$\begin{aligned} \Delta T &= \frac{T_{sys}}{\sqrt{Bt_a}}\alpha_F \\ \Delta V_{12} &= \frac{\sqrt{T_{sys1}T_{sys2}}}{\sqrt{Bt_a}}\frac{\alpha_F}{|\tilde{r}_{12}(0)|} \end{aligned} \tag{5.52}$$

and the integration time is effectively extended to cover the total time during which samples are averaged.

If the rate increases, correlation between samples must be taken into account. The procedure is not detailed here since it is fully developed in section 5.4.2 for cases two and three so only a brief summary is provided for the interested reader. In essence, it consists of using (5.66) but with the autocorrelation function of the integrated signal $\mathrm{v}(t)$ or $\Gamma(t)$ instead of that of $\mathrm{p}(t)$. The result is summarized in the plot of figure 5.5, which represents the ratio between the standard deviation of the measurement and that predicted by (5.52) as a function of the sampling rate expressed as $f_s\tau_i$. To simplify evaluation, the approximated version of (1.110) is taken to estimate the autocorrelation function of $\mathrm{v}(t)$ or $\Gamma(t)$. It implies that an *ideal* analog integrator with a relatively large value of the product $B\tau_i$ is used. In this particular example, the total time t_a dedicated to digital averaging is made equal to 300 times that of the analog integration time τ_i.

As seen in the graphic, for sampling rates below $1/\tau_i$, even though samples are uncorrelated, the number of them is low, and the standard deviation does not reduce to (5.52) for a given time t_a. On the other hand, increasing the sampling rate beyond $1/\tau_i$ does not improve the sensitivity. In any case, (5.52) is always a lower

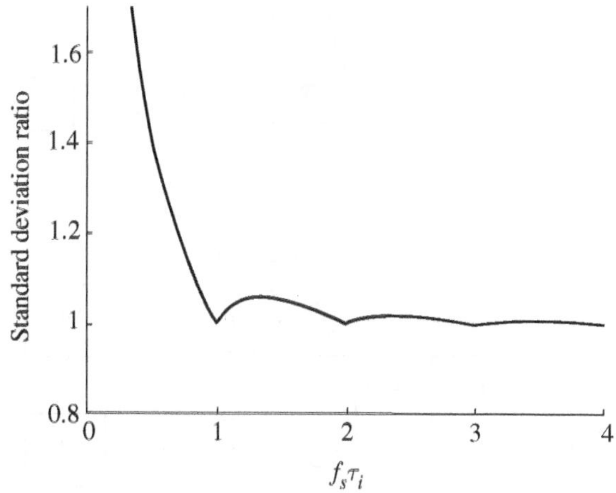

Figure 5.5 Impact of numerical averaging of the integrated output signal on measurement uncertainty.

bound of the measurement uncertainty. Using a practical low pass filter instead of an ideal integrator changes the shape of the plot but does not alter the main conclusion.

5.4.1.2 Second Case

The second case means replacing the analog integrator of figure 2.9 or 3.7 by a sampler, an analog-to-digital converter and numerical averaging to set the integration time. It carries *implicitly* the assumption that the spectrum of the signal before sampling, either p(t) or $\rho(t)$, has only low-frequency content so that the band centered at $2f_0$ (see figure 2.12 for reference) is not present. The analog-to-digital converter must thus be designed to operate in a band that goes from zero up to about the bandwidth B of the radiometric radio frequency signals.

The measurement setups are schematized in figure 5.6, on (a) for total power and on (b) for correlation measurement. The blocks labeled A/D are analog-to-digital converters governed by a clock at frequency f_s. The signals entering from the left come directly from the receivers' output ports and, in case of correlation measurement, include a quadrature filter in one of them to get the Hilbert transform. It can be typically implemented by an in-phase and quadrature mixer, as described in section 5.2. The low pass filters labeled "LPF" are formally included to visualize the cancelation of high frequency terms inherent to signal multiplication. Correlators are here analog and the symbol used is consistent with that of figure 3.9.

For correlation measurement, only two signal multipliers are needed since using four of them and combining the corresponding digital outputs is totally unnecessary. Due to the low pass filtering introduced in the correlators, the autocorrelation function of $\rho'(t)$ is only the first term of (3.119), which is equal to (3.97) so equivalent to using four multipliers and no filters. Implementing

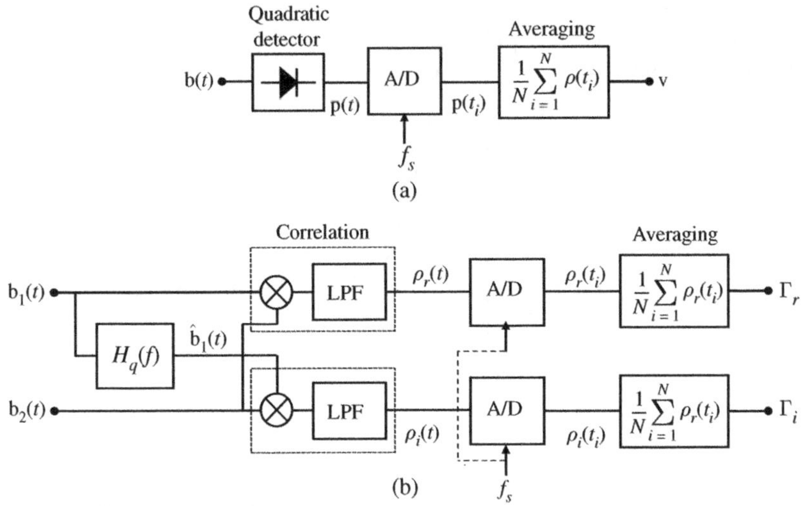

Figure 5.6 Instantaneous power (a) and correlation (b) sampling.

an analog circuit to combine four multipliers' outputs before sampling is also a theoretical option but, as the reader will easily understand, impractical. Similarly, the power measurement depicted on figure 5.6(a) uses *necessarily* a quadratic detector (not a square law device) to provide an instantaneous power with low-frequency content (see section 2.4.4). This is not an important restriction since this is the usual implementation of an analog power detector.

If N samples taken at times t_i are averaged, the outcomes of figure 5.6 are, respectively

$$\begin{aligned} \mathrm{v} &= \frac{1}{N}\sum_{i=1}^{N} \mathrm{p}(t_i) \quad \text{(power)} \\ \Gamma &= \frac{1}{N}\sum_{i=1}^{N} \rho(t_i) \quad \text{(correlation)} \end{aligned} \tag{5.53}$$

where $\Gamma = \Gamma_r + j\Gamma_i$ and $\rho(t) = \rho_r(t) + j\rho_i(t)$. Since averaging preserves the mean, for the two random variables defined above,

$$\begin{aligned} \overline{\mathrm{v}} &= \overline{\mathrm{p}(t_i)} = \mathcal{P} \\ \overline{\Gamma} &= \overline{\rho(t_i)} = \Gamma_{12} \end{aligned} \tag{5.54}$$

where $\mathcal{P}$ is the total integrated power (2.60) and Γ_{12} the cross-correlation (3.82). Their standard deviation is the subject of section 5.4.2.

5.4.1.3 Third Case

In the third case, the receivers' output signals (at RF or IF) are directly digitized. Depending on the available technology of analog-to-digital conversion, a mixer (section 5.1) may be needed in order to bring the signals to a frequency range matching the technological state-of-the-art of digital systems. In any case, the signals before sampling have a spectrum centered at f_0 and with bandwidth B. It is schematically depicted in figure 2.12(a).

Now, the power detector of figure 2.9 or the complex multiplier in figure 3.7 is also implemented by software. Power measurement is shown in figure 5.7 for two different options. Starting with the block diagram on figure 5.7(a), the instantaneous power is numerically computed from the signal samples according to (2.89). For the sake of simplicity, the analog-to-digital converter is assumed to sample the output signal b(t) precisely defined just after equation (2.59). With this convention, the output voltage is obtained by averaging N samples of (2.89)

$$\mathrm{v} = \frac{1}{N}\sum_{i=1}^{N} \mathrm{b}^2(t_i) \tag{5.55}$$

where t_i denotes the sampling times and C_d has been set to 1 since there is no need to put another value. This is obviously the digital version of a square law device (section 2.4.3) and illustrates the easiest way to measure power with numerical techniques. It suffices to average the square of N samples without the need of any other component. A possible constant $C_d \neq 1$ characterizing the proportionality between the square of the sampled signal and that of the actual instantaneous power is assumed embedded in the radiometric gain (2.73), determined by calibration. The same happens with the instrumental offset introduced in (2.71) for analog systems, which is here assumed null, although in practical systems, it may not be strictly so due to errors in the analog-to-digital converter.

The block diagram on figure 5.7(b) is a digital implementation of a quadratic detector based on (2.108), also with $C_d = 1$. The quadrature filter $H_q(f)$ can be

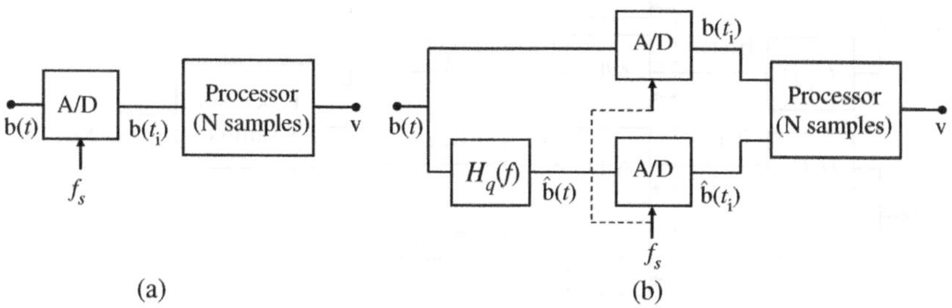

Figure 5.7 Digital implementation of a square law device (a) and a quadratic detector (b) for power measurement. For each case, the block "Processor" implements, respectively, equations (5.55) and (5.56).

made by an in-phase and quadrature mixer (section 5.2), although using a time delay (section 5.3) is an interesting option further explored below in section 5.4.5. After sampling the receiver output signal and its Hilbert transform *both at the same epoch*, the radiometric measurement is computed according to (2.108)

$$\mathrm{v} = \frac{1}{2N}\sum_{i=1}^{N}\mathrm{b}^2(t_i) + \frac{1}{2N}\sum_{i=1}^{N}\hat{\mathrm{b}}^2(t_i) \tag{5.56}$$

In general, if any of the two equations (5.55) or (5.56) is expressed in terms of samples of instantaneous power $\mathrm{p}(t_i) = \mathrm{b}^2(t_i)$, they become *mathematically identical* to the first one of (5.53), corresponding to the second case. This means that, even though the technological implications of this third case may be completely different than those of the second one, from the point of view of performance analysis, it is totally equivalent (see section 5.4.2).

The block diagram for correlation measurement in the third case, digital correlator, is depicted in figure 5.8. Two options for the analog-to-digital converters are considered: multibit and one-bit. For the first one the cross correlation Γ_{12} (3.82) is measured. If *one bit converters* are used, the *normalized* correlation M_{12} (3.135) is obtained.

Cross correlation can be computed using either two or four real-number multiplications. Assuming unnormalized correlation with two multipliers, from (3.103) (or other equivalent products)

$$\Gamma = \frac{1}{N}\sum_{i=1}^{N}\mathrm{b}_1(t_i)\mathrm{b}_2(t_i) + j\frac{1}{N}\sum_{i=1}^{N}\hat{\mathrm{b}}_1(t_i)\mathrm{b}_2(t_i) \tag{5.57}$$

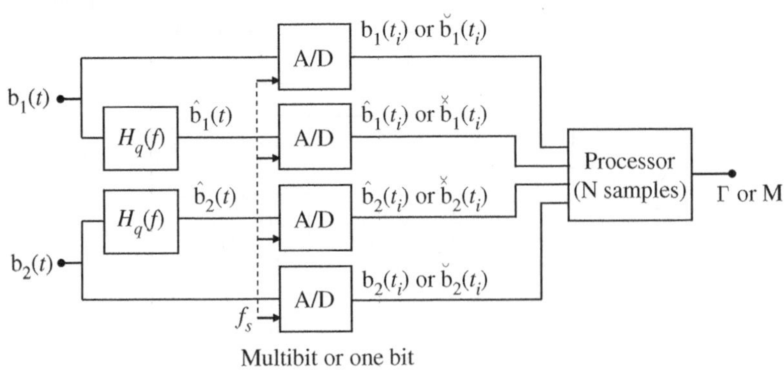

Figure 5.8 Digital correlator using either multibit or one bit analog-to-digital conversion. The processor may implement two or four multipliers.

from which it is apparent that $\overline{\Gamma} = \Gamma_{12}$. Note that hardware can be simplified by using only three analog-to-digital converters and one Hilbert filter. The four multipliers approach is based on (3.101)

$$\begin{aligned} \mathfrak{Re}[\Gamma] &= \frac{1}{2N}\sum_{i=1}^{N} \mathrm{b}_1(t_i)\mathrm{b}_2(t_i) + \frac{1}{2N}\sum_{i=1}^{N} \hat{\mathrm{b}}_1(t_i)\hat{\mathrm{b}}_2(t_i) \\ \mathfrak{Im}[\Gamma] &= \frac{1}{2N}\sum_{i=1}^{N} \hat{\mathrm{b}}_1(t_i)\mathrm{b}_2(t_i) - \frac{1}{2N}\sum_{i=1}^{N} \mathrm{b}_1(t_i)\hat{\mathrm{b}}_2(t_i) \end{aligned} \tag{5.58}$$

also satisfying that $\overline{\Gamma} = \Gamma_{12}$.

As it is demonstrated in sections 5.4.3 and 5.4.4, both approaches (two or four multipliers) differ only on the impact of the sampling rate on the standard deviation of Γ. As expected, making sampling rate arbitrarily large always reverts the measurement uncertainty to the value obtained in analog integration, which is independent on the number of multipliers, as concluded in section 3.4.2.

Note finally that both (5.57) and (5.58) can be expressed in terms of instantaneous correlation samples $\rho(t_i)$, and then they convert to the second equation (5.53), corresponding to the second case of those mentioned at the start of this section. Again, for performance analysis, the second and third cases are equivalent.

5.4.1.4 One Bit Correlation

Normalized correlation is measured with the technique introduced in section 3.4.5 based on the arc sine law. In the setup of figure 3.10 the component named "Limiter" is commonly implemented by a sampler followed by a "one bit analog-to-digital converter" that provides only the sign of the input signal (one bit). This operation results in samples of clipped signals that are then multiplied and averaged.

Using two multipliers, the normalized correlation is estimated by (3.134)

$$\mathrm{M} = \sin\left(\frac{\pi}{2N}\sum_{i=1}^{N} \breve{\mathrm{b}}_1(t_i)\breve{\mathrm{b}}_2(t_i)\right) + j\,\sin\left(\frac{\pi}{2N}\sum_{i=1}^{N} \breve{\hat{\mathrm{b}}}_1(t_i)\breve{\mathrm{b}}_2(t_i)\right) \tag{5.59}$$

and the corresponding expressions using four multipliers is

$$\begin{aligned} \mathfrak{Re}[\mathrm{M}] &= \sin\left(\frac{\pi}{4N}\sum_{i=1}^{N} \breve{\mathrm{b}}_1(t_i)\breve{\mathrm{b}}_2(t_i)\right) + \sin\left(\frac{\pi}{4N}\sum_{i=1}^{N} \breve{\hat{\mathrm{b}}}_1(t_i)\breve{\hat{\mathrm{b}}}_2(t_i)\right) \\ \mathfrak{Im}[M] &= \sin\left(\frac{\pi}{4N}\sum_{i=1}^{N} \breve{\hat{\mathrm{b}}}_1(t_i)\breve{\mathrm{b}}_2(t_i)\right) - \sin\left(\frac{\pi}{4N}\sum_{i=1}^{N} \breve{\mathrm{b}}_1(t_i)\breve{\hat{\mathrm{b}}}_2(t_i)\right) \end{aligned} \tag{5.60}$$

Disregarding the nonlinear effect (3.143) and taking into account that averaging preserves the mean; in both cases, the mean of M correctly represents the normalized correlation, $\overline{\mathrm{M}} \approx \mathrm{M}_{12}$.

In practice, since the clipped signals may take only values +1 and −1, the average of their product is actually evaluated as

$$\frac{1}{N}\sum_{i=1}^{N}\breve{\mathrm{b}}_1(t_i)\breve{\mathrm{b}}_2(t_i) = 2\frac{N_1}{N} - 1 \tag{5.61}$$

where N_1 is the number of signal coincidences, either both +1 or both −1. This is the main advantage of the technique and the reason why it is so commonly used: a correlator consists only of an exclusive NOR gate to detect signal coincidences and a counter; and this can be very efficient. In fact, the raw measurement of such systems is N_1, the number of coincidences within the total number of samples N allocated in the integration time. The normalized correlation is then computed by inserting (5.61) into (5.59) or (5.60).

Evaluating the uncertainty in the measurement of M is not straightforward in general due to the nonlinear operation inherent in the sinus functions. Nevertheless, analytical results can be deduced for small instantaneous correlation. This is consistent with the analysis of section 3.4.5, in which the assumption of uncorrelated $\mathrm{b}_1(t)$ and $\mathrm{b}_2(t)$ was needed in (3.145) to get an estimation of the variance of M. Presumably, the main conclusions based on this supposition will also be valid for higher correlation values. This premise is stressed by recalling that, for analog systems with unclipped signals, the uncertainty (3.94) depends only on the total power but not on the cross-correlation itself.

For small instantaneous correlation, the sinus can be replaced by its argument in either (5.59) or (5.60) and then

$$\mathrm{M} \approx \frac{\pi}{2}\frac{1}{N}\sum_{i=1}^{N}\breve{\rho}(t_i) \tag{5.62}$$

where the complex instantaneous correlation for clipped signals is here defined as

$$\breve{\rho}(t) = \breve{\rho}_r(t) + j\breve{\rho}_i(t) \tag{5.63}$$

with $\breve{\rho}_r(t)$ and $\breve{\rho}_i(t)$ given, respectively, by (3.142) and (3.153) for two signal multipliers or (3.157) for four signal multipliers.

Equation (5.62) is mathematically equivalent to any of (5.53), and this feature allows using a common methodology to study the measurement uncertainty in power or correlation measurement, either multibit or one bit. The standard deviation of the variables defined in (5.53), either for case 2 or 3, and (5.62) is the subject of the next section 5.4.2.

5.4.2 Impact on Measurement Uncertainty

Even though numerical averaging always provides the correct mean, for a given integration time, a numerically averaged measurement is always noisier than the one obtained by a continuous integrator. At low sampling rates, errors can be large while in the limit, when the rate tends to infinity, they coincide with those of analog averaging. Two metrics are defined in this section for assessing the impact of sampling rate on the measurement uncertainty: The *noise excess* due to numerical averaging and the *effective integration time.*

In general, equation (1.13) can be used to estimate the variance of either v or Γ expressed by (5.53) or M by (5.62). Since all these three equations are mathematically equivalent, the procedure is detailed only for the first one (instantaneous power) and then the results are particularized for the other two.

Interpreting in the covariance definition (1.4) $z = \mathrm{p}(t_i)$ and $w = \mathrm{p}(t_j)$ and assuming $\mathrm{p}(t)$ stationary so that all samples have the same mean and standard deviation, the covariance of two samples becomes

$$C_{\mathrm{p}(t_i)\mathrm{p}(t_j)} = R_{\mathrm{p}'}(t_i - t_j) \tag{5.64}$$

where $\mathrm{p}'(t) = \mathrm{p}(t) - \langle \mathrm{p}(t) \rangle$ is the instantaneous power with the mean subtracted. Equation (1.49) has been used to write the autocorrelation function of this process. At a rate f_s two consecutive samples are separated at a time interval $t_s = 1/f_s$. The number of samples taken during an integration time τ_i is then

$$N = f_s \tau_i \tag{5.65}$$

Since there are $(N - i)$ time segments of length $i\,t_s$, introducing (5.64) into (1.13) and including the division by N of (5.53) yields

$$\sigma_{\mathrm{v}}^2 = \sigma_{\mathrm{p}}^2 \left[\frac{1}{N} + \frac{2}{N^2} \sum_{i=1}^{N-1} (N - i)\, r_{\mathrm{p}'}(i\,t_s) \right] \tag{5.66}$$

where $r_{\mathrm{p}'}(\tau)$ is the normalized autocorrelation function defined in (1.55),

$$r_{\mathrm{p}'}(\tau) = \frac{R_{\mathrm{p}'}(\tau)}{R_{\mathrm{p}'}(0)} = \frac{R_{\mathrm{p}'}(\tau)}{\sigma_{\mathrm{p}}^2} \tag{5.67}$$

which in this particular case is real.

The same procedure can be applied to the correlation measurement of the second equation (5.53) with the following result

$$\sigma_{\Gamma}^2 = \sigma_{\rho}^2 \left[\frac{1}{N} + \frac{2}{N^2} \sum_{i=1}^{N-1} (N - i)\, \Re e[r_{\rho'}(i\,t_s)] \right] \tag{5.68}$$

where now the normalized autocorrelation function $r_{\rho'}(i\,t_s)$ is complex in general, so the real part must be applied in agreement with (1.13).

Finally, for normalized correlation (5.62), the same procedure yields to

$$\sigma_{\mathrm{M}}^2 = \frac{\pi^2}{4}\sigma_{\breve{\rho}}^2 \left[\frac{1}{N} + \frac{2}{N^2}\sum_{i=1}^{N-1}(N-i)\,\Re e[r_{\breve{\rho}'}(i\,t_s)]\right] \tag{5.69}$$

which differs from (5.66) and (5.68) in the multiplicative constant $\pi^2/4$.

5.4.2.1 Uncorrelated Samples

Consecutive samples are uncorrelated if the autocorrelation function inside any of the above expressions is null for all values of i. If this condition holds, only the first term remains, so the standard deviations simplify to

$$\sigma_{\mathrm{V}} = \frac{\sigma_{\mathrm{P}}}{\sqrt{N}}\ ;\ \sigma_{\Gamma} = \frac{\sigma_{\rho}}{\sqrt{N}}\ ;\ \sigma_{\mathrm{M}} = \frac{\pi}{2}\frac{\sigma_{\breve{\rho}}}{\sqrt{N}} \tag{5.70}$$

so the standard deviation is reduced by a factor equal to the square root of the number of samples. Based on this result, some references define the "effective number of samples" as the inverse of the terms inside the brackets in the above expressions for the variances. This concept, however, is not further developed in this book.

5.4.2.2 Metrics

Instead, two other metrics are defined. The first one estimates the noise excess that is generated by averaging discrete samples instead of using a continuous integration,

$$\mathcal{N}_{\mathrm{V}} = \frac{\sigma_{\mathrm{V}}}{\sigma_{\mathrm{V}a}}\ ;\ \mathcal{N}_{\Gamma} = \frac{\sigma_{\Gamma}}{\sigma_{\Gamma a}}\ ;\ \mathcal{N}_{\mathrm{M}} = \frac{\sigma_{\mathrm{M}}}{\sigma_{\mathrm{M}a}} \tag{5.71}$$

where the subscript a in all cases denotes "analog integration." This metric can be interpreted as the degradation that numerical averaging produces in the radiometric sensitivity of a radiometer. The denominators in the above expressions are directly obtained from (2.81), (3.94), and either (3.155) or (3.162), repeated here for convenience

$$\sigma_{\mathrm{V}a} = \frac{C_d\mathcal{P}}{\sqrt{B\tau_i}}\alpha_F\ ;\ \sigma_{\Gamma a} = \frac{\sqrt{\mathcal{P}_1\mathcal{P}_2}}{\sqrt{B\tau_i}}\alpha_F\ ;\ \sigma_{\mathrm{M}a} = \frac{\sqrt{2(\mathcal{I}+\mathcal{J})}}{\sqrt{B\tau_i}}\alpha_F \tag{5.72}$$

where the last one only applies when four multipliers are used. Otherwise $\mathcal{I}+\mathcal{J}$ should be substituted by $2\mathcal{I}$. The reader must refer to the original expressions to get the definitions of the parameters involved.

The second metric, particularly useful in performance analysis, is the "effective integration time" τ_{eff}, defined such that the uncertainty on the system temperature or visibility is *always* expressed as

$$\Delta T = \frac{T_{sys}}{\sqrt{B\tau_{eff}}}\ ;\ \Delta V_{12} = \frac{\sqrt{T_{sys1}T_{sys2}}}{\sqrt{B\tau_{eff}}} \tag{5.73}$$

independently of the specific implementation. In the case of one-bit correlation this definition, as well as the standard deviation of (5.72), implies that the second term of (3.140) is neglected. So, they both strictly hold only for the special case of zero visibility.

Taking into account (2.75) and (3.86) and using also (5.72), the ratio between effective and real integration times can be easily expressed in terms of the noise excess (5.71) for each case

$$\begin{aligned} \frac{\tau_{\mathit{eff}}}{\tau_i} &= \frac{1}{\alpha_F^2 \mathcal{N}_{\mathrm{V}}^2} \quad \text{(Power)} \\ \frac{\tau_{\mathit{eff}}}{\tau_i} &= \frac{1}{\alpha_F^2 \mathcal{N}_{\Gamma}^2} |\tilde{r}_{12}(0)|^2 \quad \text{(Multibit correlator)} \\ \frac{\tau_{\mathit{eff}}}{\tau_i} &= \frac{1}{\alpha_F^2 \mathcal{N}_{\mathrm{M}}^2} \frac{|\tilde{r}_{12}(0)|^2}{2(\mathcal{I} + \mathcal{J})} \quad \text{(One bit correlator)} \end{aligned} \tag{5.74}$$

where, again, $\mathcal{I} + \mathcal{J}$ should be substituted by $2\mathcal{I}$ if only two multipliers are used.

Both the noise excess and the effective integration time are function of the sampling rate. They are evaluated in sections 5.4.3 and 5.4.4 for two different scenarios defined on the basis of the instantaneous power or correlation spectrum: either having only low-frequency content or including a band at twice the center frequency.

5.4.3 Low-frequency Spectrum

Equations (5.66), (5.68), or (5.69) are first evaluated by substituting into them the autocorrelation functions (2.106), (3.97), and (3.160). They all have in common that the associated spectrum has no band at twice the center frequency and correspond in practice to any of the following situations:

- Sampling the instantaneous power or cross-correlation, that is the second case mentioned at the start of section 5.4.1, so using the setups of figure 5.6. Without loss of generality, a value of $C_d = 1$ is assumed for power measurement.
- Digital implementation of a quadratic detector with the block diagram depicted in figure 5.7(b). Total power is estimated by (5.56).
- Digital correlation measurement, either using multibit or one-bit, using four signal multipliers, that is using (5.58) or (5.60), respectively.

The two last bullets correspond to the third case of those listed at the start of section 5.4.1, so they imply sampling of RF (or IF) signals.

Directly by inspection, inserting (2.103) and (2.106) into (5.66),

$$\sigma_{\mathrm{V}}^2 = \mathcal{P}^2 \left[\frac{1}{N} + \frac{2}{N^2} \sum_{i=1}^{N-1} (N - i) |\tilde{r}(i t_s)|^2 \right] \tag{5.75}$$

and using (3.91) and (3.97) in (5.68)

$$\sigma_\Gamma^2 = \mathcal{P}_1\mathcal{P}_2 \left[\frac{1}{N} + \frac{2}{N^2} \sum_{i=1}^{N-1} (N-i)\Re e[\tilde{r}_1(i\,t_s)\tilde{r}_2^*(i\,t_s)] \right] \tag{5.76}$$

which coincides mathematically with the previous one provided that both receivers have identical frequency response, that is if $\tilde{r}_1(\tau) = \tilde{r}_2(\tau) = \tilde{r}(\tau)$. The noise excess metric (5.71) in each case is found by dividing the square root of these expressions by the corresponding standard deviations (5.72). Note that the total powers are always canceled.

One bit correlation needs more elaboration. First of all, introducing (3.157) into (5.63) and taking into account that the clipped signals may take only values +1 and −1,

$$\left\langle |\breve{\rho}(t)|^2 \right\rangle = \left\langle \breve{\rho}_r^2(t) \right\rangle + \left\langle \breve{\rho}_i^2(t) \right\rangle = 1 \tag{5.77}$$

Under the assumption of uncorrelated input signals, always contemplated in this context, $\langle \breve{\rho}(t) \rangle = 0$ and the variance of $\breve{\rho}(t)$ becomes unity:

$$\sigma_{\breve{\rho}}^2 = 1 \Rightarrow r_{\breve{\rho}'}(\tau) = R_{\breve{\rho}}(\tau) \tag{5.78}$$

On the other hand, applying the general relation (1.59) and using (3.160),

$$\Re e[R_{\breve{\rho}}(\tau)] = 2R_{\breve{\rho}_r}(\tau) = R_{\breve{b}_1}(\tau)R_{\breve{b}_2}(\tau) + R_{\check{\breve{b}}_1\breve{b}_1}(\tau)R_{\check{\breve{b}}_2\breve{b}_2}(\tau) \tag{5.79}$$

and substituting (3.147) and (3.161),

$$\begin{aligned} \Re e[r_{\breve{\rho}'}(i\,t_s)] = &\frac{4}{\pi^2} \arcsin \Re e[\tilde{r}_1(i\,t_s)e^{j2\pi f_0 i\,t_s}] \arcsin \Re e[\tilde{r}_2(i\,t_s)e^{j2\pi f_0 i\,t_s}] \\ &+ \frac{4}{\pi^2} \arcsin \Im m[\tilde{r}_1(i\,t_s)e^{j2\pi f_0 i\,t_s}] \arcsin \Im m[\tilde{r}_2(i\,t_s)e^{j2\pi f_0 i\,t_s}] \end{aligned} \tag{5.80}$$

which must be directly introduced into (5.69) to get the variance of the normalized correlation measurement. The explicit equation is not provided here to avoid burdening the text, but the reader should be able to write it with no difficulty. Be aware that, as stated in (5.79), $\sigma_{\breve{\rho}}^2 = 1$. It is interesting to observe that $\Re e[\tilde{r}_1(i\,t_s)\tilde{r}_2^*(i\,t_s)]$ in (5.76) is *identical* to (5.80) if the arc sin law is not applied to the correlation functions. The reader may check it using (A.11) and (A.12).

5.4.3.1 Results

Figure 5.9 shows the noise excess (5.71) as a function of the sampling rate normalized to the bandwidth for the three cases analyzed in this section. In all of them, the product bandwidth–integration time is fixed to $B\tau_i = 300$, high enough to ensure the validity of the approximated version of (1.111), used implicitly in all analyses (see also figure 2.11). Furthermore, from (5.65) the number of samples

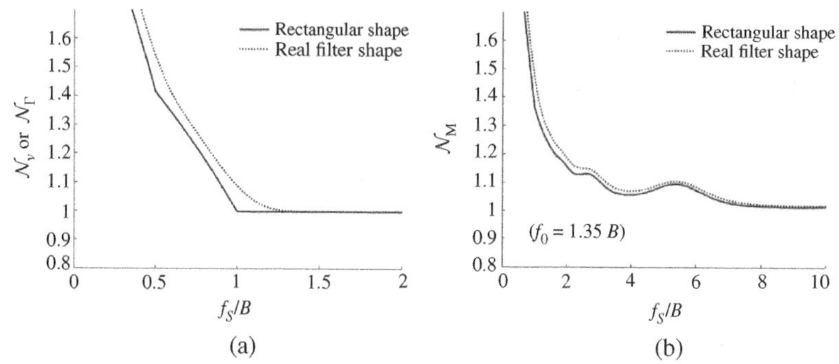

Figure 5.9 Noise excess due to numerical averaging with respect to analog integration as defined by (5.71). Plot (a) holds for both total power and multibit correlation measurement. Plot (b) is for one-bit correlation.

becomes in this case $N = 300f_s/B$ so the horizontal axis reflects also the number of samples. For comparison, in SMOS this product is (20 MHz × 1.2 s = 2.4x10^6), four orders of magnitude larger.

Two identical receivers are assumed in (5.76) so figure 5.9(a) holds either for power or multibit correlation measurement. If both filters differ, the shape changes slightly, but the overall conclusions may still be valid if the filter differences are not too large. Note that this plot is independent of the center frequency since this parameter does not appear in (5.75) nor in (5.76).

On the other hand, the center frequency does have an impact when one bit correlation is used. The plot of figure 5.9(b) illustrates this case and assumes, as an example, $f_0 = 1.35B$. Higher center frequencies translate into more ripples and also an increase in the frequency at which this metric saturates to one. To visualize this saturation the horizontal axis scale is in this case extended to a higher value with respect to figure 5.9(a).

As indicated in the legends, in all plots, the solid line corresponds to an ideal rectangular filter, for which $\tilde{r}(i\,t_s) = \text{sinc}\; B\,i\,t_s$ (2.70). The dotted line shows the results for the measured frequency response drawn also as a dotted line in figure 2.10.

The effective integration time (5.74) is illustrated in figure 5.10. The observed increase of effective integration time associated with using a nonideal filter is due to having the parameter α_F lower than one for the example used. This same effect is also observed in figure 2.11. The better performance in terms of noise of a real filter is here associated with an increase of effective integration time.

As expected, oscillations are also observed in the plot for one-bit correlation. The important feature to highlight in these plots is that, in general, the effective integration time for one-bit correlation is always lower than one, depends on the sampling rate and stabilizes at about 0.7 depending on the filter shape, in consistency with (3.164).

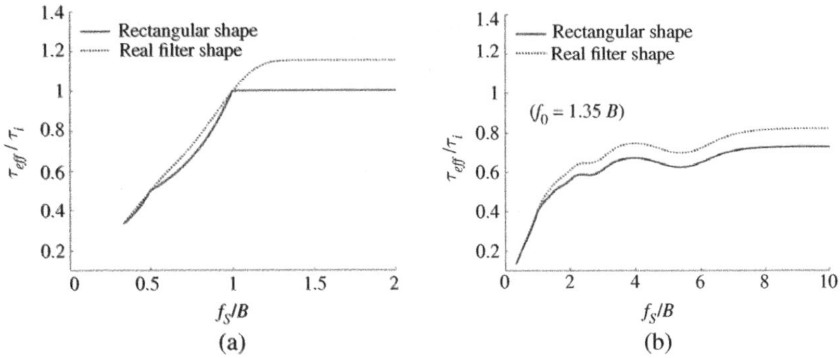

Figure 5.10 Normalized effective integration time as defined by (5.74). Plot (a) holds for both total power and multibit correlation measurement. Plot (b) is for one-bit correlation.

5.4.3.2 Uncorrelated Samples

Samples are uncorrelated if the second term of (5.75) or (5.76) [or (5.80) in case of one bit] vanishes for all values of the index i. Assuming, in correlation measurement, that both filters are identical, this translates in all cases in having null samples of the fringe washing function.

Therefore, for ideal rectangular filters (2.70), consecutive samples are uncorrelated if sinc $Bit_s = 0 \, \forall i$, which is translated into a sampling rate of

$$f_s = B \quad \text{(uncorrelated samples)} \tag{5.81}$$

For nonideal filters the fringe washing function may be not identically zero at this rate but should be sufficiently small to consider samples also uncorrelated (see plot of figure 2.10(b)). Introducing this sampling rate into (5.65) it turns out that the number of samples is $N = B\tau_i$, which substituted into (5.70) results in this case

$$\sigma_V = \frac{\mathcal{P}}{\sqrt{B\tau_i}} \; ; \; \sigma_\Gamma = \frac{\sqrt{\mathcal{P}_1\mathcal{P}_2}}{\sqrt{B\tau_i}} \; ; \; \sigma_M = \frac{\pi}{2}\frac{1}{\sqrt{B\tau_i}} \tag{5.82}$$

and then the noise excess (5.71) becomes

$$\mathcal{N}_V = \mathcal{N}_\Gamma = \frac{1}{\alpha_F} \; ; \; \mathcal{N}_M = \frac{\pi}{2}\frac{1}{\alpha_F\sqrt{2(\mathcal{I}+\mathcal{J})}} \tag{5.83}$$

which corresponds to an effective integration time (5.74) of

$$\frac{\tau_{eff}}{\tau_i} = \begin{cases} 1 & \text{(for power or multibit correlation)} \\ \dfrac{4}{\pi^2} & \text{(for one-bit correlation)} \end{cases} \tag{5.84}$$

where, since both receivers are assumed identical, it has been assumed $\tilde{r}_{12}(0) = 1$. The reader may want to check that these results are consistent with the plots in figures 5.9 and 5.10. The above value for one-bit correlation is evaluated to $1/2.467$, which is a result reported in some articles, for example [5] and also in [6]. It is interpreted as that one-bit correlation reduces integration time by a factor of about 2.467. The reader should be aware of the specific conditions in which this is true.

Note that, in case of unclipped signals and ideal rectangular filter, sampling at a rate $f_s = B$ or beyond is totally equivalent to a continuous integration. This is clearly seen in both figures 5.9 and 5.10. Only in this case the uncorrelated condition can be considered as an optimum value in the sense that sampling beyond this frequency does not change the effective integration time. For a nonideal filter or one-bit correlation, oversampling with respect to uncorrelated condition is needed to bring standard deviation down to the analog case. Note also that at $f_s = 0.5B$ samples are also uncorrelated and this is why there is a change of slope at this point. The same happens at other integer ratios. These cases fail to be relevant since the number of samples drops and the associated standard deviations become larger.

5.4.4 Spectrum with High-frequency Content

Different results are obtained if the equations corresponding to a square law device (2.95), two signal multipliers (3.119) and its clipped version (3.148) [together with a similar one for the imaginary part] are introduced, respectively, into (5.66), (5.68) and (5.69). In these three cases the spectrum of the instantaneous power or correlation contains a high-frequency band at twice the center frequency and this has an impact on noise reduction through averaging. This case corresponds to the following practical situations, which are the most common in digital radiometers and both belong to the *third case* of those mentioned at the beginning of section 5.4.1, implying a direct sampling of the RF (or IF) signals.

- Digital implementation of a square law detector. In this case power is measured using the block diagram on figure 5.7(a) and according to (5.55).
- Digital correlation measurement, either multi- or one-bit, using only two real multiplications, so applying (5.57) or (5.59), respectively (or equivalent products).

For the first bullet above (total power measurement), instead of directly inserting (2.95) into (5.66), a more compact expression for $r_{p'}(i\,t_s)$ is obtained from (2.91) after substitution of (2.64). Introducing the result of this substitution into (5.66)

$$\sigma_V^2 = 2\mathcal{P}^2 \left[\frac{1}{N} + \frac{2}{N^2} \sum_{i=1}^{N-1} (N - i)\Re e[\tilde{r}(i\,t_s) e^{j2\pi f_0 i\,t_s}]^2 \right] \tag{5.85}$$

Similarly, for the second bullet, the real part of the first two terms of (3.119) can be written as twice the product of the real parts of the autocorrelation functions [see (A.11)]. Normalizing and taking the real part,

$$\mathfrak{Re}[r_{\rho'}(i\,t_s)] = \frac{1}{2+|M_{12}|^2}\Big[2\mathfrak{Re}[\tilde{r}_1(i\,t_s)e^{j2\pi f_0 i\,t_s}]\mathfrak{Re}[\tilde{r}_2(i\,t_s)e^{j2\pi f_0 i\,t_s}] + |M_{12}|^2\mathfrak{Re}[\overline{\tilde{R}}_{12}(i\,t_s)e^{j4\pi f_0 i\,t_s}]\Big] \tag{5.86}$$

where M_{12} is the normalized correlation (3.135) and $\overline{\tilde{R}}_{12}(\tau)$ is defined in (3.118). This expression, together with (3.108), must now be introduced into (5.68) to obtain the variance of the correlation measurement.

To simplify the analysis, only the particular case of independent input signals $b_1(t)$ and $b_2(t)$ is considered, which means making $\Gamma_{12} = M_{12} = 0$. For analog integration, this restriction has no impact on the variance of $\Gamma(t)$ [see (3.94)] but for numerical averaging, it may have some influence. How much, it is difficult to evaluate as it depends on the brightness temperature of the external source, but it is expected to be minor and, in any case, not considered here. Making then $\Gamma_{12} = M_{12} = 0$, and using (3.108),

$$\begin{aligned}\mathfrak{Re}[r_{\rho'}(i\,t_s)] &= \mathfrak{Re}[\tilde{r}_1(i\,t_s)e^{j2\pi f_0 i\,t_s}]\mathfrak{Re}[\tilde{r}_2(i\,t_s)e^{j2\pi f_0 i\,t_s}]\\ \sigma_\rho^2 &= 2\mathcal{P}_1\mathcal{P}_2\end{aligned} \tag{5.87}$$

which largely simplifies the expression of σ_Γ^2 when inserted into (5.68). Furthermore, if both receivers have identical frequency response, then $\tilde{r}_1(\tau) = \tilde{r}_2(\tau) = \tilde{r}(\tau)$ and σ_Γ^2 becomes *mathematically identical* to (5.85). Therefore, as in the case of section 5.4.3, both total power and multibit correlation share the same results.

For one-bit correlation, from (3.142) and (3.153)

$$\left\langle|\breve{\rho}(t)|^2\right\rangle = \left\langle\breve{\rho}_r^2(t)\right\rangle + \left\langle\breve{\rho}_i^2(t)\right\rangle = 2 \tag{5.88}$$

and, under the assumption of uncorrelated input signals $\langle\breve{\rho}(t)\rangle = 0$, the variance of $\breve{\rho}(t)$ now becomes equal to two.

$$\sigma_{\breve{\rho}}^2 = 2 \Rightarrow r_{\breve{\rho}'}(\tau) = \frac{1}{2}R_{\breve{\rho}}(\tau) \tag{5.89}$$

Applying the general relation (1.59) and using now (3.148), knowing that $R_{\breve{\rho}_i'}(\tau) = R_{\breve{\rho}_r'}(\tau)$,

$$\mathfrak{Re}[r_{\breve{\rho}'}(i\,t_s)] = \frac{4}{\pi^2}\arcsin\mathfrak{Re}[\tilde{r}_1(i\,t_s)e^{j2\pi f_0 i\,t_s}]\arcsin\mathfrak{Re}[\tilde{r}_2(i\,t_s)e^{j2\pi f_0 i\,t_s}] \tag{5.90}$$

which is then substituted into (5.69) along with $\sigma_{\breve{\rho}}^2 = 2$. The reader can easily notice that this equation reverts to (5.87) when the arc sin law is not applied to the correlation functions.

5.4.4.1 Results

Figure 5.11 shows the noise excess (5.71) that results from the above expressions as a function of sampling rate. As in figure 5.9, a bandwidth–integration time product of $B\tau_i = 300$ is assumed. The left plot holds for either total power or multibit correlation while the one at right for one-bit correlation. In all correlation measurements two identical receivers are considered as well as independent signals at antenna level. The same two filter shapes (see figure 2.10) are included with the same conclusion that the main impact of considering a nonideal rectangular filter is smoothing the curves.

Now, even for unclipped signals, there is a clear dependency with the center frequency and the same value $f_0 = 1.35\,B$ is taken as an example in both plots. As expected, after some transient behavior, at high sampling rates the standard deviation always saturates to the final value corresponding to analog integration.

The main difference between these plots and those of figure 5.9 is that now two distinct zones, separated by a vertical dashed line, are observed. An oscillatory behaviour dominates the left one while it is monotonic at the right. This is especially marked in the plot on the left since in the other one the ripples also present in figure 5.9 are superimposed. The sampling rate that delimits both zones, taken from the plot corresponding to a rectangular shape, is

$$f_s(\text{lim}) = 2\left(f_0 + \frac{B}{2}\right) \tag{5.91}$$

which, taking a look at the spectrum of b(t) (see for example figure 2.12), turns out to be equal to twice the maximum frequency of the signal. Interestingly, this value coincides with the Nyquist sampling rate. The reader is invited to read section 5.4.6 for a more detailed discussion.

Figure 5.12 shows the effective integration time associated with this case. As expected, the "two-zone" behavior is also apparent. Again, at high sampling rates the effective integration time is larger for a real filter due to having a value of α_F

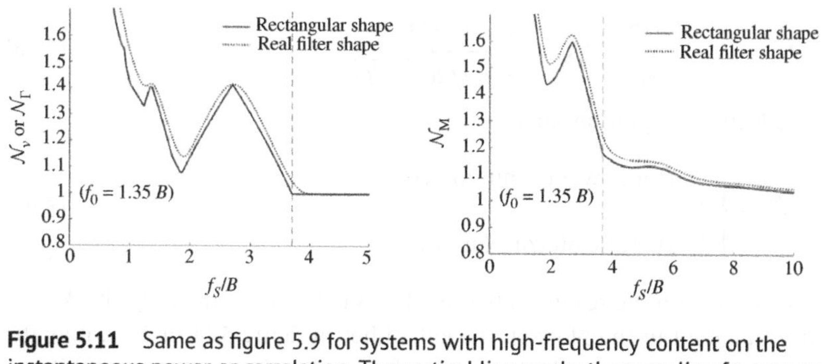

Figure 5.11 Same as figure 5.9 for systems with high-frequency content on the instantaneous power or correlation. The vertical line marks the sampling frequency that delimits two different behaviors (5.91).

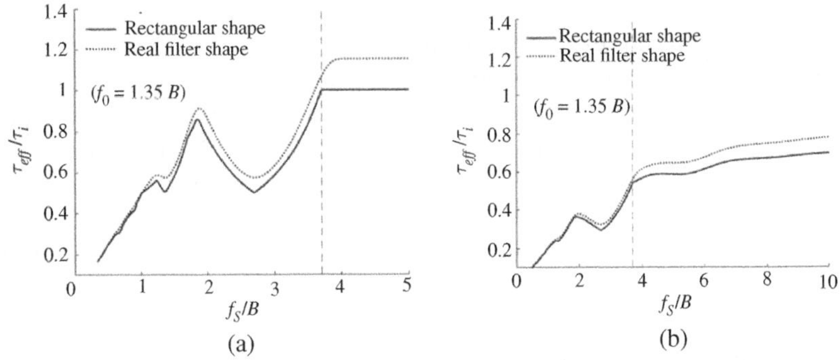

Figure 5.12 Same as 5.10 for systems with high-frequency content on the instantaneous power or correlation. The dashed vertical line has the same meaning as that of figure 5.11.

lower than unity. Globally, the same conclusions as those related to figure 5.10 apply. Particularly, for a one-bit correlation system, increasing the frequency rate beyond f_s(lim) always increases the effective integration time. On the other hand, for multibit correlation it saturates to a constant value. These results, particularly those related to one-bit correlation, were originally presented in [7].

5.4.4.2 Uncorrelated Samples

Obviously, by taking a glance at (5.85), (5.87), or even (5.90), it is apparent that the condition (5.81), needed to have uncorrelated samples, still holds. Therefore, at $f_s = B$, the standard deviations become

$$\sigma_V = \frac{\sqrt{2}P}{\sqrt{B\tau_i}} \; ; \; \sigma_\Gamma = \frac{\sqrt{2P_1P_2}}{\sqrt{B\tau_i}} \; ; \; \sigma_M = \frac{\pi}{2}\frac{\sqrt{2}}{\sqrt{B\tau_i}} \tag{5.92}$$

which differ from (5.82) by a factor of $\sqrt{2}$ in all cases. In consequence

$$\mathcal{N}_V = \mathcal{N}_\Gamma = \frac{\sqrt{2}}{\alpha_F} \; ; \; \mathcal{N}_M = \frac{\pi}{2}\frac{\sqrt{2}}{\alpha_F\sqrt{2(\mathcal{I}+\mathcal{J})}} \tag{5.93}$$

and the effective integration time is

$$\frac{\tau_{eff}}{\tau_i} = \begin{cases} \dfrac{1}{2} & \text{(for power or multibit correlation)} \\ \dfrac{2}{\pi^2} & \text{(for one bit correlation)} \end{cases} \tag{5.94}$$

fully consistent with the results of figure 5.12, as the reader can easily check.

As it is apparent in the plots on the left, either in figure 5.11 or 5.12, now the minimum sampling rate at which numerical averaging (in power or multibit correlation) is equivalent to a continuous integration is not $f_s = B$, as deduced from figure 5.10, but f_s(lim) given in (5.91). In other words, sampling at a rate beyond the condition for uncorrelated samples does reduce in this case the standard deviation, which in the limit converges always to the same value attained with an analog integrator.

5.4.5 I/Q Alternate Sampling

An interesting option consists of implementing digitally the time delay quadrature filter of section 5.3. It can be done in a very simple way by fixing the sampling rate to four times the center frequency $f_s = 4f_0$ so consecutive samples become delayed a quarter of a period and alternate samples correspond, respectively, to the signal and its Hilbert transform approximately (or i and q outputs in the terminology of section 5.3). This concept can be applied to both power and correlation measurement.

5.4.5.1 Power Measurement

For a given signal $b(t)$ sampled at this rate, even samples are now interpreted as "signal" and odd samples as "Hilbert transform of the signal" (or the other way around) and introduced in (5.56). The sampling rate of each component $\mathrm{b}(t)$ and $\hat{\mathrm{b}}(t)$ becomes then half the total sampling rate and, consequently, the graphs of figures 5.9 and 5.10 can still be used as long as f_s is divided by two: $f_s \rightarrow f_s/2$. For an ideal rectangular filter, the minimum sampling rate needed to saturate the effective integration time to 1 is then equal to $2B$ instead of B; and a little bit larger for a real filter. In this case, the one-bit correlation is not applicable.

This same result can be obtained with a different interpretation of the same equation (5.56). Averaging even and odd samples separately and computing the mean is totally equivalent to averaging them all, so this equation is *identical* to (5.55). Therefore, the setup of figure 5.7(a) is applicable and also the subsequent analysis leading to expressing the variance of the output measurement as (5.85). Substituting $t_s = 1/f_s$ with f_s equal to four times the center frequency results in $2\pi f_0 i t_s = i\pi/2$ and therefore, using (A11), it can be written as

$$\sigma_V^2 = 2\mathcal{P}^2 \left[\frac{1}{N} + \frac{1}{N^2} \sum_{i=1}^{N-1} (N-i) \left(|\tilde{r}(i t_s)|^2 + (-1)^i \Re e[\tilde{r}^2(i t_s)] \right) \right] \tag{5.95}$$

Now, according to property 3 of the fringe washing function (section 3.3.2), if the frequency response is symmetrical around the center frequency, the fringe washing function is real. Then the second term in the above equation vanishes for odd i, meaning that consecutive samples are uncorrelated, consistently with the fact that they represent the signal and its Hilbert transform. Otherwise, if the band is not symmetrical, a center frequency error appears, as duly analyzed in section 5.3.2.

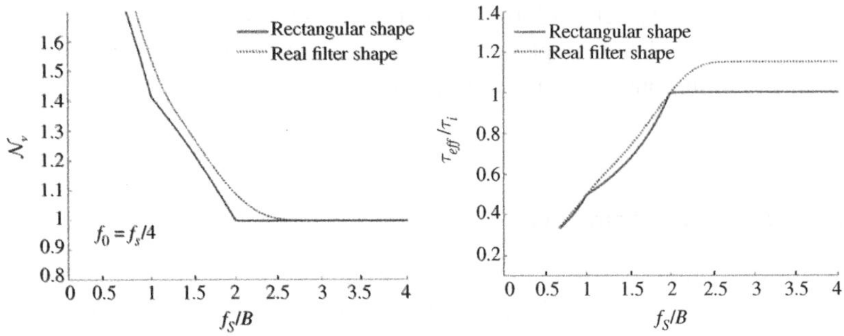

Figure 5.13 Noise excess metric and effective integration time for total power measurement sampling at four times the center frequency.

Introducing $f_s = 4f_0$ into (5.91) shows that the condition $f_s > f_s(\text{lim})$ is always met if $f_s > 2B$, which corresponds to having $f_0 > B/2$, assumed always implicitly, as it is the minimum position of the band to avoid overlapping with the negative part of the spectrum.

Figure 5.13 shows the noise excess (5.71) and effective integration time (5.74) as a function of the sampling rate obtained from (5.95). As expected, these plots are identical to those of figures 5.9(a) and 5.10(a) for half the sampling rate. Consistently with the assertion of the previous paragraph, now no oscillatory zone is seen.

5.4.5.2 Cross-correlation Measurement

Sampling at four times the center frequency to implement the Hilbert transform is also an option for correlation measurements. Figure 5.14 shows the setup for this case. It is very attractive since it is highly simplified with respect to that of figure 5.8, although its practical application can be limited due to the restriction pointed out in relation to (5.38).

As mentioned earlier, in this case, the sampling rate of each component is halved, so the results of section 5.4.3 (four signal multipliers) or 5.4.4 (two signal

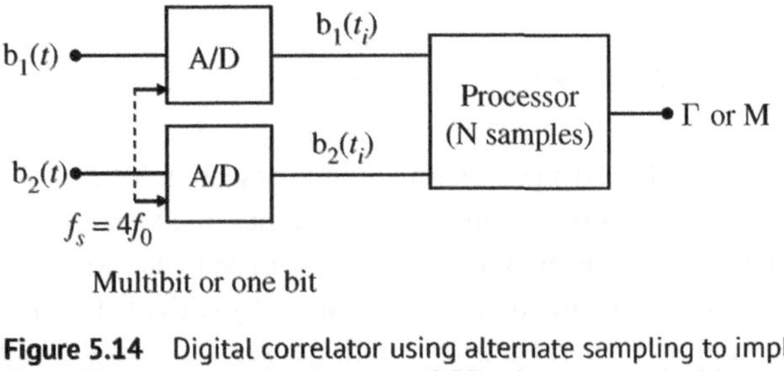

Figure 5.14 Digital correlator using alternate sampling to implement the quadrature filters. The processor implements (5.58) with even and odd samples for signal and Hilbert transform.

multipliers) are valid with only dividing the sampling rate by two. In the first case, the complex correlation is computed by appropriately substituting into (5.58) the even and odd samples. Plots of figures 5.9 and 5.10 are then applicable after the mentioned sampling rate reduction. As pointed out before in relation to power measurement, the sampling rate must be larger than $2B$ (equivalent to making $f_0 > B/2$) to saturate the effective integration time to one. And this condition is *always* met. Oversampling is then beneficial only for nonideal filters and, especially, for one bit correlation.

If two multipliers are used, alternate samples are substituted into (5.57) to get the complex correlation and the analysis of section 5.4.4 applies. In consequence, for optimum performance, the sampling rate *of each component* must be larger than (5.91). Substituting $f_s \to f_s/2$ and making $f_0 = f_s/4$

$$\frac{f_s}{2} > 2\left(f_0 + \frac{B}{2}\right) = \frac{f_s}{2} + B \tag{5.96}$$

which turns out to be *impossible* (it would need negative bandwidth). In conclusion, if alternate samples are selected for the signal and its Hilbert transform, correlation must *always* be computed with four multipliers. Otherwise, using only two multipliers results in higher noise, which cannot be improved by oversampling since the rate is fixed to four times the center frequency.

5.4.6 Nyquist Zones

In digital receivers for telecommunications, it is fundamental to select the sampling frequency according to the Nyquist criterion (see appendix A.8). Otherwise, the spectrum suffers aliasing and this is translated into signal distortion. This effect can be understood with the help of figure 5.15, which shows a signal spectrum replicated due to sampling at frequency f_s. Both the positive and the negative parts of the spectrum are translated at multiples of the sampling rate. The semicircles drawn in dashed lines are included only to help understanding the spectra replication. As it is apparent from the figure, a necessary condition for avoiding aliasing is having the signal bandwidth smaller than $f_s/2$, which is expressed as

$$f_s > 2B \tag{5.97}$$

but this is not enough. Additionally, the band must be allocated in one of the so-called "Nyquist zones," numbered as 1 to 5 in the figure. No matter which

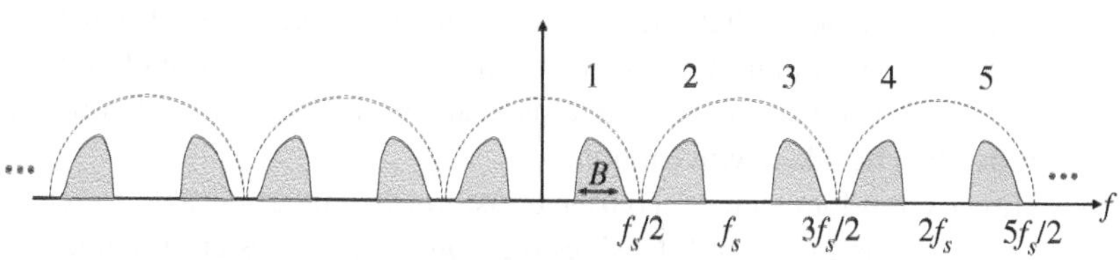

Figure 5.15 Spectrum replication due to sampling.

one of these numbered bands is the original unsampled signal, the spectrum of the sampled signal is always the one shown in the figure. Therefore the time domain sampled signal is also the same. A proper low-pass filter can recover the very same signal without distortion if it was originally in zone 1. For any other zone (number k), sampling plus low pass filtering is equivalent to mixing with a local oscillator with frequency $kf_s/2$ for even k or $(k-1)f_s/2$ for odd k. Note that for even Nyquist zones, the low-frequency band is reversed with respect to the original one, which may be an issue in some systems.

The bounds of the Nyquist zone number k are

$$\begin{aligned} &\text{Minimum frequency:} \quad (k-1)\tfrac{f_s}{2} \\ &\text{Maximum frequency:} \quad k\tfrac{f_s}{2} \end{aligned} \tag{5.98}$$

so, if the original band is allocated in zone k, the allowed center frequency excursion to keep always the band within the zone is

$$(k-1)\frac{f_s}{2} + \frac{B}{2} < f_0 < k\frac{f_s}{2} - \frac{B}{2} \tag{5.99}$$

which implies (5.97), as this one is also obtained from the left- and right-hand side terms of this inequality. In practice, some margin should be added to allow for a slow decay of the spectrum outside the nominal bandwidth, but this subtlety is not considered here. The above inequalities can be rewritten as

$$\frac{2}{k}\left(f_0 + \frac{B}{2}\right) < f_s < \frac{2}{k-1}\left(f_0 - \frac{B}{2}\right) \tag{5.100}$$

from which, using the left- and right-hand side terms, the following restriction for f_0 is found

$$f_0 > (2k-1)\frac{B}{2} \tag{5.101}$$

In the first zone ($k = 1$), (5.100) states that the maximum sampling rate is infinite. If, furthermore, $f_0 = B/2$ then the left inequality coincides with (5.97).

In summary, given a spectrum with bandwidth B centered at f_0, to avoid aliasing the sampling rate must satisfy (5.100) with k, a positive integer fulfilling (5.101).

5.4.6.1 Application to Radiometry

The above concepts are *not* directly applicable to random signals of the kind found in radiometers, particularly thermal noise. Or, at least, a discussion is needed. To start with, stationary random processes have infinite energy and therefore they do not have Fourier transform. Any analysis made on the basis of the frequency domain must then be taken with caution.

For stationary random signals, Fourier transform is only applicable to relate correlation functions to spectral densities (section 1.2.4). A system intended to retrieve the spectral density of a signal from correlation measurements between samples at different time intervals needs indeed to consider (5.100) to avoid aliasing. But this is not the case of a microwave radiometer, an instrument dedicated to measure the total power of a signal, or the cross-correlation of two signals in the case of an interferometer or in polarimetry. This statement is apparent by taking a glance at (1.51) and noting that it is totally independent on the sampling rate: for a given process, the mean value of the average of N samples of the instantaneous power is always equal to its mean power. And this is true without regard to the condition (5.100). Aliasing has then no impact on total power or correlation.

However, sampling rate does have an impact on the measurement uncertainty when the integration time is finite (radiometric sensitivity). It has been properly studied in section 5.4.2 through the analysis of the correlation between samples and without needing any additional condition based on signal spectrum replicas. Nevertheless, for illustration purposes, it is interesting to visualize the position of the spectral densities in the three cases mentioned at the beginning of section 5.4.1 and their compatibility with the Nyquist zones.

Figure 5.16 shows schematically the spectrum in each case. For the first one it is copied from that of the right of figure 1.2 and assumes that the spectrum of the instantaneous power or correlation is flat within the bandwidth of the analog integration filter. The second case, in figure 5.16(b), is the spectrum of p(t) or $\rho(t)$ filtered to keep only the low-frequency band. Ideal rectangular filters are assumed, so it is the low-frequency triangle depicted in figure 2.12(b). Finally, for the third case, the spectrum corresponds to the receivers' output signals before power detection or correlation. Therefore, it has a band-pass of bandwidth B centered around f_0. Also ideal rectangular filters are assumed.

From the strict point of view of aliasing, figure 5.16(a) and (b) should be sampled at a rate higher than $2/\tau_i$ (disregarding secondary lobes) or $2B$, respectively. However, according to the plots of figures 5.5 and 5.9(a) the minimum rate is half this value, namely $1/\tau_i$ and B. In both cases, sampling faster does not improve the performance in terms of noise. In consequence, for these two cases, avoiding

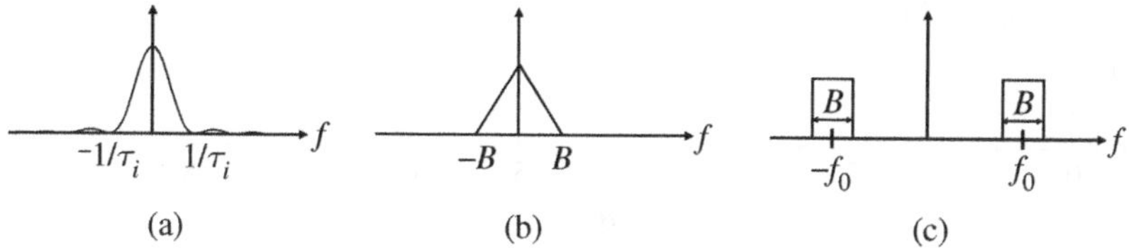

Figure 5.16 Power spectral density of the signal before sampling. The three cases stated at the beginning of section 5.4.1 are illustrated from (a) to (c).

aliases is not a good design criterion. It is now pertinent to remind that the spectrum in figure 5.16(b) corresponds to the setups in figure 5.6 for both power and correlation measurements.

In the third case, in order to avoid aliasing the sampling frequency must satisfy (5.100), so it depends on the Nyquist zone, where the bandwidth is allocated. On the other hand, for ideal rectangular filters, samples are uncorrelated at the same rate $f_s = B$ independently on the band position. If power is measured with a quadratic detector, as in figure 5.7(b), or correlation is measured with that of figure 5.8 with four multipliers, then the plot of figure 5.9 predicts that the minimum sampling rate is still $f_s = B$ *independently of the Nyquist zone.* Strong aliasing may be present without compromising the quality of the measurements so again, avoiding aliasing is not a design criterion.

However, different conclusions are obtained if a square law detector (figure 5.7(a)) or two real multipliers are used for power and correlation measurements, respectively. In these two cases, even though samples are still uncorrelated at $f_s = B$, oversampling reduces noise and the optimum sampling rate becomes (5.91), equivalent to (5.100) for the first Nyquist zone. Moreover, the spectrum can be allocated in any zone and still getting the minimum noise. This is illustrated in figure 5.17, which is the same as in figure 5.11 but for

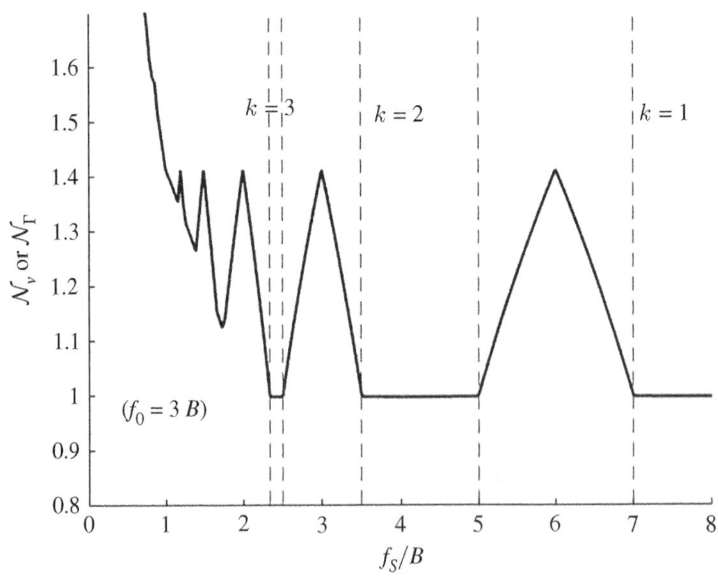

Figure 5.17 Same as figure 5.11 for a higher value of center frequency to illustrate the compatibility with the Nyquist zones. Only ideal rectangular filter is shown.

$f_0 = 3B$. According to (5.101), this center frequency supports Nyquist zones 1 to 3. Consistently, when the sampling rate is within the bounds (5.100) for each zone, marked in the figure by dashed lines, the standard deviation is minimized and is equal to that of an analog system. In this case, which is the most common, a design based on avoiding aliasing *is* a good option. Be aware, however, that the plot of figure 5.17 applies to ideal rectangular filters. As illustrated in figure 5.11, actual filters have the effect of smoothing the curve and this may have impact if the Nyquist zone is narrow, for example the third one.

From the above discussion, it follows that the Nyquist sampling rate is *not*, in general, the frequency at which consecutive samples are uncorrelated. These are two different concepts, the first one related to the spectrum repetition and the other one to the zeroes of the correlation function.

5.4.7 Correlation in the Frequency Domain

A digital radiometer (third case of those listed at the beginning of section 5.4.1) can also be implemented in the frequency domain by applying a discrete Fourier transform to the time domain samples and using the identity (A.55) demonstrated in appendix A.8. This option may be useful to apply filters in the frequency domain in order to reject spurious signals, although it implies an important increase in computational complexity.

5.4.7.1 Power

Applying directly (A.55) to (5.55),

$$\mathrm{v} = \frac{1}{N^2} \sum_{i=1}^{N} |\mathrm{DFT}\ (\mathbf{b})_i|^2 \tag{5.102}$$

where DFT $(\mathbf{b})_i$ is the ith element of the discrete Fourier transform of the vector formed by the sampled time domain real signal $\mathrm{b}_i = \mathrm{b}(t_i)$, $\mathbf{b} = [\mathrm{b}_1, \mathrm{b}_2, \dots, \mathrm{b}_N]$.[2]

Implementation wise, it may be interesting to reduce the number of sums taking into account the symmetry of the Fourier transform of real signals. For example, if N is odd, assuming that the DFT vector is not shifted, the above equation is equivalent to

$$\mathrm{v} = \frac{\mathrm{DFT}(\mathbf{b})_1^2}{N^2} + \frac{2}{N^2} \sum_{i=2}^{(N+1)/2} |\mathrm{DFT}\ (\mathbf{b})_i|^2 \tag{5.103}$$

Otherwise, for even N and/or shifted DFT, appropriate equivalent formulas can be easily found.

2 Consistent with the rest of this chapter, indexing starts at 1.

If no more processing is done in the frequency domain, (5.102) and (5.103) are both *identical* to (5.55) so the conclusions of section 5.4.4 regarding this equation apply. As it is apparent from figure 5.11, for an ideal filter, the sampling rate must be larger than $2(f_0 + B/2)$ in order to have the same standard deviation of the output signal as an analog system.

The same concept can be adopted in the digital implementation of a quadratic detector, depicted on figure 5.7(b). Samples of the analytic signal are first computed as

$$b(t_i) = \mathrm{b}(t_i) + j\hat{\mathrm{b}}(t_i) \tag{5.104}$$

and then (5.56) in the frequency domain becomes

$$\mathrm{v} = \frac{1}{2N^2}\sum_{i=1}^{N} |\mathrm{DFT}\ (\boldsymbol{b})_i|^2 \tag{5.105}$$

where now DFT $(\boldsymbol{b})_i$ is the ith element of the discrete Fourier transform of the vector of analytic signal samples (5.104). In this case, since the time domain vector is complex, no reduction similar to (5.103) is applicable.

Since (5.105) is numerically equivalent to (5.56), the properties derived in section 5.4.3 for this one are valid. Using for example figure 5.9, it follows that, assuming ideal filter, the standard deviation of v becomes equal to that of an analog system only if the sampling rate is larger than the signal bandwidth B with no other requirement, independent thus of the center frequency.

As explained in section 5.4.5, the Hilbert transform can be implemented by sampling at four times the center frequency and selecting alternate samples. Obviously, this option is compatible with the frequency domain processing. Equation (5.104) now selects even samples for the real part and odd samples for the imaginary part (or the other way around, of course) and (5.105) is applied to this signal.

Independently of the above statement, for any sampling rate, the Hilbert transform of (5.56) can be implemented in the frequency domain and applied to the very same samples. Evaluating (5.56) in the frequency domain using (A.55) results in

$$\mathrm{v} = \frac{1}{2N^2}\sum_{i=1}^{N} (|\mathrm{DFT}\ (\mathbf{b})_i|^2 + |\mathrm{DFT}\ (\hat{\mathbf{b}})_i|^2) \tag{5.106}$$

and, since $\hat{\mathrm{b}}(t)$ is the output of a quadrature filter, with frequency response (1.123), when $\mathrm{b}(t)$ is the input signal, the discrete Fourier transform of the Hilbert transform can be substituted by

$$\mathrm{DFT}\ (\hat{\mathbf{b}})_i = (-j\mathrm{sgn}\, f_i)\mathrm{DFT}\ (\mathbf{b})_i \tag{5.107}$$

where f_i is the frequency of the corresponding sample. Introducing this result into (5.106) recovers *exactly* the same equation (5.102), so no advantage is obtained with this method.

5.4.7.2 Cross-correlation

Frequency domain is also applicable to correlation measurements. Using four signal multipliers, equation (5.58) is first rewritten in a more compact form

$$\Gamma = \frac{1}{2N}\sum_{i=1}^{N} b_1(t_i)b_2^*(t_i) \tag{5.108}$$

where $b_1(t_i)$ and $b_1(t_i)$ are samples of the analytic signals of $\mathrm{b}_1(t)$ and $\mathrm{b}_2(t)$,

$$\begin{aligned} b_1(t_i) &= \mathrm{b}_1(t_i) + j\hat{\mathrm{b}}_1(t) \\ b_2(t_i) &= \mathrm{b}_2(t_i) + j\hat{\mathrm{b}}_2(t) \end{aligned} \tag{5.109}$$

and then (5.108) is evaluated in the frequency domain using (A.55)

$$\Gamma = \frac{1}{2N^2}\sum_{i=1}^{N} \mathrm{DFT}\,(\boldsymbol{b}_1)_i\ \mathrm{DFT}(\boldsymbol{b}_2)_i^* \tag{5.110}$$

where $\boldsymbol{b}_1$ and $\boldsymbol{b}_1$ are the vectors of analytic signal samples (5.109). Since four signal multipliers are assumed, the analysis of section 5.4.3 applies, and the noise excess and effective integration time are given in figures 5.9 and 5.10, respectively. Consequently, to achieve a standard deviation of Γ equal to the analog counterpart, the sampling rate must be larger than the signal bandwidth. This processing is fully compatible with the approach of section 5.4.5 where the Hilbert transform is obtained from alternate samples, and the sampling rate is equal to four times the center frequency.

If using only two real multipliers, the starting point is equation (5.57) in the frequency domain

$$\Gamma = \frac{1}{N^2}\sum_{i=1}^{N} \mathrm{DFT}\,(\mathbf{b}_1)_i\ \mathrm{DFT}(\mathbf{b}_2)_i^* + j\frac{1}{N^2}\sum_{i=1}^{N} \mathrm{DFT}\,(\hat{\mathbf{b}}_1)_i\ \mathrm{DFT}(\mathbf{b}_2)_i^* \tag{5.111}$$

and, using (5.107),

$$\Gamma = \frac{1}{N^2}\sum_{i=1}^{N} \mathrm{DFT}\,(\mathbf{b}_1)_i\ \mathrm{DFT}(\mathbf{b}_2)_i^*(1 + \mathrm{sgn}\, f_i) \tag{5.112}$$

which is null for negative frequencies and twice the value of the DFT product for positive frequencies. This result is compatible with the spectrum of analytic signals, limited to the positive side of the frequency axis. Assuming odd number of samples and no DFT shifting applied, this equation can be expressed as (5.103)

$$\Gamma = \frac{\mathrm{DFT}\,(\mathbf{b}_1)_1\ \mathrm{DFT}\,(\mathbf{b}_2)_1}{N^2} + \frac{2}{N^2}\sum_{i=2}^{(N+1)/2} \mathrm{DFT}\,(\mathbf{b}_1)_i\ \mathrm{DFT}(\mathbf{b}_2)_i^* \tag{5.113}$$

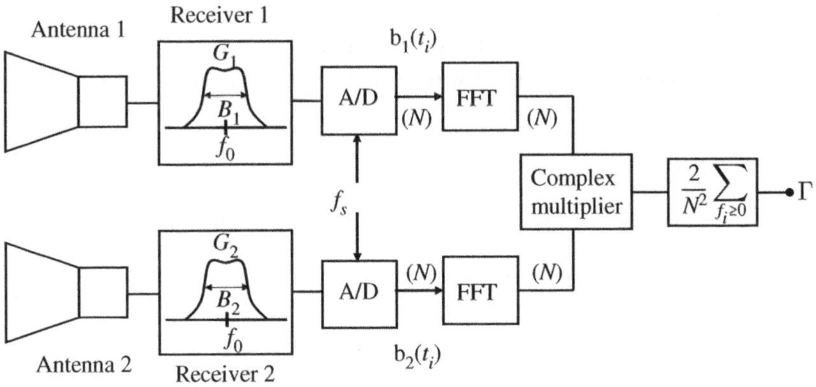

Figure 5.18 Conceptual block diagram of an interferometer that uses a correlator in the frequency domain according to equation (5.113). The complex multiplier multiplies one input by the conjugate of the other. In the sum of spectral lines, the DC is included with half amplitude.

This result provides the correct value of correlation as it is equivalent to (5.57) with the Hilbert transform obtained directly in the frequency domain. An odd number of samples is required to ensure that the spectrum of the discrete Fourier transform is symmetric in frequency.

A possible implementation of a frequency domain correlator based on (5.113) is shown in figure 5.18. The receivers' output signals are directly converted to digital numbers at a sampling rate f_s compatible with (5.91). Discrete Fourier transform is computed using N samples of each digitized signal, and the result is entered in equation (5.113). The procedure is very straightforward as no windowing, normalization, or any other processing is applied to the input signals. Just a discrete Fourier transform.

The number of points of the DFT must be odd, but its concrete value is not a critical parameter. For a given sampling frequency, it fixes the *single snapshot* integration time $\tau_i = N/f_s$, but noise can be further reduced by averaging several individual snapshots, effectively increasing the overall integration time. From the point of view of computing time, averaging many short discrete Fourier transforms may be slower than averaging a few number of long ones, so it is maybe preferable to have relatively long sequences. Also, a small systematic error may appear for very short sequences due to heavy undersampling in the frequency domain.

Obviously, this schema is also valid for clipped signals (section 3.4.5) as property (A.55) holds for any kind of signals, including these ones. The analysis for one-bit correlation of section 5.4.1 applies.

References

1 Stephen A. Maas. *Microwave Mixers*. Artech House, Inc., Norwood, MA, 1986. ISBN 0-89006-171-8.

2 Ignasi Corbella, Francesc Torres, Adriano Camps, Andreas Colliander, Manuel Martín-Neira, Serni Ribó, Kimmo Rautiainen, Nuria Duffo, and Mercè Vall-llossera. MIRAS end-to-end calibration: Application to SMOS L1 processor. *IEEE Transactions on Geoscience and Remote Sensing*, 43(5):1126–1134, May 2005.

3 Ignasi Corbella, Manuel Martin-Neira, Roger Vilaseca, Albert Catal Francesc Torres, and Martin Suess. A novel digital IQ demodulation for interferometric radiometers. *Remote Sensing*, 13(6):1156, March 2021. doi: 10.3390/rs13061156.

4 Alan V. Oppenheim and Ronald W. Shafer. *Digital Signal Processing*. Prentice Hall, Inc., New Jersey, 1975. ISBN 0-13-214635-5.

5 J. B. Hagen and D. T. Farley. Digital-correlation techniques in radio science. *Radio Science*, 8(8, 9):775–784, August-September 1973.

6 A. Richard Thomson, James M. Moran, and George W. Swenson Jr. *Interferometry and Synthesis in Radio Astronomy*. John Wiley & Sons, Inc., 1986.

7 Ignasi Corbella, Francesc Torres, Adriano Camps, Javier Bará Nuria Duffo, and Mercè Vall-llossera. L-band aperture synthesis radiometry: Hardware requirements and system performance. In *International Geoscience and Remote Sensing Symposium, IGARSS 2000*, volume 7, pages 2975 –2977, Honolulu, HW, USA, 24–28 July 2000. IEEE.

6

Calibration and Characterization

Radiometers are calibrated by measuring known sources, designated as "calibration standards," to determine a set of parameters of a predefined model intended to describe the instrument operation. For interferometric or polarimetric radiometers this process involves both complex correlation and total power measurements [1–3]. Therefore both of them are treated in this chapter, which deals with methods and techniques for characterizing radiometric instruments in general and calibrating the raw measurements acquired.

6.1 Calibration Standards

Calibration standards are a special kind of sources for which the response of an ideal, perfectly calibrated instrument, is supposedly known. Deviations of the actual measurements from this ideal response provide the calibration parameters. To calibrate a radiometer, standards should be placed, in principle, externally in order to include all antenna effects, particularly loss and phase shift. In total power radiometers antenna loss uncertainty has a very important impact on antenna temperature measurement. On the other hand, antenna relative phase has impact on correlation measurements.

External targets used as standards can be difficult to characterize and, depending on the instrument size, they can be impractical, particularly if frequent calibration is needed. Internal calibration, consisting of injecting noise to the receivers' input ports, may be advantageous in this case since it can be done efficiently and fast. In practice, a combination of external and internal calibration is often the best option. Frequent internal calibration is carried out regularly while external calibration is used at some selected times to correct for hardware not included in the previous process.

Principles of Interferometric and Polarimetric Radiometry, First Edition. Ignasi Corbella.

6.1.1 Antenna and Calibration Planes

The "antenna plane" denotes the conventional port where both antenna temperature (2.46) and visibility (3.9) are defined. It corresponds to the antenna output port without including the associated ohmic efficiency [see figure 1.19]. The "calibration plane" is the point in the receiver chain where noise is injected during internal calibration. Clearly, calibrating using internal noise injection fails to characterize all components located between antenna and calibration planes. Therefore, to minimize calibration errors, noise should be injected as close as possible to the antenna. But, in practice, some physical distance is unavoidable so finite time delay and power loss are to be expected even in the best scenario.

The situation is modeled in figure 6.1 where two ideal lossless antennas labeled k and j are connected to respective receivers through passive networks to account for loss and delay. This model is compatible with the one defined in section 1.6.11 but including also a time delay and additional loss in transmission lines or other components inserted. The total *available power gain* (1.184) of the passive networks that connect lossless antennas and receivers are here denoted by η_k and η_j, assumed frequency independent at least within the receivers' bandwidth. They include both antenna ohmic efficiency and loss in physical connections to receivers. The corresponding time delays are t_k and t_j. If, for whatever reason, a given system cannot be modeled by a loss and a delay, then the subsequent analysis is not valid.

With these conventions, the transducer power gain of, for example, receiver k is $G_{Tk}(f) = \eta_k G^C_{Tk}(f)$ where the superscript C indicates "calibration plane." Furthermore, according to property 4 of the fringe washing function (section 3.3.2),

$$\tilde{r}^C_{kj}(t) = \tilde{r}_{kj}(t - \Delta t)e^{-j2\pi f_0 \Delta t} \tag{6.1}$$

where $\Delta t = t_k - t_j$. From the definition of the correlator gain (3.83), taking into account that the symbol G stands for transducer power gain at the center frequency

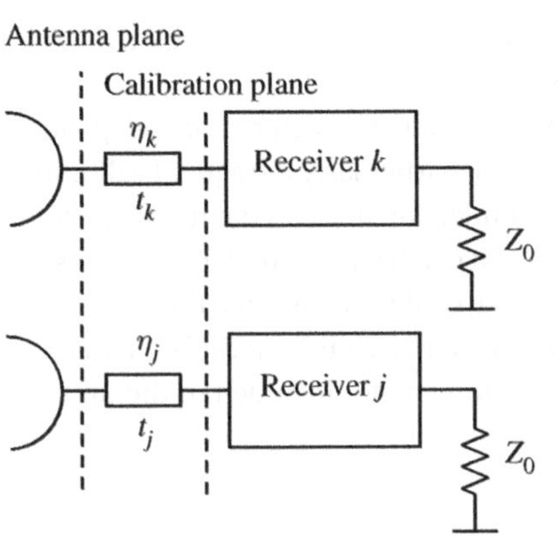

Figure 6.1 Definition of antenna and calibration planes. Connection of both planes is modeled by passive networks with available power gain $\eta_{k,j}$ and time delay $t_{k,j}$.

and that the bandwidth is unchanged by a time delay

$$G_{kj} = G_{kj}^C \sqrt{\eta_k \eta_j}\, \tilde{\bar{r}}_{kj}^C(t_j - t_k)\, e^{j(\phi_k - \phi_j)} \approx G_{kj}^C \sqrt{\eta_k \eta_j} e^{j(\phi_k - \phi_j)} \tag{6.2}$$

where $\phi_{k,j} = -2\pi f_0 t_{k,j}$ are the corresponding phases of the frequency responses particularized at the center frequency f_0. If the connections between planes can be modeled by transmission lines, then $t_{k,j} = \ell_{k,j}/c$ where $\ell_{k,j}$ are their respective lengths and c the velocity of wave propagation in the transmission line. The approximation indicated at the right-hand side of the above equation holds for sufficiently small time difference so that $\tilde{\bar{r}}_{kj}^C(t_j - t_k) \approx 1$. In practice this is a reasonable approximation.

Equation (6.2) allows estimating the correlator gain at antenna plane from the one calibrated internally, provided the elements in between are previously characterized in terms of available gain and phase (see section 6.2.2). The visibility *at calibration plane* is directly (3.86)

$$V_{kj}^C = \frac{\Gamma_{kj}}{G_{kj}^C} \tag{6.3}$$

where Γ_{kj} is the raw correlation measurement. Conversion to antenna plane is then straightforward from (6.2)

$$V_{kj} = \frac{\Gamma_{kj}}{G_{kj}} = \frac{V_{kj}^C}{\sqrt{\eta_k \eta_j} e^{j(\phi_k - \phi_j)}} \tag{6.4}$$

On the other hand, for each single antenna, the system temperature is the total equivalent noise temperature at a given plane. Directly from (1.218) and taking into account that η_k is also the ratio of available power gains between planes

$$T_{sysk} = \frac{T_{sysk}^C}{\eta_k} \tag{6.5}$$

where T_{sysk}^C is the system temperature referenced at the calibration plane. Note the consistency between this equation and (6.4) due to the fact that, as pointed out after equation (3.60), the system visibility coincides with the system temperature when both antennas collapse into one.

6.1.1.1 Normalized Correlation

If the instrument is measuring *normalized* correlation, then the correlator gain is equal to the fringe washing function at the origin (3.137). In this case (6.2) reduces to

$$G_{kj} = G_{kj}^C \tilde{\bar{r}}_{kj}^C(t_j - t_k)\, e^{j(\phi_k - \phi_j)} \approx G_{kj}^C e^{j(\phi_k - \phi_j)} \tag{6.6}$$

where now $G_{kj}^C = \tilde{r}_{kj}^C(0)$ and all other symbols have the same meaning. Introducing this result together with (6.5) into (3.136)

$$V_{kj} = \frac{M_{kj}\sqrt{T_{sysk}^C T_{sysj}^C}}{G_{kj}^C\sqrt{\eta_k\eta_j}e^{j(\phi_k-\phi_j)}} \tag{6.7}$$

which is fully consistent with (6.4). The reader must be aware, though, that the definition of G_{kj}^C in both expressions is different. This equation is used in SMOS data processing to get the visibility at antenna plane after calibrating both total power and correlation at calibration plane [4].

6.1.2 Plane Change in Total Power Radiometers

The situation is much more difficult for total power radiometers designed to measure antenna temperature and calibrated using internal noise injection. This is so because the noise contribution of all subsystems inserted between antenna and calibration planes must be correctly taken into account. There are several methods to do that, but one particularly elegant consists of transferring the equivalent noise temperature of the internal loads to the antenna plane.

Consider for example the setup of figure 6.2 consisting of a lossy antenna and a connection to the calibration plane with some extra loss, for example a coaxial cable. Two physical temperatures are assumed in this model, one in the antenna itself and other in the cable, but more elaborated models with additional temperatures can easily be accommodated.

Assume that the noise temperature of a given internal calibration standard is T_S^C. Typically, it would consist of a matched load or noise diode connected through a switch (see section 6.1.5 for details). Using the flux diagram method of section 1.5.8, the equivalent noise temperature at the antenna plane that would produce *exactly* the same value of T_S^C at the output port is straightforwardly computed

$$T_s = T_S^C\frac{L}{\eta_a} - T_{p_a}\left(\frac{1}{\eta_a} - 1\right) - T_{p_l}\frac{1}{\eta_a}(L-1) \tag{6.8}$$

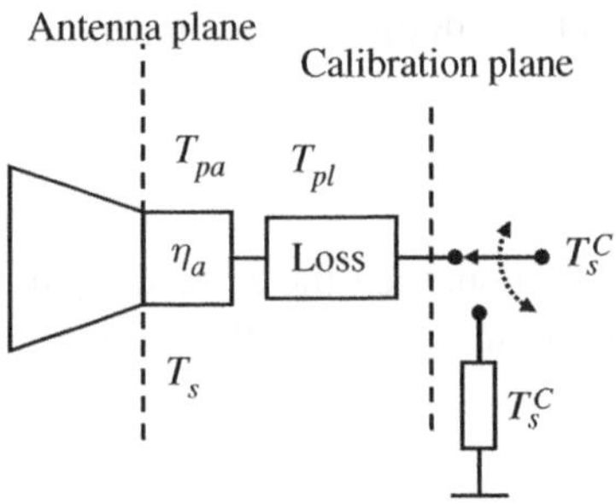

Figure 6.2 Antenna front-end model used to describe the transfer of an internal calibration standard to antenna plane. T_{p_a} and T_{pl} are physical temperatures.

where L is the inverse of the available power gain of the lossy network and η_a the antenna ohmic efficiency (1.285). If more components in cascade are present at different temperatures, more terms are simply added, but the procedure is the same. Injecting T_s at antenna plane is totally equivalent to switching the calibration plane to the internal standard. In other words, no matter the physical plane where the noise is actually injected, it can always be made equivalent to an external target with (6.8) as equivalent antenna temperature. By this procedure, the internal calibration standards are said to be *transferred* to the antenna plane.

> **Remark**: Even though switching the receiver input port to the internal load at T_s^C or to the antenna pointing to an external target at T_s are equivalent in terms of input noise temperature, they may be different in matching. If it is not the same in both switch positions, calibration parameters (i.e., gain and offset, see section 6.2.6) retrieved from internal loads differ from those that would be obtained from the equivalent external targets. This is so because both gain and noise equivalent temperature depend on the matching at the receiver input port. In a practical system this effect should be minimized or corrected by proper characterization.

In the particular case of having the whole front-end at a the same temperature, then $T_{p_a} = T_{p_l} = T_p$ and (6.8) simplifies to

$$T_s = (T_s^C - T_p)\frac{L}{\eta_a} + T_p \quad \text{(Homogeneous temperature)} \tag{6.9}$$

which reduces simply to T_p if the internal load has also the same temperature ($T_s^C = T_p$). Therefore, if the front end is kept at a homogeneous temperature, an internal standard at this temperature is totally equivalent to an external standard at the same temperature, independently of the loss between planes.

In general, the switch inserted to connect the internal load must be placed as close as possible to the antenna to minimize errors. Nevertheless, even with a good characterization of all the elements in the chain, calibrating a radiometer based *only* on internal standards transferred to antenna plane may be risky and can generate errors beyond what is acceptable. An important source of error is the unavoidable mismatch between adjacent elements. As pointed out at the end of section 1.5.8 in reference to equation (1.219), mismatch plays a role in the attenuation L, which is the inverse of the available power gain (1.184), in turn a function of the source reflection coefficient and S-parameters.

For example, assuming in figure 6.2 that the antenna is perfectly matched (null reflection coefficient), then the attenuation of the cable (connection) is

$$L = \frac{1 - |S_{22}|^2}{|S_{21}|^2} \tag{6.10}$$

where the S-parameters are those of the cable. Of course, if this one is perfectly matched ($S_{22} = 0$), then the attenuation is just the inverse of $|S_{21}|^2$, but otherwise the numerator departs from unity and must be included.

If the antenna is not perfectly matched, its reflection coefficient both in amplitude and in phase must be known and substituted into (1.184) to get the available gain of the cable. A complete characterization of all components in terms of *all four S-parameters* is then essential. This is even more important if additional elements, for example orthomode transducers or filters, are included between antenna and switch. An incomplete characterization of these components may give rise to errors on noise temperature and gain that can accumulate making internal calibration useless. On the other hand, internal calibration provides an easy and fast way to calibrate, and it is very convenient for operational radiometers.

6.1.2.1 Internal Standard Characterization

An alternative approach to avoid precise measurement of all individual components goes through characterizing the internal calibration loads using the same radiometer calibrated, from time to time or even only once, with well-known external calibration targets.

If two targets at different known temperatures are placed in front of the antenna of figure 6.2, then equation (6.54) can be used to estimate the total gain G and offset v_0' referenced to antenna plane. Once this is done, when the switch is connected to the internal load, the equivalent noise temperature at antenna plane is

$$T_s = \frac{v - v_0'}{G} \tag{6.11}$$

where v is the raw radiometric measurement (voltage). Note that the above noise temperature T_s is a measurement of (6.8). Therefore, to calibrate the radiometer during normal operation, it suffices to activate the switch to the internal load and use the value of T_s obtained once from this equation to estimate the injected temperature at antenna plane.

This method is more robust than using (6.8) with individual component characterization since the actual measurement takes into account all mismatches and losses. However, it fails to include the dependence of T_s on the physical temperatures, which is apparent in (6.8). To do so, the procedure must be carried out for different combinations of controlled physical temperatures in each component and, from this set of measurements, corresponding sensitivity coefficients must be derived from a linear fit. For example, in the case of the two subsystems of figure 6.2

$$S_a = \frac{\partial T_s}{\partial T_{p_a}}\ ;\ S_l = \frac{\partial T_s}{\partial T_{p_l}} \tag{6.12}$$

and then these coefficients are used at any internal calibration epoch (t_c) to estimate the noise temperature of the internal standard

$$T_s(t_c) = T_s(t_0) + S_a(T_{p_a}(t_c) - T_{p_a}(t_0)) + S_l(T_{p_l}(t_c) - T_{p_l}(t_0)) \tag{6.13}$$

where t_0 indicates the time at standard characterization. This equation is fully consistent with the theoretical model (6.8) if the sensitivity coefficients are directly the terms multiplying the physical temperatures. However, in this approach both the sensitivity coefficients and the noise temperature of the standard at t_0 are measured instead of estimated from different component parameters. Possible dependencies of losses with temperature are here automatically included. Needless to say, this procedure can be applied to any number of subsystems in the chain.

In many radiometric systems, the antenna is large and cumbersome so it is difficult, if not impossible, to include it in the temperature characterization procedure. In this case, a good option is just using the theoretical value of the sensitivity, given directly from (6.8)

$$S_a = \left(\frac{1}{\eta_a} - 1\right) \tag{6.14}$$

in which η_a is a measurement or a theoretical estimation of the antenna ohmic efficiency, carried out independently.

In any case, antenna plane change relies on having precise physical temperature measurements at different points of the front end, particularly at the antenna. This is done by using appropriate thermal sensors precisely located and properly calibrated.

6.1.3 External Passive Targets

The two most common natural targets for radiometer calibration are the cold sky and an absorbing material at known temperature. Both of them are used to provide a reference antenna temperature in total power measurements, but their usage in interferometric radiometers is more restricted. An example of calibration of polarimetric radiometers using external targets can be found in [5]. Section 6.2.4 describes the use of the cold sky to retrieve the flat target response.

The nominal antenna temperature of an external target is estimated using (2.50) assuming that both brightness temperature and antenna pattern are known. This last one may be measured in a suitable facility or evaluated from the antenna geometry. Using the concept of main beam efficiency (2.56) and assuming constant brightness temperature T_s within the main beam, the antenna temperature becomes

$$T_a = T_s\eta_b + T_{SL} \tag{6.15}$$

where η_b is the main beam efficiency and T_{SL} is the contribution of the side lobes. Depending on the setup used in the calibration process, this last term may be either neglected or estimated by a computer simulation.

Total power radiometers using narrow beam antennas are good candidates for using this approach, but otherwise, wide antenna patterns or with relatively high side and/or back lobes may be responsible for errors due to the contribution of unexpected objects outside the calibration target. This is especially important when the brightness temperature of the calibration standard is very low, for example the cold sky or an absorbing material immersed in liquid nitrogen. Surrounding objects at much larger temperature may contribute to the radiometric measurements in a similar or even larger amount than that of the target, invalidating the calibration. In this case, the complete expression (2.50) must be used.

The brightness temperature of the cold sky is equal to the cosmic microwave background at 2.725 K plus the galactic contribution, obtained independently from available radio astronomy data. For ground-based radiometers, care must be taken to include all atmospheric contributions, which depend on water content, aerosols, and others, sometimes difficult to assess. Use of standard atmosphere is a first-guess, but to increase accuracy, a complete radiative transfer model is needed. This topic is out of the scope of the present book and the interested reader may wish to consult more specific works about it. In the case, radiometers onboard satellites using the cold sky as a calibration target, the impact of changing the attitude of the platform is also important and may give rise to large errors in the estimation of calibration parameters.

Absorbing materials are commonly used in total power radiometers equipped with highly directive scanning antennas. Calibration is carried out then by periodically pointing the antenna to the calibration target. The brightness temperature of the absorbing material is inferred from (2.11) based on their estimated emissivity and the physical temperature measured by dedicated sensors. Depending on the frequency and the kind of absorber, the emissivity can be approximated to unity (black body), and then the brightness temperature is just the measured physical temperature. To have an absolute temperature calibration, the absorber can be immersed in liquid nitrogen at a known temperature of 77 K. This option is valid only as a single instrument characterization on ground but not operationally as a periodic calibration strategy. In satellite-based or airbone instruments, it is totally impractical.

A complete study of absorbing materials as calibration targets for radiometry falls beyond the scope of this book. As a matter of fact, their usage in interferometric radiometers, and especially those oriented to Earth observation from space, is very limited, if not null.

6.1.4 Probe Antenna

A known source for interferometric calibration can be implemented by a probe antenna emitting noise toward the instrument in the setup shown schematically in figure 6.3. Applying the linearity property, the visibility of a given baseline can be expressed as the sum of both contributions, one from the power emitted by the probe and another from the thermal noise of the chamber itself

$$V_{kj} = V_{kj}^{s}(T_s - T_{ch}) + V_{kj}^{ch}(T_{ch}) \tag{6.16}$$

consistent with the fact that, if the source temperature matches that of the chamber ($T_s = T_{ch}$), the probe does not affect the measurements since all elements inside the chamber are kept at the same temperature.

The visibility of the empty chamber is directly given by (3.58) with $T_B(\theta, \phi) = T_{ch}$. Assuming that the whole setup is kept at a constant ambient temperature, then the receivers physical temperatures are also equal to T_{ch} and the visibility vanishes $V_{kj}^{ch} = 0$. The contribution to the visibility of the antenna interaction, given by the second term of (3.44), is thus included in the chamber visibility and should not be applied also to the visibility of the point source.

The antenna visibility of a point source consisting of a radiating antenna can be computed by inserting in (3.6) the signals collected by the antennas, in this case, evaluated in (1.271). Consistently with (6.16), the equivalent temperature of the source is now ($T_s - T_{ch}$). According to the results of section 3.3.1, the measurable system visibility (3.58) has the same expression as the antenna visibility but including the $-T_r$ term and the fringe washing function and using the wave number at the center frequency. Now, as already pointed out, the $-T_r$ term should not

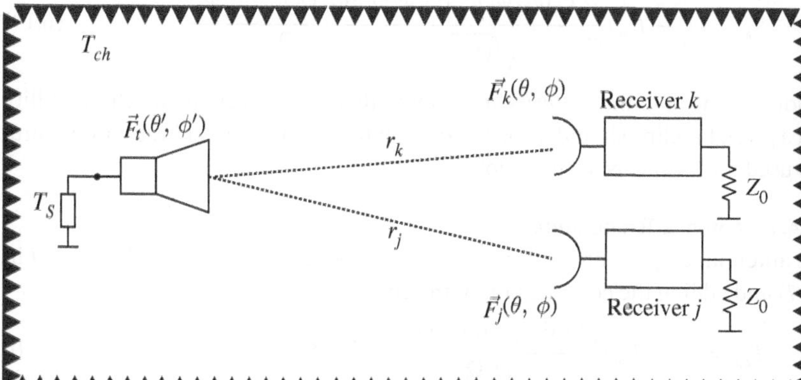

Figure 6.3 A probe antenna as an external calibration standard for interferometric measurements.

be included as it is part of the chamber visibility. The visibility (6.16) becomes then

$$V_{kj} = (T_s - T_{ch})\frac{\lambda^2}{r_k r_j}\frac{F_k(\theta_k,\phi_k)F_j^*(\theta_j,\phi_j)}{\sqrt{\Omega_k\Omega_j}}\frac{F_t(\theta_k',\phi_k')F_t^*(\theta_j',\phi_j')}{\Omega_t}c_k c_j^*$$

$$\ldots\tilde{r}_{kj}\left(\frac{-(r_k-r_j)}{c}\right)e^{-jk_0(r_k-r_j)} \tag{6.17}$$

where c_k and c_j are, respectively, the non-Hermitian dot products (A.2) between the corresponding polarization vectors (1.235) of the receiving antennas and that of the transmitting antenna

$$\begin{aligned} c_k &= \hat{e}_{ak}(\theta_k,\phi_k) \odot \hat{e}_{at}(\theta_k',\phi_k') \\ c_j &= \hat{e}_{aj}(\theta_j,\phi_j) \odot \hat{e}_{at}(\theta_k',\phi_k') \end{aligned} \tag{6.18}$$

in which the angular coordinates without prima (θ,ϕ) refer to the receiving antennas positions and those primed (θ',ϕ') to the transmitting antenna. Other symbols in (6.17) are defined in the context of the visibility equation in chapter 3. To simplify, perfect matching has been assumed by making $C = 1$.

The equivalent noise temperature of the source T_s in (6.17) must account for the transmitting antenna loss using the model of section 1.6.11. For example, if the source is implemented as a resistor at temperature T_r followed by an amplifier of available gain G_A, then $T_s = T_r G_A \eta_a$ where η_a is the antenna ohmic efficiency.

Equation (6.17) is greatly simplified if the probe is centered with respect to the interferometer so that $r_k = r_j = r$. Moreover, assuming that the polarization of the probe matches those of the receiving antennas, the polarization mismatch coefficients (6.18) become $c_k = c_j = 1$ and a further simplification is applicable. In this case

$$V_{kj} = (T_s - T_{ch})\frac{\lambda^2}{r^2}\frac{F_k(\theta_k,\phi_k)F_j^*(\theta_j,\phi_j)}{\sqrt{\Omega_k\Omega_j}}\frac{F_t(\theta_k',\phi_k')F_t^*(\theta_j',\phi_j')}{\Omega_t} \tag{6.19}$$

As long as the antenna patterns are known, this equation estimates the visibility that a given baseline should measure and can thus be used as calibration standard. It is used for this purpose in section 6.2.2.

6.1.4.1 Antenna Temperature

The antenna temperature of each receiver can be written by just making $k = j$ in (6.17) and adding the contribution of the chamber,

$$T_{ak} = (T_s - T_{ch})\frac{\lambda^2}{r_k^2}\frac{t(\theta_k,\phi_k)t_t(\theta_k',\phi_k')}{\Omega_k\Omega_t}|c_k|^2 + T_{ch} \tag{6.20}$$

which is fully compatible with (1.273) taking into account that $|c_k|^2$ is the polarization mismatch coefficient C_p. Again, this equation may serve as an external calibration standard for total power measurements provided, the antennas are well characterized.

6.1.5 Internal Load

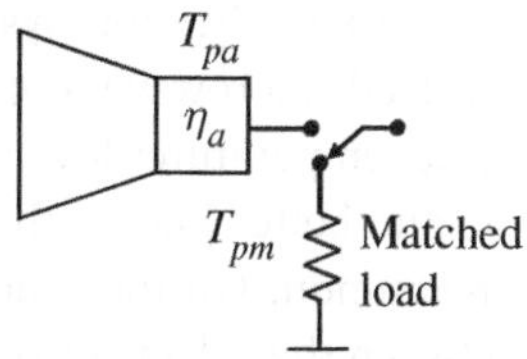

Figure 6.4 Matched load as internal calibration standard. Its equivalent noise temperature is always equal to the physical temperature T_{p_m}.

The most simple internal calibration standard is just a matched load (figure 6.4). Since the noise equivalent temperature of a resistor is, by definition, equal to its physical temperature, a thermal sensor suffices to characterize it. This is independent of the actual impedance of the load, but matching is needed to account for the mismatch error discussed in the remark of section 6.1.2

In general, the plane change (6.8) must be applied to transfer the noise temperature of the load to the antenna plane. In the special case of a load at ambient temperature, if additionally, all front-end temperatures are equal, switching to the load is equivalent to inserting the whole antenna into an anechoic chamber at the same temperature, as in figure 2.6. In both cases, the equivalent noise temperature at the antenna output port is equal to the common physical temperature, as predicted by (2.52). This is true even if the antenna is lossy and it is consistent with (6.9) when all temperatures coincide, as already pointed out in the text following this equation.

This is a very interesting result since it implies that the equivalent temperature of the calibration standard is insensitive to front-end characterization errors. Even having slightly different temperatures and using the full equation (6.8) the possible uncertainty in the knowledge of the different losses is multiplied by the temperature difference, presumingly very small. In the context of a calibration procedure with several calibration standards, this option allows having at least one of them with virtually no uncertainty.

A matched load at a different temperature than ambient can be implemented by electric heaters, peltier devices or cryogenic setups. In any case, a thermal sensor is required to determine its physical temperature.

6.1.5.1 Noise Diodes and Active Cold Loads

Another commonly used internal calibration standard is a noise diode with a known Excess Noise Ratio defined as

$$\mathrm{ENR} = 10\ \log\left(1 + \frac{T_s}{T_0}\right) \tag{6.21}$$

where T_s is the equivalent noise temperature of the diode and $T_0 = 290\,K$ is the reference temperature. The noise temperature of a diode is, generally, much larger than the antenna temperature the radiometer is designed to measure. Therefore, the dynamic range of the receiver must be extended to account for this higher input temperature during calibration. System specifications must ensure linearity for both nominal and calibration operations.

Alternatively, some systems use "active cold loads" which are devices based on reversed microwave amplifiers producing, under certain conditions, equivalent noise temperatures lower than ambient [6]. In general, noise diodes are more stable and better known than active cold loads, so the use of these last ones is still uncommon. On the other hand, observed antenna temperature usually lays in between that of an active cold load and ambient, so the dynamic range needs not to be extended.

6.1.6 Noise Distribution

Correlated signals for calibration are generated by a power splitter driven by a common source. This is illustrated in figure 6.5, which shows the same two receivers of figure 3.5 (section 3.2.5) but with antennas replaced by a noise source and a power divider in order to inject correlated noise into both of them. The noise distribution network is kept at a homogeneous temperature T_n and is characterized by a three-port S-parameter matrix $[S]$ (see section 1.5.4). The source is a noise diode with an equivalent temperature T_s and is connected to port number 0 in order to match the other port numbers with those of the receivers. The intrinsic noise waves c_0, c_1, and c_2 are function of the S-parameters and physical temperature through the Bosma theorem (section 1.5.10).

Waves b_{n1} and b_{n2} are defined *exactly* as b_{a1} and b_{a2} in figure 3.5. They are the *intrinsic* noise waves of the complete two-port network that includes both the noise source and the power splitter. From the definition of intrinsic waves (1.189), they can thus be computed by replacing both receivers with matched loads at absolute zero temperature. Then

$$b_{n1} = a_0 S_{10} + c_1 \ ; \ b_{n2} = a_0 S_{20} + c_2 \tag{6.22}$$

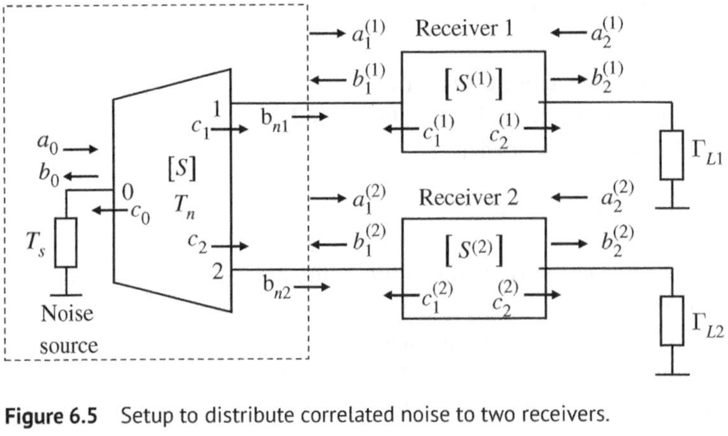

Figure 6.5 Setup to distribute correlated noise to two receivers.

where a_0 is the total inward wave at port 0, function of the reflection coefficients and intrinsic waves in the interconnection. Specifically, it is equal to the expression for b in (1.196) with the adequate identification of terms. To simplify the analysis, the source will be assumed perfectly matched to the reference impedance, so a_0 equals the source intrinsic wave and so it has power equal to $\overline{|a_0|^2} = kT_s$. Furthermore, in this case, this wave is uncorrelated with others in the system. With this assumption,

$$\begin{aligned} \overline{|b_{ni}|^2} &= k|S_{i0}|^2 T_s + kT_n \left(1 - \sum_{l=0}^{2} |S_{il}|^2\right) \quad (i = 1, 2) \\ \overline{b_{n1} b_{n2}^*} &= kS_{10} S_{20}^* (T_s - T_n) - kT_n \sum_{l=1}^{2} S_{1l} S_{2l}^* \end{aligned} \tag{6.23}$$

where the expanded form of the Bosma theorem (1.234) has been used to write the power and correlation products of c_1 and c_2.

The setup of figure 6.5 can be analyzed with the same methodology followed in section (3.2.5). Under the approximations (3.39), equation (3.40) hold just by replacing b_{a1} and b_{a2} by b_{n1} and b_{n2}, respectively. This is equivalent to substituting in (3.41) and (3.44) the antenna temperature T_{a1} and the visibility $\mathcal{V}_{12}$ by the corresponding expressions that result of introducing (6.23) into (3.31) and (3.32) with the replacements of intrinsic waves. The procedure is straightforward and the reader would be willing to work out the details. The result for (3.44) is a "system visibility" $\mathbb{V}_{12}$ that in this case, since there is no external target, will be referred to as *correlation temperature* and designated by T_{12}.

Instead of writing the expression for T_{12}, the formulation is extended to N receivers connected to a single source through an $(N + 1)$-port noise distribution network. The method outlined in the previous paragraph results in a correlation temperature for a receiver pair kj

$$T_{kj} = \frac{S_{k0} S_{j0}^* (T_s - T_n) + \sum_{l=1}^{N} S_{kl} S_{jl}^* (\mathcal{T}_1^{(l)} - T_n)}{\sqrt{1 - |S_{kk}|^2} \sqrt{1 - |S_{jj}|^2}} \tag{6.24}$$

where $\mathcal{T}_1^{(l)}$ is the definition of (1.213) for receiver l. It represents the backward noise of this receiver and, in practice, can be replaced by its physical temperature T_{rl} if the receiver includes an input isolator.

The temperature of the distribution network T_n is expected to be similar to that of the receivers, so the second term in the numerator can often be neglected. This is even more justified if the splitter has isolated outputs so that the S-parameters relating them vanish $S_{kl} = S_{jl} = 0$. It is a reasonable assumption if it is implemented by an N-way Wilkinson divider or if the different branches include long cables with fairly high loss. If, moreover, the splitter ports are matched

($S_{kk} = S_{jj} = 0$), then the correlation temperature can be approximated by

$$T_{kj} \approx S_{k0} S_{j0}^* (T_s - T_n) \tag{6.25}$$

which vanishes whenever the temperature of the noise source becomes equal to that of the power splitter. To synthesize a relatively high value of T_{kj}, the equivalent temperature of the source T_s should then be much larger than the physical temperature of the distribution network T_n. If the source is very strong so that $T_s \gg T_n$ then $T_{kj} \approx S_{k0} S_{j0}^* T_s$ which is a rather intuitive expression.

The same analysis allows writing the system temperature in (3.41) after replacing the antenna temperature as indicated before, that is taking into account (6.23). For receiver number k, the result is

$$T_{sysk} = \frac{|S_{k0}|^2 T_s + T_n(1 - |S_{k0}|^2) + \sum_{l=1}^{N} (\mathcal{T}_1^{(l)} - T_n)|S_{kl}|^2 + \mathcal{T}_2^{(k)}}{1 - |S_{kk}|^2} \tag{6.26}$$

where $\mathcal{T}_2^{(k)}$ is the receiver noise temperature of receiver k for matched input, as defined in (1.213). To check this equation the reader must be aware that approximation (3.39) is being applied, so $\mathcal{T}_{12}^{(k)} = 0$ and $\Lambda_k = 1$.

Four different contributions to the total noise are easily identified in the above expression: source, network, cross-coupling, and receiver. With the same approximations leading to equation (6.25), the system temperature becomes

$$T_{sysk} \approx |S_{k0}|^2 T_s + T_n(1 - |S_{k0}|^2) + T_{Rk} \tag{6.27}$$

where now $\mathcal{T}_2^{(k)} = T_{Rk}$, the receiver noise temperature. This equation is consistent with the intuitive addition of attenuated source power, contribution of the network expressed as in (1.204) and receiver noise.

The results presented in this section were, in a slightly different format, initially derived in [7]. A distributed noise distribution network was implemented in the MIRAS instrument for phase and amplitude calibration [8].

6.2 Parameter Retrieval

The key measurement of an interferometric or polarimetric radiometer is the cross-correlation between the output signals of two receivers. Calibration is carried out by injecting correlated noise at the input port of both receivers and retrieving the correlator gain in amplitude and in phase. Additionally, any signal leakage between receivers produces a correlation offset that is estimated by injecting uncorrelated noise.

Interferometric radiometers always include at least one total power radiometer dedicated to measure the visibility at the origin, which is the antenna temperature.

Also, if normalized correlation is measured, total power radiometers are needed to measure the system temperature of each receiver. Total power radiometer calibration is thus a fundamental part of the interferometer calibration and therefore is dealt with also in this section.

6.2.1 Correlator Gain

The correlator gain at calibration plane can be calibrated using two levels ("hot" and "warm") of noise injection using the setup of figure 6.5. Subtracting the correlation measurement (3.82) for these two levels taking into account that the visibility in this case is equal to the correlation temperature (6.24),

$$\Gamma_{kj}^{H} - \Gamma_{kj}^{W} = G_{kj}^{C} \frac{S_{k0} S_{j0}^{*}}{\sqrt{1 - |S_{kk}|^2} \sqrt{1 - |S_{jj}|^2}} (T_s^H - T_s^W) \tag{6.28}$$

where the superscripts H and W refer to hot and warm, respectively. Note that all terms of (6.24) not related to the noise source are canceled. The correlator gain G_{kj}^{C} can be directly solved from this equation provided the S parameters of the distribution network and the difference between hot and warm noise source temperatures are known.

A perfectly valid option for the warm noise level is ambient temperature, which means connecting a matched resistor in port 0 of the distribution network (see figure 6.5) with no heating or cooling. The warm noise temperature is then directly measured by a thermal sensor. The hot level can be implemented by a noise diode of known ENR so that the equivalent temperature is (6.21). Switching between both levels can be done just by turning on and off the bias of the noise diode assuming that in off status it is a matched resistor. Another option is inserting a switch to select either the biased noise diode or the resistor.

If the warm level is close to the noise injection network physical temperature and also to those of the receivers, then (6.24) predicts very low correlation temperature, so Γ_{kj}^{W} is also very small, probably negligible with respect to Γ_{kj}^{H}. Also, if a noise diode with high ENR is used, then T_s^W would be negligible with respect to T_s^H. With these approximations,

$$G_{kj}^{C} \approx \frac{\Gamma_{kj}^{H}}{T_s^H} \frac{\sqrt{1 - |S_{kk}|^2} \sqrt{1 - |S_{jj}|^2}}{S_{k0} S_{j0}^{*}} \tag{6.29}$$

Nevertheless, even in this situation it may be safer to use the whole equation (6.28) with no approximations. This helps reducing calibration errors due to temperature differences across the whole calibration system.

If the difference $T_s^H - T_s^W$ is unknown, it can be estimated with a well calibrated total power radiometer connected to one specific branch of the noise distribution network. The detected output voltage of this *reference radiometer* is (2.72) with the system temperature (6.26). Therefore, if it is located at port number i and two levels of noise are injected,

$$\mathrm{v}^H - \mathrm{v}^W = G\frac{|S_{i0}|^2}{1-|S_{ii}|^2}(T_s^H - T_s^W) \tag{6.30}$$

so that the difference in noise source temperature $(T_s^H - T_s^W)$ can be solved if the reference radiometer gain G and the network S-parameters are known by other means. It is important to point out that the reference radiometer must be calibrated independently using other known calibration standards.

This approach is implemented in MIRAS [4] using a Noise Injection Radiometer (NIR) as reference [9]. However (6.30) is only applied to calibrate the power measurement systems of individual receivers since normalized correlation is measured.

6.2.1.1 Normalized Correlation

The uncertainties in the knowledge of the reference radiometer gain and the S-parameter amplitude are not present when measuring normalized correlation. If both receivers in the baseline have associated individual total power radiometers, combining (3.136) and (2.72), the following equation follows

$$\mathcal{M}_{kj} = M_{kj}\sqrt{(\mathrm{v}_k - \mathrm{v}_{ok})(\mathrm{v}_j - \mathrm{v}_{oj})} = \sqrt{G_k G_j} G_{kj}^C T_{kj} \tag{6.31}$$

where the visibility has been replaced by the correlation temperature (6.24) and G_{kj}^C is now just the fringe washing function at the origin in accordance to (3.137). The symbol $\mathcal{M}_{kj}$ used in the left-hand side is self-defined and used to simplify expressions. It can be interpreted as *denormalized* M_{kj} with respect of the voltages measured in the associated total power radiometers.

On the other hand, if (6.30) is applied to both receivers, the geometric mean of the measurement of both receivers is

$$\begin{aligned}&\sqrt{(\mathrm{v}_k^H - \mathrm{v}_k^W)(\mathrm{v}_j^H - \mathrm{v}_j^W)}\\ &\quad = \sqrt{G_k G_j}\frac{|S_{k0}||S_{j0}|}{\sqrt{1-|S_{kk}|^2}\sqrt{1-|S_{jj}|^2}}(T_s^H - T_s^W)\end{aligned} \tag{6.32}$$

Now, subtracting both noise injection levels of (6.31) and using the above result together with (6.24), the correlator gain can be solved,

$$G_{kj}^C = \frac{\mathcal{M}_{kj}^H - \mathcal{M}_{kj}^W}{\sqrt{(\mathrm{v}_k^H - \mathrm{v}_k^W)(\mathrm{v}_j^H - \mathrm{v}_j^W)}}\frac{|S_{k0}||S_{j0}|}{S_{k0}S_{j0}^*} \tag{6.33}$$

This is a very robust equation since there is only one parameter to be known a priori from instrument characterization, namely the relative phase between the transmission S-parameters. The rest of the terms are direct measurements, both normalized correlation and detected voltages, and the instrumental offset independently characterized. It is important to recall that the retrieved G^C_{kj} is the fringe washing function at the origin $\tilde{r}_{kj}(0)$ both in amplitude and in phase. The above equation was originally derived in [10] and is currently implemented in SMOS processing.

6.2.2 Inter-element Phase and Amplitude

Calibration of the correlator gain using noise injection fails to characterize all elements situated between antenna and calibration planes. In section 6.1.1 (figure 6.1), these subsystems are described by two parameters, a gain and a time delay or phase, which have to be calibrated using necessarily an external target. A good option is using a probe antenna inside an anechoic chamber, described in section 6.1.4. Since there is no need to make things complicated, the preferred location for the probe antenna is centered between both interferometer antennas, pointing the boresight to them and with aligned polarization. In these conditions, the expected visibility is (6.19) where all parameters are assumed known. In particular, the antenna patterns must be measured in advance in a suitable facility.

With the probe antenna activated, the visibility at calibration plane V^C_{kj} is obtained through (6.3) with G^C_{kj} retrieved as explained in section 6.2.1. Then, from (6.4), the following equation can be written for the phases

$$\phi_{kj} = \arg\left(\frac{V^C_{kj}}{V_{kj}}\right) = \phi_k - \phi_j \tag{6.34}$$

where ϕ_k and ϕ_j are the phases defined in (6.2) and V_{kj} is the visibility at antenna plane, estimated in (6.19). Therefore, from the left-hand part of the above equation,

$$\phi_{kj} = \arg(V^C_{kj}) - (\alpha_{tk} - \alpha_{tj}) - (\alpha_k - \alpha_j) \tag{6.35}$$

in which the α's are the phases of the corresponding antenna patterns at the directions stated in (6.19). The subscript t indicates transmitting antenna. By definition (see section 1.6.2), they are all equal to zero at boresight but have some evolution at different spatial directions. In any case, these pattern phases are assumed to be known from the independent antenna characterization. The phase of the internally calibrated visibility $\arg(V^C_{kj})$ is also known, so ϕ_{kj} is determined from (6.35).

From the right-hand side of (6.34) applied to all baselines, the following linear system of equations can be formed

$$\begin{bmatrix} \phi_{12} \\ \phi_{13} \\ \vdots \\ \vdots \\ \phi_{n-1\,n} \end{bmatrix} = \begin{bmatrix} 1 & -1 & 0 & 0 & \cdots & 0 & 0 & 0 \\ 1 & 0 & -1 & 0 & \cdots & 0 & 0 & 0 \\ \vdots & \vdots & \vdots & \vdots & \ddots & \vdots & \vdots & \vdots \\ 0 & 0 & 0 & 0 & \cdots & 1 & 0 & -1 \\ 0 & 0 & 0 & 0 & \cdots & 0 & 1 & -1 \end{bmatrix} \begin{bmatrix} \phi_1 \\ \phi_2 \\ \vdots \\ \phi_n \end{bmatrix} \tag{6.36}$$

from which the individual phases ϕ_k are retrieved by an appropriate inversion algorithm. The above matrix has $n(n-1)/2$ rows (one for each baseline) and n columns where n is the total number of antennas. Nevertheless, the system of equations can be reduced so as to include only the minimum number of rows needed to have all receiver phases represented at least once; for example, including the $n-1$ pairs formed by the first antenna and each one of the rest. In this case, assigning zero phase to the first antenna, those of the others are immediately found. In general, the system is well conditioned and is efficiently solved by the Moore–Penrose pseudo-inverse algorithm. In any case, a constant added to all phases of a given solution is also a solution.

A similar equation can be written from (6.4) for the separable amplitudes, but in this case logarithms must be used to convert the multiplication to a sum:

$$2\log\left(\frac{|V_{kj}^C|}{|V_{kj}|}\right) = \log\eta_k + \log\eta_j \tag{6.37}$$

The amplitude of the expected visibility $|V_{kj}|$ is also determined through (6.19) so the left-hand side of the above equation is known for all baselines. Then a system of equations similar to (6.36) can be formed and solved for all η_i. In this case the matrix is the same but replacing the -1 by $+1$. Again, a Moore–Penrose inverse provides an efficient method to solve it.

Nevertheless, applying this method to get the inter-element amplitudes is prone to errors due to uncertainties in the knowledge of $|V_{kj}|$. Improved results may be obtained from total power measurements using (6.5) after calibrating the system temperature with well known external and internal standards.

6.2.2.1 Normalized Correlations

If the instrument measures normalized correlation, only the inter-element phases have to be calibrated as the amplitudes η_k and η_j are obtained through external calibration of the system temperatures. Combining (6.5) and (6.7) the visibility becomes

$$V_{kj} = \frac{M_{kj}\sqrt{T_{sysk}T_{sysj}}}{G_{kj}^C e^{j(\phi_k-\phi_j)}} \tag{6.38}$$

where the phases are to be determined. They are retrieved using the same procedure explained before, particularly by solving the system of equations (6.36) with $\arg(V_{kj}^C)$ in (6.35) substituted by the phase of M_{kj}/G_{kj}^C. This approach was successfully applied for on-ground SMOS payload characterization [11].

6.2.3 Correlator Offset

Any signal path existing between the two receivers of a baseline, for example, local oscillator or calibration signal distribution circuitry, can be responsible of generating a nonzero correlation through a common signal arriving to both. In this eventuality, the total output signal of a given receiver becomes the sum of two terms, the nominal one coming from the antenna and other(s) generated internally in the system. The cross-correlation (3.82) between the total analytic signals out of two receivers k and j is then

$$\Gamma_{kj} = \frac{1}{2}\left\langle (b_k(t) + c_k(t))(b_j(t) + c_j(t))^* \right\rangle = \Gamma_{kj}^A + \Gamma_{kj}^U \tag{6.39}$$

where $c_k(t)$ and $c_j(t)$ are the internally leaked signals, assumed uncorrelated with the ones coming from the antenna $b_k(t)$ and $b_j(t)$. The term $\Gamma_{kj}^U = (1/2)\left\langle c_k(t)c_j^*(t)\right\rangle$ can be estimated by injecting uncorrelated noise at antenna ports, so forcing $\Gamma_{kj}^A = 0$.

Uncorrelated signals are generated for this purpose by individual resistors, typically at ambient temperature, located at each receiver input port. In this situation the intrinsic waves b_{n1} and b_{n2} in figure 6.5 are uncorrelated, hence also the output signals are uncorrelated. The measurement in this condition is only the "U-noise correlation" Γ_{kj}^U, which depends only on internal signals and so it is independent of the visibility.

Once Γ_{kj}^U has been measured, it must be subtracted from all correlation measurements to remove the offset. The visibility is then retrieved using (3.82) taking into account (6.39)

$$V_{kj} = \frac{\Gamma_{kj} - \Gamma_{kj}^U}{G_{kj}} \tag{6.40}$$

For this correction to be effective, the uncertainty in the measurement of Γ_{kj}^U must be much lower than that of Γ_{kj}. Otherwise, the standard deviation of the visibility is affected and in consequence the radiometric sensitivity is degraded. In practice the integration time used for measuring uncorrelated noise must be roughly an order of magnitude larger than the one used in antenna measurements.

Even though it can be corrected, the design of receivers' layout should be optimized for minimizing correlation offset. It can be reduced if the interfering signals are kept to a much lower amplitude than the nominal ones, so the effect is larger if the signals are coupled near the antenna, where the incoming waves are lower. According to this statement, if correlation offset is known to be produced in the local oscillator distribution of a heterodyne design, then the RF gain must be large enough to minimize correlation offset. In other words, from the point of view of reducing this error, given a total specified gain, it should be split in a higher value at RF and lower at IF.

6.2.3.1 Normalized Correlation

Introducing the sum of both antenna and interference signals into the normalized correlation (3.135), this one becomes

$$M_{kj} = \frac{\langle (b_k(t) + c_k(t))(b_j(t) + c_j(t))^* \rangle}{\sqrt{\langle |b_k(t) + c_k(t)|^2 \rangle \langle |b_j(t) + c_j(t)|^2 \rangle}}$$
$$= \frac{\left\langle b_k(t)b_j^*(t) \right\rangle + \left\langle c_k(t)c_j^*(t) \right\rangle}{\sqrt{(\langle |b_k(t)|^2 \rangle + \langle |c_k(t)|^2 \rangle)(\langle |b_j(t)|^2 \rangle + \langle |c_j(t)|^2 \rangle)}} \tag{6.41}$$

In contrast with (6.39), this equation cannot be written as the sum of the *nominal* normalized correlation plus an offset independent of antenna signals, as these ones appear in the common denominator. However, denormalizing this expression using (3.136), the result can indeed be expressed as the sum of two terms, each one depending on different signal pair correlations:

$$\frac{M_{kj}\sqrt{T_{sysk}T_{sysj}}}{G_{kj}} = V_{kj} + V_{kj}^U \tag{6.42}$$

where V_{kj} is the visibility (3.57) and V_{kj}^U the equivalent one for U-noise injection, defined with the same expression (3.57) but replacing the nominal signals $b_{k,j}(t)$ by the interfering ones $c_{k,j}(t)$. Note that, even though they are originated internally somewhere in the system, the parameters involved in the expression of V_{kj}^U derived from (3.57) –namely gains, bandwidths and fringe washing function– correspond to the *full receiver chains*. The system temperatures appearing in the above expression represent the total noise of the system, including the signals collected by the antennas, the noise generated by the receivers and the leaked signals.

When uncorrelated noise is injected, the visibility V_{kj} vanishes and the system temperatures include the noise generated by the input resistors instead of that of the signals coming from the antennas. Therefore, the "U-noise visibility" is

$$V_{kj}^U = \frac{M_{kj}^U\sqrt{T_{sysk}^U T_{sysj}^U}}{G_{kj}} \tag{6.43}$$

where M_{kj}^U, T_{sysk}^U, and T_{sysj}^U are, respectively, the normalized correlation and system temperatures measured while injecting uncorrelated noise.

Once the equivalent visibility for U-noise injection is obtained, the corrected visibility is retrieved from (6.42):

$$V_{kj} = \frac{M_{kj}\sqrt{T_{sysk}T_{sysj}}}{G_{kj}} - V_{kj}^{U} \tag{6.44}$$

The same comments about the uncertainty in the measurement of Γ_{kj}^{U} stated after equation (6.40) apply to V_{kj}^{U}. U-noise measurements must be performed with large integration time in order to avoid degrading the radiometric sensitivity.

6.2.4 Flat Target Response

Cross-coupling of noise generated by one receiver into the other through the antennas within a baseline is analyzed in section 3.2.5. Under some suitable approximations, section 3.2.6 shows that it can be accounted for by the "$-T_r$" term. For any polarimetric combination, equation (3.49) applied to the measurable system visibility [see equation (3.58)] states that, for a baseline formed by receivers k and j, this one can be written as

$$V_{kj} = V'_{kj} - T_r FTR_{kj} \tag{6.45}$$

where T_r is the physical temperature of the receivers, V'_{kj} is the visibility without the "$-T_r$" term and FTR_{kj} is the Flat Target response (3.50). Superscripts denoting polarization are not included for simplicity.

The flat target response can be estimated from the actual measurement of the visibility of a flat target. If the brightness temperature $T_B(\theta,\phi)$ is replaced by that of a flat target (2.51) $T_B(\theta,\phi) = T_{FT}$, constant by definition, then, directly from (3.12) $V'_{kj} = T_{FT}\,FTR_{kj}$. The flat target response can thus be retrieved as

$$FTR_{kj} = \frac{V_{kj}(FT)}{T_{FT} - T_r(FT)} \tag{6.46}$$

where $V_{kj}(FT)$ is the visibility measured when the instrument is surrounded by the flat target and $T_r(FT)$ the receiver's physical temperature during this measurement. Introducing now this equation into (6.45) the *corrected* visibility V'_{kj} for any arbitrary situation is readily solved

$$V'_{kj} = V_{kj} + \frac{T_r}{T_{FT} - T_r(FT)} V_{kj}(FT) \tag{6.47}$$

This visibility is the one used in the inversion algorithms of section 4.4. As stated, its relation to the scene brightness temperature is (3.77) but without the "$-T_r$" term, or the polarimetric versions derived from (3.25).

In principle, different options can be chosen for the flat target. A possible one is an anechoic chamber at a known temperature. However, this is prone to large errors since T_{FT} is then similar to $T_r(FT)$ and the denominator of (6.47) tends to vanish while $V_{kj}(FT)$ is also close to zero.

The preferred choice is the sky, implemented by pointing the whole instrument to a galactic pole. The existence of some stars should be insignificant, so for this attitude, the brightness temperature can be considered constant in the front hemisphere. Then T_{FT} is equal to the microwave cosmic background, at 2.726 K, plus other galactic contributions that are responsible of additional thermal emission. For example, the brightness temperature of the sky at L-band is plotted in figure 6.6. It was derived by N. Floury (unpublished) for the SMOS mission based on data taken from available radioastronomic measurements [12]. The black line shows the unit circle defined after equation (3.77) for an instrument pointing to the north Galactic pole. It marks the border between the front and back hemispheres. Although there is some influence from the Galaxy, most of the incoming radiation corresponds to a fairly constant brightness temperature. The median in the whole front hemisphere is estimated from this figure as $T_{FT} = 3.52$ K. This is the value that should be introduced into (6.47).

The visibility correction just described is systematically applied to all brightness temperature measurements of the SMOS satellite, where the flat target response is derived from observations of the sky spanning about 15 minutes.

In the back hemisphere, the presence of the Earth makes the brightness temperature depart from a flat target. But, if the antenna back lobes are small,

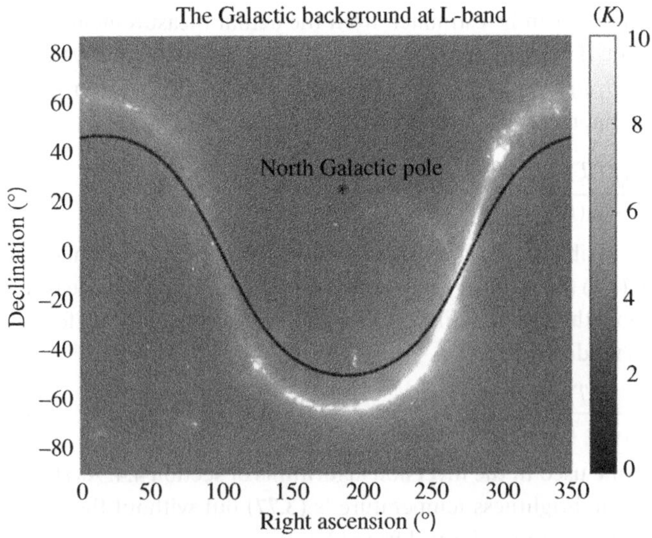

Figure 6.6 Brightness temperature of the sky. The black line delimits the front and back hemispheres of an instrument pointing to the north Galactic pole.

this contribution may be negligible, although this premise should be checked by simulation or experimentally.

6.2.5 Fringe Washing Function Shape

The term "fringe washing function shape" refers to the ratio of the fringe washing function and its value at the origin (3.56). It appears inside the integral of the visibility equation (3.77) and in consequence forms part of the G-matrix elements (4.59) in the visibility inversion algorithm.

This term is important when imaging large sources for which at some spatial directions (ξ, η), its value inside the integral may have quite low amplitude, especially for large instruments. Also, if the receivers' center frequencies have relative large dispersion, in application of property 3 (see section 3.3.2), the phase of the normalized fringe washing function may have a significant excursion in the range of times of interest.

There are essentially two ways of estimating the fringe washing function shape. The first one is measuring the receivers' frequency responses with a vector network analyzer and using (3.61) to compute numerically the fringe washing function of any receiver pair. The frequency responses must include all components of the receiver chains from antenna to correlator, particularly the RF and IF sections, in case they use frequency conversion. The definition of frequency response of (5.1) must be used in this case.

Figure 6.7 shows an example of fringe washing function computed from two measured frequency responses taken from the MIRAS instrument, similar to the one used in figure 2.10. As seen, the amplitude corresponds very well to a sinc

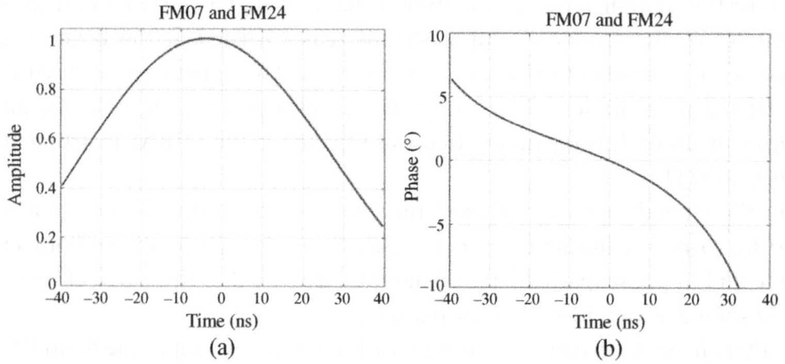

Figure 6.7 Normalized fringe washing function computed from two measured frequency responses. Filters are flight models of MIRAS with references indicated in the titles. (a) Amplitude and (b) phase.

function, somewhat displaced, and the phase is quite linear for small time span around the origin, although getting nonlinear for higher times.

The second method for retrieving the fringe washing function shape consists of measuring the correlator gain G_{kj} through noise injection (section 6.2.1)[1] while inserting appropriate time delays in the receivers' chains. It is based on (3.83), which relates the correlator gain to the fringe washing function at the origin. When a time delay is included in one of the two receivers, property 4 of the fringe washing function applies. Given thus, a baseline formed by receivers k and j, if a delay t_d is inserted either into one or the other, the corresponding correlator gains become

$$\begin{aligned}\text{If receiver } k \text{ is delayed: } & G_{kj}^{-} = k\sqrt{B_k B_j}\sqrt{G_k G_j}\,\tilde{r}_{kj}(-t_d)e^{-j2\pi f_0 t_d} \\ \text{If receiver } j \text{ is delayed: } & G_{kj}^{+} = k\sqrt{B_k B_j}\sqrt{G_k G_j}\,\tilde{r}_{kj}(t_d)e^{j2\pi f_0 t_d}\end{aligned} \tag{6.48}$$

from which the fringe washing function shape is easily derived

$$\bar{\tilde{r}}_{kj}(-t_d) = \frac{G_{kj}^{-} e^{j2\pi f_0 t_d}}{G_{kj}}\ ;\ \bar{\tilde{r}}_{kj}(t_d) = \frac{G_{kj}^{+} e^{-j2\pi f_0 t_d}}{G_{kj}} \tag{6.49}$$

In this method, all three parameters G_{kj}, G_{kj}^{-} and G_{kj}^{+} are measured with the procedure explained in section 6.2.1 in these three scenarios: zero delay, delay in receiver k and delay in receiver j, respectively. Additional delays can be inserted to get more points of the fringe washing function. Needless to say, the correlator gain for the normalized case (6.33) is also valid in (6.49).

The delay can be implemented either in RF or IF and the main difference between both options (besides hardware issues) is that the center frequency f_0 in the above expressions is the RF or the IF, respectively. In general, equation (5.2) applies.

An interesting option for implementing a delay in digital systems that produce samples of the receivers' output signals (case 3 listed at the beginning of section 5.4.1) consists of using previous samples. The delay implemented is then restricted to multiples of the sampling period, but this is not usually an important constraint. It is the option that was selected in the MIRAS instrument characterization [13].

Due to the smooth behavior of the fringe washing function, the two points provided by (6.49) should be enough for estimating the whole shape, both in amplitude and in phase, particularly taking into account that the range of times of interest should always be within the main lobe.

If the filters have sharp edges, the fringe washing function amplitude is similar to a sinc function (see figure 6.7). The phase, in turn, can be approximated by a

1 The superscript C is omitted for simplicity.

polynomial. Therefore, the following approximations are reasonable

$$\begin{aligned} |\bar{\tilde{r}}_{kj}(t)| &\approx A \text{ sinc } (B(t - C)) \\ \arg \bar{\tilde{r}}_{kj}(t) &\approx Dt^3 + Et \end{aligned} \tag{6.50}$$

where A through E are parameters retrieved by curve fitting of the measurements (6.49). Since $|\bar{\tilde{r}}_{kj}(0)| = 1$, the following relation must hold

$$A \text{ sinc } BC = 1 \tag{6.51}$$

so only two parameters are to be solved for the amplitude, consistent with having only two equations (6.49).

Besides the mathematical usage of these parameters for getting an estimation of the fringe washing function shape, they also have physical meaning. Assume that the two filters involved have ideal rectangular shape with the same bandwidth centered both at $f_c \neq f_0$ and nonzero time delay difference between both $\Delta t = t_k - t_j$. Combining properties 4 and 5 and 6 (section 3.3.2), the fringe washing function with respect to f_0 is

$$\tilde{r}_{kj}(t) = \text{sinc } (B(t - \Delta t))\, e^{-j2\pi f_c \Delta t} e^{j2\pi(f_c - f_0)t} \tag{6.52}$$

from which, after dividing by $\tilde{r}_{kj}(0)$, the following interpretation is apparent by comparing with (6.50)

- B in units of Hz is an estimation of the receivers bandwidth.
- C in units of seconds is an estimation of the difference in group delay between receivers $(t_k - t_j)$.
- $E/(2\pi)$ in units of Hz is an estimation of the center frequency error with respect to the reference frequency $E = 2\pi(f_c - f_0)$.

A simple interpretation for D cannot be found with this simple model. Note that the exponent of t associated to this term has been set equal to 3 in (6.50). This is consistent with the shape of the phase seen in figure 6.7, but other choices are possible depending on the actual behavior of the phase for the selected filters.

6.2.6 Receiver Gain and Offset

The basic equation that describes the operation of a total power radiometer is (2.72), which is now rewritten as

$$\text{v} = GT_a + \text{v}_0' \tag{6.53}$$

with $\text{v}_0' = GT_R + \text{v}_0$. From this expression, it is apparent that voltage and temperature are related by a simple linear relationship, so calibration is reduced to estimating two parameters: a slope and a bias. The general approach, sometimes referred to as "two-point calibration," consists of measuring two known standards

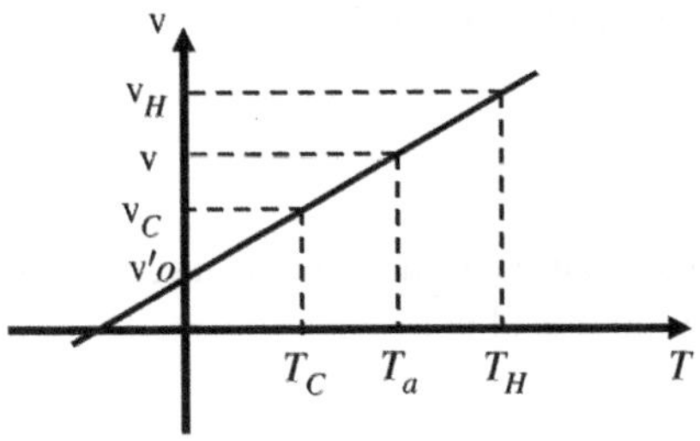

Figure 6.8 Linear input–output radiometer response to interpret calibration.

(often called "hot" and "cold") and solve for both unknowns using (6.53) along with the corresponding measurements. If T_H and T_C are the expected antenna temperatures of the standards, then (see drawing of figure 6.8)

$$G = \frac{v_H - v_C}{T_H - T_C} \qquad v'_o = \frac{T_H v_C - T_C v_H}{T_H - T_C} \tag{6.54}$$

where v_H and v_C are the corresponding output voltages. The method can be applied either with external or internal standards or even mixing both kinds. When internal loads are used, the remark about matching conditions of section 6.1.2 applies.

The key for an accurate calibration is the correct choice of the standards and also the quality of their characterization. Not surprisingly, any uncertainty in their expected antenna temperature is directly propagated to errors in the estimation of gain and offset and then in the final measurement. A simple analysis of the calibration equations (6.54) allows estimating the sensitivity of the measured antenna temperature with respect to T_C or T_H. Substituting (6.54) into (6.53) and solving for T_a provides the estimate of antenna temperature for a given set of calibration and nominal voltage measurements

$$\hat{T}_a = v\frac{T_H - T_C}{v_H - v_C} - \frac{T_H v_C - T_C v_H}{v_H - v_C} \tag{6.55}$$

from which the following sensitivities are found

$$\frac{\partial \hat{T}_a}{\partial T_C} = \frac{T_H - T_a}{T_H - T_C} \qquad \frac{\partial \hat{T}_a}{\partial T_H} = \frac{T_a - T_C}{T_H - T_C} \tag{6.56}$$

As expected, the uncertainty depends inversely on the difference between hot and cold temperatures so the larger the difference between the standards' temperatures, the better the accuracy of the measurement. For example, assume a radiometer measuring a scene of $T_a = 200$ K calibrated using a hot load at $T_H = 1400$ K and a cold load $T_C = 370$ K (values taken from SMOS [11]). An error in the estimation of the cold temperature of 1 K translates into a measurement error of −0.16 K while an uncertainty of also 1 K in the hot temperature produces 1.16 K error in the measurement. Calibration standards have thus to be previously characterized in a controlled environment before their use in radiometric measurements. Any unexpected or uncontrolled change in them induces necessarily a calibration error.

Although it is not always possible, in order to avoid nonlinearity issues due to extrapolation, the hot standard temperature should be preferably larger than the maximum expected antenna temperature and the cold standard lower (as drawn at the left of figure 6.8).

6.2.7 Instrumental Offset

The total offset in (6.53) is the sum of two components with very different origin: the receiver thermal noise and the instrumental offset defined in (2.71). In systems where predetection signals are directly converted to digital (figure 5.7) this offset may be negligible since detection is made by software. In analog detectors it depends on the postdetection amplifiers and often has an important impact on power measurements.

By introducing an additional attenuation just before detection and assuming that the receiver gain is large enough so that this attenuator does not influence the receiver noise figure, it is possible to measure separately the instrumental offset. The methodology, proposed in [14], is based on the setup sketched in figure 6.9 and is sometimes referred to as "four-point technique."

Depending on the position of the two switches, four different voltages are measured, from which the following system of equations is obtained:

$$\begin{aligned} v_1 &= G(T_C + T_R) + v_o \\ v_2 &= G(T_H + T_R) + v_o \\ v_3 &= G/L(T_C + T_R) + v_o \\ v_4 &= G/L(T_H + T_R) + v_o \end{aligned} \tag{6.57}$$

Its solution provides estimates for the four unknowns, namely G, T_R, L, and v_o. The one for the offset is

$$v_o = \frac{v_2 v_3 - v_1 v_4}{(v_2 - v_4) - (v_1 - v_3)} \tag{6.58}$$

which has the interesting feature that it is independent on the hot and cold temperatures injected, as it depends only on the measured voltages. It is then a very

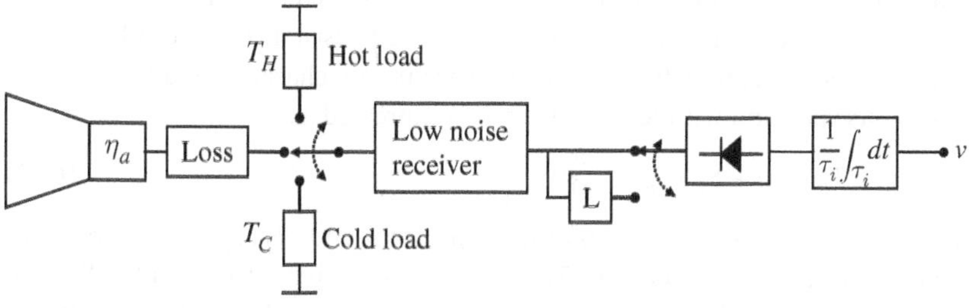

Figure 6.9 Setup for measuring the instrumental offset.

robust measurement that provides a very good estimation of the instrumental offset and only depends on the linearity of the detector. The amount of attenuation inserted L needs not be known since it is obtained as a side solution of the system.

$$L = \frac{v_2 - v_1}{v_4 - v_3} \tag{6.59}$$

Furthermore, both the gain and the receiver noise temperature can also be retrieved by solving (6.57)

$$G = \frac{v_2 - v_1}{T_H - T_C} \qquad T_R = \frac{T_C(v_2 - v_4) + T_H(v_3 - v_1)}{v_1 - v_2 - v_3 + v_4} \tag{6.60}$$

These retrievals depend on the hot and cold temperatures, so they are less robust than the ones for instrumental offset and attenuation. Note that the equation for the gain above is the same as (6.54) since $v_1 = v_C$ and $v_2 = v_H$.

In principle, this technique can be implemented either with internal or external calibration standards, but is usually done with internal ones. All comments about the accuracy of the calibration stated in the above sections still apply.

6.2.7.1 One-point Calibration

The independent knowledge of the instrumental offset together with the assumption that the receiver noise temperature is very stable, which is usually true in solid-state receivers, allows simplifying calibration by reducing it to just a gain retrieval [15].

If the receiver input port is switched to a matched load at ambient temperature (figure 6.4), according to (2.72) the measured output voltage becomes

$$v_m = G(T_{P_m} + T_R) + v_o \tag{6.61}$$

where T_{P_m} is the physical temperature of the matched load and v_m the voltage when the load is connected. The gain can thus be retrieved as

$$G = \frac{v_m - v_o}{T_{P_m} + T_R} \tag{6.62}$$

In this equation, the instrumental offset is assumed to be known, either null or through the four-point technique described before. One of the two loads (hot or cold) may be used in this case as the matched resistor. On the other hand, the receiver noise temperature T_R must be accurately characterized beforehand, including its variation with temperature. It is expressed as

$$T_R = T_R(T_{p0}) + S_{T_R}(T_p - T_{p0}) \tag{6.63}$$

where T_p is the physical temperature of the front-end, $T_R(T_{p0})$ is the receiver noise temperature at temperature T_{p0} and S_{T_R} a sensitivity coefficient. All three parameters, namely $T_R(T_{p0})$, S_{T_R} and T_{p0}, are measured in advance after a

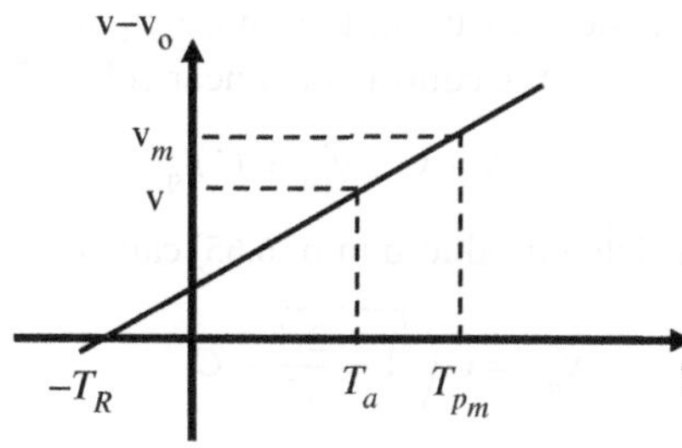

Figure 6.10 Graphical interpretation of the one-point method to retrieve the radiometric gain.

careful characterization of the receivers in a thermal chamber. Receiver noise temperature measurement is a highly standardized procedure that can provide low systematic uncertainties.

A graphical interpretation of the one-point calibration technique is provided in figure 6.10. As seen, it can be viewed as a two-point method with one of the calibration standards equal to the receiver noise temperature and the second one a matched load.

Any uncertainty on the receiver noise temperature, either due to unexpected drifts or poor characterization, is obviously translated into an error in the calibrated gain. From (6.62), assuming no error in the instrumental offset, the relative gain and noise temperature errors relate to each other by

$$\frac{\Delta G}{G} = -\frac{\Delta T_R}{T_R} \frac{T_R}{T_{p_m} + T_R} \tag{6.64}$$

so gain errors are always attenuated with respect to those of T_R since $T_R/(T_{p_m} + T_R) < 1$.

The remark of section 6.1.2 about different matching conditions in the two states of the switch is also relevant in this case. If needed, the receiver noise temperature used in (6.63) can be corrected to take into account this effect.

6.3 Nonlinearity

Calibration of a total power radiometer is implicitly based on the linear relationship stated in (2.72). In practice, depending on the particular design of the receiver and the power detector, a second-order term may be nonnegligible, so the response is better described by

$$\text{v} = aT_{sys}^2 + GT_{sys} + \text{v}_o \tag{6.65}$$

where v_o is the instrumental offset, T_{sys} is the system temperature, and a is a coefficient in units of V/K^2, ideally equal to zero.

Instead of using the above equation, nonlinearity is preferably described in terms of the equivalent "linear voltage" by defining

$$\mathrm{v}' = \mathrm{v} - \mathrm{v}_o \; ; \; \mathrm{v}'_{lin} = G\,T_{sys} \tag{6.66}$$

which, introduced into (6.65) can be solved to get

$$\mathrm{v}'_{lin} = C\sqrt{1 + \frac{2\mathrm{v}'}{C}} - C \tag{6.67}$$

where

$$C = \frac{G^2}{2a} \tag{6.68}$$

is a constant in units of Volts that characterizes the nonlinearity. The larger this constant, the more linear the system is. Once known, (6.67) provides the voltage of an equivalent linear system as a function of the actually measured voltage.

6.3.1 Deflection Ratio

A robust method to measure a radiometer's nonlinear behavior is schematized in figure 6.11. The radiometer is driven by a noise source followed by a stepped attenuator to generate a number of different input noise power levels. At each one, extra noise power is injected during a short period of time in order to measure the slope of the input–output characteristic curve. For a given input level labeled "i," the "deflection ratio" is defined as

$$D_i = \frac{\Delta \mathrm{v}_i}{\Delta \mathrm{v}_1} \tag{6.69}$$

where $\Delta \mathrm{v}_i$ is the voltage increment when the extra noise is injected in the corresponding level. Level number 1 corresponds to the lowest input power. For an ideal linear detector, this parameter is equal to one for all input levels ($D_i = 1\,\forall i$), but in real detectors, it may depart from unity at high input powers. With this approach,

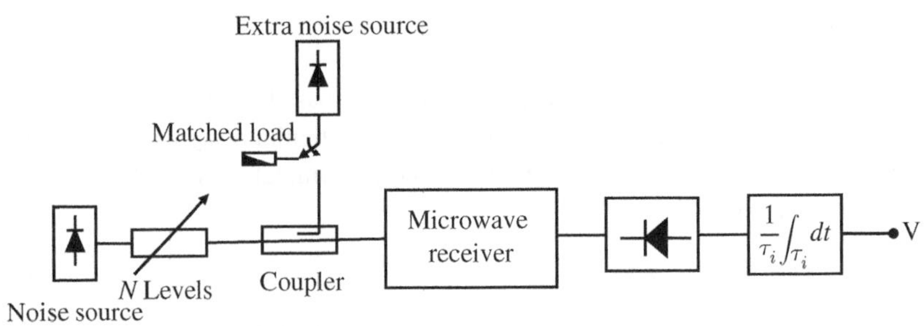

Figure 6.11 Setup for measuring the receiver's nonlinearity.

nonlinearities are observed as deviations of the extra noise deflection as the input noise temperature changes.

To estimate parameter C, the equivalent linear deflection ratio is first expressed as (6.69) but with the equivalent linear voltages (6.67)

$$D_{lin_i}(C) = \frac{\Delta v'_{lin_i}}{\Delta v'_{lin_1}} \tag{6.70}$$

so that, for the correct value of C, the deflection ratio should be 1 for all levels. Starting with a first guess of C, $D_{lin_i}(C)$ is computed for all levels and the following error function is defined

$$\varepsilon(C) = \frac{1}{N-1}\sum_{i=2}^{N}(D_{lin_i}(C) - 1)^2 \tag{6.71}$$

where N is the total number of noise levels. This function is then optimized by sweeping the value of C until a minimum of $\varepsilon(C)$ is obtained. The process should converge in relatively few iterations. An example of the application of this method for MIRAS can be found in [16, 17].

6.3.2 Impact on Instrumental Offset

Equation (6.67) applies only to voltages without instrumental offset. Otherwise it must be first subtracted from the raw detected voltage. On the other hand, the four-point method (6.58) is fundamentally based on a linear relationship between detected voltage and system temperature, so its application in this case results in some errors.

Figure 6.12 illustrates this issue using a simulated scenario. A known second order polynomial is selected to model the voltage to system temperature response

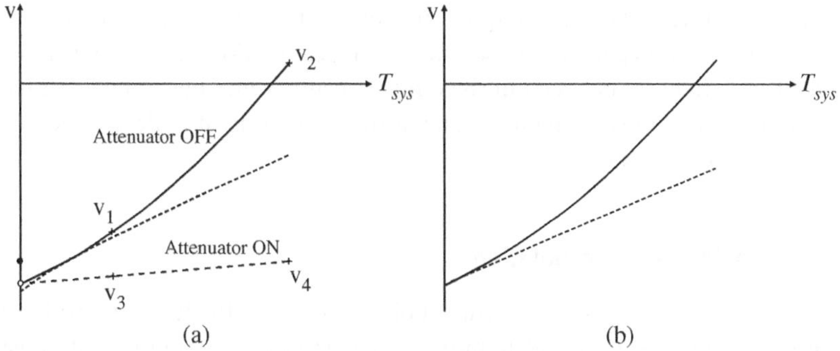

Figure 6.12 Example of the linearity correction with simulated data. (a) First offset subtraction and linearization. (b) Original and linearized voltages after a second iteration.

as (6.65) with reasonable values of gain (first-order term) and offset (zero order term) but with an exaggerated second-order term for illustration purposes. The nonlinear parameter C is then computed from (6.68).

The plot on figure 6.12(a) shows qualitatively the detected voltage when the attenuator needed in the four-point method is activated (dashed line) or not (solid line). The four-point voltages v_1 through v_4 are indicated on the lines. Due to the nonlinear behavior, use of these points in (6.58) fails to estimate the actual offset: the result is the black dot in the vertical axis, quite far away from the voltage at $T_{sys} = 0$.

The linearized voltage (6.67) after adding the offset just retrieved is shown in dash-dotted line in the same plot. As seen, it has a much more linear behavior, although with a slightly lower value at the origin with respect to the original voltage. A second application of the four-point method with these already linearized voltages results in a second offset estimate, shown in the figure as a circled dot. In this case, the solution is much closer to the actual offset. Finally, the right-hand side plot of figure 6.12 represents again the detected voltage (attenuator OFF) together with the result of linearizing (6.67) and adding the second estimate of the offset. As seen, the result is a highly linear voltage with an offset correctly estimated.

6.4 Calibration Rate

During a radiometer operation, calibration events are periodically launched to acquire measurements of the calibration standards and infer from them the corresponding parameters. The rate at which calibration operations are to be performed is an important aspect of designing radiometric systems. Parameters such as inter-element phase or flat target response may need only one calibration in the mission lifetime. Others may require frequent updates, for example correlator gain phase or receivers gain and offset. In the end, calibration rate is determined as a result of a trade-off between measurement accuracy and percentage of time dedicated to science measurements. General rules are difficult to derive, as they depend on the specific hardware.

6.4.1 Averaging and Interpolation

To reduce uncertainty due to thermal noise, several individual retrievals of calibration parameters, each one nominally in a single integration period, must be averaged before their usage in correcting science measurements. Since the uncertainty varies inversely with the square root of the number of measurement,

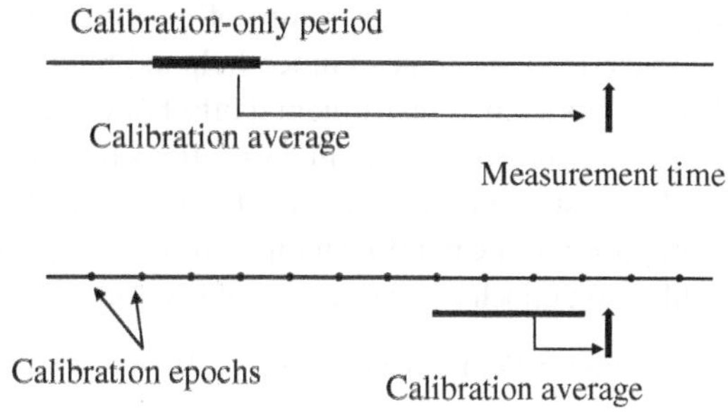

Figure 6.13 Two averaging strategies for calibration.

for a reduction of an order of magnitude with respect to science measurements, 100 individual calibration retrievals are needed.

Figure 6.13 shows two different strategies to manage calibration events. In the first one, a certain period of time is dedicated exclusively to calibrate, with no science measurements. The corresponding averaged parameters are then used in all subsequent science epochs. In the second case, very short calibration events are launched relatively often and a certain number of them are averaged and used later in measurement time. Both methods may have the same percentage of time dedicated to calibration although they have pros and cons. In the first approach, the instrument is more stable between individual calibrations so they are more consistent to each other when they are averaged. On the other hand, it may result in important gaps in science data. This is mitigated in the second approach at the expense of averaging more dissimilar individual calibrations. Again, no general rule can be derived about the best strategy as it depends on the application and hardware.

In any case, parameters retrieved during calibration events have to be interpolated to the measurement epochs. Several strategies are possible and the most simple one consists of just using the parameters that have been measured in the most recent calibration event, previous to the current epoch. It is the usage implicit in figure 6.13. This approach must be necessarily applied in systems that provide continuous measurements and cannot be based on future data. Otherwise, a better choice consists of using the nearest event, either before or after the current epoch. This is typically applied when reprocessing data from a given set of raw measurements already acquired. A third option is just using a linear interpolation between the closer calibration events.

6.4.2 Temperature Correction

Regardless of using the latest or the nearest set of calibration parameters, interpolation is improved if the dependency with physical temperature is taken into account. Temperature variation is known to be the major cause of drift in the gain and noise temperature of electronic circuitry, so an accurate thermal characterization can be of much help in predicting their values during operation. Parameters such as instrumental offset or correlator gain amplitude or others can be more insensitive to thermal variations but still a correction can be useful.

The most common model for temperature characterization is a linear dependency around the nominal temperature. Given a parameter P measured at certain calibration epoch t_c, its value at a given time t is

$$P(t) = P(t_c) + S_P(T_p(t) - T_p(t_c)) \tag{6.72}$$

where T_p is the physical temperature and S_P a sensitivity coefficient, which has to be previously characterized using a thermal chamber. This approach requires the installation of a temperature sensor properly located to be representative of the thermal evolution. In case of temperature gradients, it could be necessary to have several sensors and use a different sensitivity coefficient for each one by extending the above equation with more terms.

If both the previous and the next calibration data are available, interpolation between them improves the estimation of the parameter at the science epoch. Assume, for example that two estimates $P(t_1)$ and $P(t_2)$ are available at respective temperatures $T_p(t_1)$ and $T_p(t_2)$, then the estimated parameter at an epoch t between t_1 and t_2 is

$$P(t) = P_i(t) + S_P(T_p(t) - T_{p_i}(t)) \tag{6.73}$$

where $P_i(t)$ and $T_{p_i}(t)$ are, respectively, the linear interpolation of $P(t_1)$ and $P(t_2)$ and of $T_p(t_1)$ and $T_p(t_2)$ at the science epoch t:

$$\begin{aligned} P_i(t) &= P(t_1) + \frac{P(t_2) - P(t_1)}{t_2 - t_1}(t - t_1) \\ T_{p_i}(t) &= T_p(t_1) + \frac{T_p(t_2) - T_p(t_1)}{t_2 - t_1}(t - t_1) \end{aligned} \tag{6.74}$$

A parameter with no temperature dependence has $S_P = 0$ and then (6.73) becomes $P(t) = P_i(t)$ with $P_i(t)$ given by the first of the equations above, so it is equivalent to a linear interpolation between retrievals.

References

1 Michael Brown, Francesc Torres, Ignasi Corbella, and Andreas Colliander. SMOS calibration. *IEEE Transactions on Geoscience and Remote Sensing*, 46(3):646–658, March 2008.

2 Ignasi Corbella, Francesc Torres, Nuria Duffo, Vera González-Gambau, Miriam Pablos, Israel Duran, and Manuel MartNeira. MIRAS calibration and

performance: Results from the SMOS in-orbit commissioning phase. *IEEE Transactions on Geoscience and Remote Sensing*, 49(9):3147–3155, September 2011.

3 Ignasi Corbella, Francesc Torres, Nuria Duffo, Israel Durán, Verónica González-Gambau, Roger Oliva, Josep Closa, and Manuel Martin Neira. Calibration of the MIRAS radiometers. *IEEE Journal of Selected Topics in Applied Earth Observations and Remote Sensing*, 12(6):1633–1646, June 2019. doi: 10.1109/JSTARS.2018.2885576.

4 Ignasi Corbella, Francesc Torres, Adriano Camps, Andreas Colliander, Manuel Martín-Neira, Serni Ribó, Kimmo Rautiainen, Nuria Duffo, and Mercè Vall-llossera. MIRAS end-to-end calibration: Application to SMOS L1 processor. *IEEE Transactions on Geoscience and Remote Sensing*, 43(5):1126–1134, May 2005.

5 Janne Lahtinen, Albin J. Gasiewski, Marian Klein, and Ignasi Corbella. A calibration method for fully polarimetric microwave radiometers. *IEEE Transactions on Geoscience and Remote Sensing*, 41(3):588–602, March 2003. doi: 10.1109/TGRS.2003.810203.

6 Sten Schmidl Sobjaerg, Niels Skou, and Jan E. Balling. Measurements on active cold loads for radiometer calibration. *IEEE Transactions on Geoscience and Remote Sensing*, 47(9):3134–3139, September 2009. doi: 10.1109/TGRS.2009.2017015.

7 Ignasi Corbella, Adriano Camps, Francesc Torres, and Javier Bará. Analysis of noise injection networks for interferometric radiometer calibration. *IEEE Transactions on Microwave Theory and Techniques*, 48(4):545–552, April 2000.

8 Francesc Torres, Adriano Camps, Javier Bará, Ignasi Corbella, and Roberto Ferrero. On-board phase and modulus calibration of large aperture synthesis radiometers: Study applied to MIRAS. *IEEE Transactions on Geoscience and Remote Sensing*, GRS-34(4):1000–1009, July 1996.

9 Juha Kainulainen, Andreas Colliander, Josep Closa, Manuel Martin-Neira, Roger Oliva, Guillermo Buenadicha, Pilar Rubiales Alcaine, Anssi Hakkarainen, and Martti T. Hallikainen. Radiometric performance of the SMOS reference radiometers–assessment after one year of operation. *IEEE Transactions on Geoscience and Remote Sensing*, 50(5):1367–1383, May 2012.

10 Francesc Torres, Ignasi Corbella, Adriano Camps, Nuria Duffo, Mercè Vall-llossera, Santiago Beraza, Carlos Gutierrez, and Manuel Martín-Neira. Denormalization of visibilities for in-orbit calibration of interferometric radiometers. *IEEE Transactions on Geoscience and Remote Sensing*, 44(10):2679–2686, October 2006.

11 Ignasi Corbella, Francesc Torres, Nuria Duffo, Manuel Martin_Neira, Verónica González-Gambau, Adriano Camps, and Mercè Vall-llossera. On-ground characterization of the SMOS payload. *IEEE Transactions on Geoscience and Remote Sensing*, 47(9):3123–3133, September 2009.

12 J. C. Testori, P. Reich, J. A. Bava, F. R. Colomb, E. E. Hurrel, J. J. Larrarte, W. Reich, and A. J. Sanz. A radio continuum survey of the southern sky at 1420 MHz. *Astronomy and Astrophysics*, 368(3):1123–1132, March 2001. doi: 10.1051/0004-6361:20010088.

13 Adriano Camps, Francesc Torres, Ignasi Corbella, Javier Bará, and F. Monzón. Automatic calibration of channels frequency response in interferometric radiometers. *Electronics Letters*, 35(2):115–116, January 1999.

14 Petri Piironen. PMS offset determination using an IF attenuator. Technical Note 14629/00/NL/SF, ESA-ESTEC, Noordwijk - The Netherlands, June 2002.

15 Ignasi Corbella, Francesc Torres, Josep Closa, Nuria Duffo, Israel Durán, Verónica González-Gambau, and Manuel Martin-Neira. One-point microwave radiometer calibration. *IEEE Geoscience and Remote Sensing Letters*, 17(3):461–463, March 2020. doi: 10.1109/LGRS.2019.2923539.

16 Francesc Torres, Nuria Duffo, Ignasi Corbella, Adriano Camps, Mercè Vall-llossera, and Lluís Sagués. Dynamic range and linearity tradeoff in detectors for interferometric radiometers. *IEE Electronics Letters*, 39(25):1852–1854, December 2003.

17 Cristina Gonzalez-Haro, Francesc Torres, Nuria Duffo, Ignasi Corbella, Roger Vilaseca, Pedro de Paco, and Manuel MartNeira. Linearity characterization of detectors for interferometric radiometers. In *International Geoscience and Remote Sensing Symposium, IGARSS 2009*, volume 4, pages 657–660, Cape Town, South Africa, 12–17 July 2009. IEEE.

A

Definitions and Concepts

A.1 Complex Vectors

The reader is assumed already familiarized with vectors in general and only some subtleties regarding complex vectors (those that have complex components) are highlighted in this appendix. A vector is here represented by a capital letter with an over arrow and, for simplicity, only two components are considered, understanding that definitions apply to any number of them.

Given two complex vectors $\vec{A}$ and $\vec{B}$, the following two *dot products* are defined:

$$\vec{A} \cdot \vec{B} = A_p B_p^* + A_q B_q^* \quad \text{Hermitian} \tag{A.1}$$

$$\vec{A} \odot \vec{B} = A_p B_p + A_q B_q \quad \text{Non-Hermitian} \tag{A.2}$$

where A_p and A_q are the components of $\vec{A}$ while those of $\vec{B}$ are similarly defined. The Hermitian dot product is not commutative since $\vec{A} \cdot \vec{B} = (\vec{B} \cdot \vec{A})^*$. The non-Hermitian dot product is commutative: $\vec{A} \odot \vec{B} = \vec{B} \odot \vec{A}$. Both of them are distributive and associative. Needless to say, for real vectors, both products coincide. By default, when no adjective is applied, the Hermitian version is assumed.

Two vectors are said to be orthogonal if the Hermitian dot product vanishes

$$\vec{A} \cdot \vec{B} = 0 \quad \text{Orthogonal vectors} \tag{A.3}$$

The amplitude squared of a complex vector is the Hermitian dot product with itself

$$|\vec{A}|^2 = \vec{A} \cdot \vec{A} = |A_p|^2 + |A_q|^2 \tag{A.4}$$

so any vector can be represented as

$$\vec{A} = A\,\hat{u} \tag{A.5}$$

where A is a complex number and $\hat{u}$ a *unit vector*, denoted by a hat substituting the arrow and usually represented by a lowercase letter. It is complex in general

Principles of Interferometric and Polarimetric Radiometry, First Edition. Ignasi Corbella.

and fulfills the following property

$$\hat{u} \cdot \hat{u} = 1 \quad \text{unit vector} \tag{A.6}$$

from which it is apparent that $|\vec{A}| = |A|$. On the other hand, the phase of $\vec{A}$, or of any vector, in general, is undefined.

Independently of the above, any vector can be expressed as

$$\vec{A} = A_p\hat{p} + A_q\hat{q} \tag{A.7}$$

where $\hat{p}$ and $\hat{q}$ are *orthogonal* unit vectors. The Hermitian dot product of a vector with each of them is the corresponding component

$$A_p = \vec{A} \cdot \hat{p} \; ; \; A_q = \vec{A} \cdot \hat{q} \tag{A.8}$$

as can be easily deduced by substituting (A.7), taking into account (A.3) and the distributive property of the dot product. Moreover, if the unit vector of (A.5) is decomposed into its components,

$$\hat{u} = u_p\hat{p} + u_q\hat{q} \tag{A.9}$$

where $|u_p|^2 + |u_q|^2 = 1$, then, substituting (A.5) into (A.7) follows that

$$A_p = A\,u_p \; ; \; A_q = A\,u_q \tag{A.10}$$

A.2 Useful Complex Number Identities

Given two arbitrary complex numbers z_1 and z_2, the following identities are easily demonstrated

$$\mathfrak{Re}[z_1]\mathfrak{Re}[z_2] = \frac{1}{2}\mathfrak{Re}[z_1z_2^*] + \frac{1}{2}\mathfrak{Re}[z_1z_2] \tag{A.11}$$

$$\mathfrak{Im}[z_1]\mathfrak{Im}[z_2] = \frac{1}{2}\mathfrak{Re}[z_1z_2^*] - \frac{1}{2}\mathfrak{Re}[z_1z_2] \tag{A.12}$$

$$\mathfrak{Im}[z_1]\mathfrak{Re}[z_2] = \frac{1}{2}\mathfrak{Im}[z_1z_2^*] + \frac{1}{2}\mathfrak{Im}[z_1z_2] \tag{A.13}$$

which are obviously also applicable when $z_1 = z_2$.

A.3 Energy Conservation and Unitary Matrix

Consider an N-port microwave network characterized by S-parameters (see figure 1.13). In the absence of noise, $\boldsymbol{b} = \mathbf{S}\boldsymbol{a}$ where $\boldsymbol{a}$ and $\boldsymbol{b}$ are column vectors containing, respectively, the inward and outward waves at each port. If the network has no loss or gain of energy, due to energy conservation the sum of input powers in all ports, defined as in (1.176), must vanish.

$$\sum_{i=1}^{N}\overline{|a_i|^2} - \sum_{i=1}^{N}\overline{|b_i|^2} = 0 \Rightarrow \overline{\boldsymbol{a}^\dagger\boldsymbol{a}} - \overline{\boldsymbol{b}^\dagger\boldsymbol{b}} = 0 \tag{A.14}$$

where † means Hermitian transpose. The above identity can be expanded to

$$\overline{\boldsymbol{a}^{\dagger}\boldsymbol{a}} - \overline{\boldsymbol{a}^{\dagger}\mathbf{S}^{\dagger}\,\mathbf{S}\boldsymbol{a}} = 0 \Rightarrow \overline{\boldsymbol{a}^{\dagger}(\mathbf{I} - \mathbf{S}^{\dagger}\mathbf{S})\boldsymbol{a}} = 0 \tag{A.15}$$

where **I** is the identity matrix. Therefore

$$\mathbf{S}^{\dagger}\mathbf{S} = \mathbf{S}\mathbf{S}^{\dagger} = \mathbf{I} \tag{A.16}$$

A matrix satisfying the above conditions is said to be unitary.

The above derivation is not limited to the S-parameters. Any matrix relating waves complying with energy conservation must be unitary.

A.4 Spherical Coordinates and Solid Angle

Figure A.1 shows the convention for the spherical system of coordinates used all along this book. The co-elevation angle θ goes from 0 to π and the azimuthal angle ϕ from 0 to 2π. Conversion from spherical to rectangular coordinates is

$$x = r\sin\theta\cos\phi\ ;\ y = r\sin\theta\sin\phi\ ;\ z = r\cos\theta \tag{A.17}$$

and the inverse conversion

$$r = \sqrt{x^2 + y^2 + z^2}\ ;\ \theta = \arccos\frac{z}{r}\ ;\ \phi = \arctan\frac{y}{x} \tag{A.18}$$

A.4.1 Differential Surface

The drawing also illustrates the computation of the differential surface in spherical coordinates. Directly from the figure:

$$dS = r^2\sin\theta d\theta d\phi \tag{A.19}$$

This result can also be demonstrated mathematically using the Jacobian of the transformation (A.17).

A.4.2 Solid Angle

Solid angle is the ratio between the area of a spherical cap and the square of the radius of the sphere, just in the same way as an angle is the ratio between the

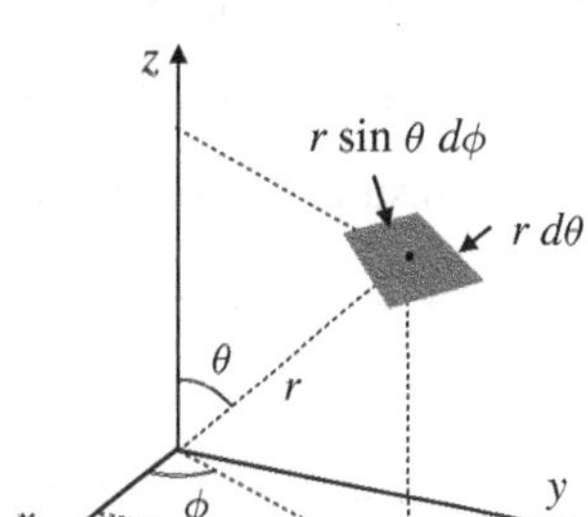

Figure A.1 Definition of spherical system of coordinates and differential surface.

arc length and the radius of the circumference. Solid angles have no units, but they are specified in steradian (denoted *sr*). Since the area of a sphere is $4\pi r^2$, the corresponding solid angle is 4π *sr*.

The differential solid angle is the ratio between the differential surface (A.19) and the square of the radius

$$d\Omega = \sin\theta d\theta d\phi \tag{A.20}$$

For example, the solid angle subtended by a *total* angle Θ is equal to the integral of $d\Omega$ in the corresponding region. Assuming the axis of the spherical cap pointing towards the z-axis and the coordinates convention of figure A.1,

$$\Omega = \iint_\Omega d\Omega = \int_0^{2\pi}\int_0^{\Theta/2} \sin\theta d\theta d\phi = 2\pi\left(1 - \cos\frac{\Theta}{2}\right) \tag{A.21}$$

The area of the spherical cap defined by this solid angle is then $r^2\Omega$ and substituting the above result produces the known expression for the surface of a spherical cap.

A.5 Quadrature Equation Inversion

Given the following equation

$$Z = \Re e[A\,V] + j\Im m[B\,V] \tag{A.22}$$

the objective is to solve for V provided A and B are known complex parameters. The above equation is first converted to matrix form

$$\begin{bmatrix} Z_r \\ Z_i \end{bmatrix} = \begin{bmatrix} A_r & -A_i \\ B_i & B_r \end{bmatrix} \begin{bmatrix} V_r \\ V_i \end{bmatrix} \tag{A.23}$$

where the subscripts r and i refer to the real and imaginary parts, respectively. The solution is immediate using matrix inversion

$$\begin{bmatrix} V_r \\ V_i \end{bmatrix} = \frac{1}{\Delta} \begin{bmatrix} B_r & A_i \\ -B_i & A_r \end{bmatrix} \begin{bmatrix} Z_r \\ Z_i \end{bmatrix} \tag{A.24}$$

where Δ is the determinant of the matrix:

$$\Delta = A_r B_r + A_i B_i = \Re e[A\,B^*] \tag{A.25}$$

The result (A.24) can in turn be converted to complex number notation

$$V = \frac{1}{\Delta}\left(\Re e[A'\,Z] + j\Im m[B'\,Z]\right) \tag{A.26}$$

where

$$A' = B_r - jA_i\ ;\ B' = A_r - jB_i \tag{A.27}$$

which can also be written as

$$\begin{aligned} A' &= \frac{1}{2}(A+B)^* - \frac{1}{2}(A-B) \\ B' &= \frac{1}{2}(A+B)^* + \frac{1}{2}(A-B) \end{aligned} \tag{A.28}$$

A.6 Special Functions

The following special functions are frequently used in this book:

$$\text{sinc:} \quad \mathrm{sinc}\,\mathrm{x} = \frac{\sin \pi \mathrm{x}}{\pi \mathrm{x}} \tag{A.29}$$

which equals one at the origin, sinc 0 = 1.

$$\text{signum:} \quad \mathrm{sgn}\,\mathrm{x} = \begin{cases} +1 & \mathrm{x} > 0 \\ 0 & \mathrm{x} = 0 \\ -1 & \mathrm{x} < 0 \end{cases} \tag{A.30}$$

$$\text{unit step:} \quad \mathrm{u(x)} = \frac{1}{2}(1 + \mathrm{sgn}\,\mathrm{x}) = \begin{cases} +1 & \mathrm{x} > 0 \\ 0.5 & \mathrm{x} = 0 \\ 0 & \mathrm{x} < 0 \end{cases} \tag{A.31}$$

The value of this function at the discontinuity in $\mathrm{x} = 0$ has no impact on integration. For any function $f(\mathrm{x})$ *continuous* at $\mathrm{x} = 0$,

$$\int_{-\infty}^{\infty} f(\mathrm{x})\mathrm{u(x)}\,d\mathrm{x} = \int_{0}^{\infty} f(\mathrm{x})\,d\mathrm{x} \tag{A.32}$$

$$\text{Rectangle:} \quad \Pi(\mathrm{x}) = \begin{cases} +1 & |\mathrm{x}| < 0.5 \\ 0 & |\mathrm{x}| > 0.5 \end{cases} \tag{A.33}$$

$$\text{Triangle:} \quad \Lambda(\mathrm{x}) = \begin{cases} 1 - |\mathrm{x}| & |\mathrm{x}| < 1 \\ 0 & |\mathrm{x}| > 1 \end{cases} \tag{A.34}$$

Dirac delta function $\delta(\mathrm{x})$: it is defined through an integration. Given any function $f(\mathrm{x})$ continuous at $\mathrm{x} = 0$,

$$\int_{\mathrm{x}_1}^{\mathrm{x}_2} f(\mathrm{x})\delta(\mathrm{x})\,d\mathrm{x} = \begin{cases} f(0) & \mathrm{x}_1 < 0 < \mathrm{x}_2 \\ 0 & \text{otherwise} \end{cases} \tag{A.35}$$

which results in the following properties

$$\delta(\mathrm{x}) = 0 \quad \mathrm{x} \neq 0$$

$$\int_{-\infty}^{\infty} \delta(\mathrm{x})\, d\mathrm{x} = 1 \tag{A.36}$$

$$f(\mathrm{x}) * \delta(\mathrm{x}) = f(\mathrm{x})$$

where the operator $*$ is the convolution (A.40). If the variable x has units (for example seconds), then $\delta(\mathrm{x})$ has the inverse of these units (for example Hertz).

A.7 Fourier Transform

The Fourier transform of a complex function of time $v(t)$ is a complex function of frequency $V(f)$ defined by[1]

$$V(f) = \mathscr{F}\,[v(t)] = \int_{-\infty}^{\infty} v(t)e^{-j2\pi ft}dt \tag{A.37}$$

Conversely, the inverse Fourier transform of $V(f)$ recovers the original signal

$$v(t) = \mathscr{F}^{-1}[V(f)] = \int_{-\infty}^{\infty} V(f)e^{j2\pi ft}df \tag{A.38}$$

The variables "time" and "frequency" are given this name for historical reasons but the above relations constitute a mathematical transformation between two domains in general.

Given a second signal $w(t)$ with Fourier transform $W(f)$, the following identity holds

$$\int_{-\infty}^{\infty} w(t)v^*(t)dt = \int_{-\infty}^{\infty} W(f)V^*(f)dt \tag{A.39}$$

which is easily demonstrated by substituting the conjugate of (A.38) into the left-hand side and changing the order of integrals. When applied to a single signal $[w(t) = v(t)]$ this result is known as Rayleigh's energy theorem. The Parseval's power theorem is a similar relation applicable to periodic signals.

A.7.1 Convolution

The convolution operator is

$$v(t) * w(t) = \int_{-\infty}^{\infty} v(t')w(t-t')dt' \tag{A.40}$$

1 Sign convention in exponentials is swapped in some references.

with the following properties

$$\begin{aligned} \text{Commutative:} &\quad w(t) * v(t) = v(t) * w(t) \\ \text{Associative:} &\quad z(t) * [w(t) * v(t)] = [z(t) * w(t)] * v(t) \\ \text{Distributive:} &\quad z(t) * [v(t) + w(t)] = z(t) * v(t) + z(t) * w(t) \end{aligned} \qquad \text{(A.41)}$$

The convolution between to functions of time is also a function of time but assignment applies to the operation, not to each function individually. In other words, writing $v(t_0) * w(t_0)$ with t_0 a constant is incorrect.

For completion, the convolution of frequency domain signals is also written

$$V(f) * W(f) = \int_{-\infty}^{\infty} V(f')W(f - f')df' \qquad \text{(A.42)}$$

A.7.2 Properties

Some useful properties of the Fourier transform are given in table A.1.

A.7.3 Transform Pairs

A handful of transform pairs is provided in table A.2. Functions involved are defined in section A.6.

A.7.4 Real Signals

A signal in the frequency domain includes both positive and negative frequencies. Denoting $V_N(f)$ and $V_P(f)$ the portions of the function for positive and for negative

Table A.1 Fourier transform properties.

Operation	Function	Transform
Linearity	$\alpha v(t) + \beta w(t)$	$\alpha V(f) + \beta W(f)$
Time delay	$v(t - t_d)$	$V(f)e^{-j2\pi f t_d}$
Scale change	$v(at)$	$\frac{1}{\lvert a\rvert} V\left(\frac{f}{a}\right)$
Sign change	$v(-t)$	$V(-f)$
Conjugation	$v^*(t)$	$V^*(-f)$
Duality	$V(t)$	$v(-f)$
Frequency translation	$v(t)e^{j2\pi f_c t}$	$V(f - f_c)$
Modulation	$v(t)\cos(2\pi f_c t + \theta)$	$\frac{1}{2}V(f - f_c)e^{j\theta} + \frac{1}{2}V(f + f_c)e^{-j\theta}$
Convolution	$v(t) * w(t)$	$V(f)W(f)$
Multiplication	$v(t)w(t)$	$V(f) * W(f)$

Source: Carlson [1]/McGraw-Hill.

Table A.2 Fourier transform pairs.

Function	Transform	Function	Transform
$B \operatorname{sinc} Bt$	$\Pi\left(\frac{f}{B}\right)$	$B \operatorname{sinc}^2 Bt$	$\Lambda\left(\frac{f}{B}\right)$
$\Pi\left(\frac{t}{\tau}\right)$	$\tau \operatorname{sinc} f\tau$	$\Lambda\left(\frac{t}{\tau}\right)$	$\tau \operatorname{sinc}^2 f\tau$
$\frac{1}{\pi t}$	$-j \operatorname{sgn} f$	$\delta(t)$	1
		1	$\delta(f)$

Source: Carlson [1]/McGraw-Hill.

frequency, respectively, any function $V(f)$ can be written in general as

$$V(f) = V_N(f)\mathrm{u}(-f) + V_P(f)\mathrm{u}(f) \tag{A.43}$$

where $\mathrm{u}(f)$ is the step unit function (A.31) and, by convention $V_N(0) = V_P(0) = V(0)/2$. Using now the conjugation property, the Fourier transform of a *real signal* $\mathrm{v}(t)$ satisfies that $V^*(-f) = V(f)$, which using (A.43) translates into $V_N(f) = V_P^*(-f)$. Then,

$$V(f) = V_P^*(-f)\mathrm{u}(-f) + V_P(f)\mathrm{u}(f) \tag{A.44}$$

and, using the conjugation property, the signal, expressed as the inverse Fourier transform of $V(f)$, becomes

$$\mathrm{v}(t) = v_P^*(t) + v_P(t) = 2\Re e[v_P(t)] \tag{A.45}$$

where $v_P(t)$ is the inverse Fourier transform of $V_P(f)$. This result is compatible with the concept of analytic signal defined in section 1.3 since $v_P(t)$ is easily recognized as one half the analytic signal of $\mathrm{v}(t)$.

A.7.5 Two-dimensional Fourier Transform

The definitions (A.37) and (A.38) are easily extended to functions of n variables. Particularly, a two-dimensional Fourier transform is

$$V(f_1, f_2) = \mathcal{F}\,[v(t_1, t_2)] = \int_{-\infty}^{\infty}\int_{-\infty}^{\infty} v(t_1, t_2)e^{-j2\pi(f_1t_1+f_2t_2)}dt_1dt_2 \tag{A.46}$$

which has the inverse

$$v(t_1, t_2) = \mathcal{F}^{-1}[v(f_1, f_2)] = \int_{-\infty}^{\infty}\int_{-\infty}^{\infty} V(f_1, f_2)e^{j2\pi(f_1t_1+f_2t_2)}df_1df_2 \tag{A.47}$$

A.8 Discrete Fourier Transform

The sampling theorem states that the Fourier transform of a signal $v(t)$ sampled at regular intervals Δt is a periodic repetition of $V(f)$ at frequencies multiple of $f_s = 1/\Delta t$. Similarly, the inverse Fourier transform of $V(f)$ sampled at regular frequency intervals, Δf is a periodic signal with period $T = 1/\Delta f$. Sampling in both time and frequency results in periodic signals in both domains. The *principal periods* in each one are defined as

$$\begin{aligned} &\text{Time domain: } -\frac{T}{2} \leq t < \frac{T}{2} \quad \text{with} \quad T = \frac{1}{\Delta f} \\ &\text{Frequency domain: } -\frac{f_s}{2} \leq f < \frac{f_s}{2} \quad \text{with} \quad f_s = \frac{1}{\Delta t} \end{aligned} \tag{A.48}$$

The periodicity of the signals implies that samples in both domains at $t + T$ or $f + f_s$ are identical, respectively, to the samples at t and f. Therefore, if $V(f)$ extends beyond $\pm f_s/2$, the periodic repetitions of the Fourier transform of the sampled signal overlap with each other and produce a phenomenon known as "aliasing." For bandlimited signals, it can be avoided by selecting a sampling rate f_s larger than $2W$ with W the bandwidth of the signal, defined from $f = 0$ to the maximum frequency at which the signal spectrum has nonzero values. This condition is known as "Nyquist sampling rate." A similar aliasing effect is produced if $v(t)$ extends beyond the principal period.

Assuming N samples in the principal period, then $T = N\Delta t = 1/\Delta f$ so

$$\Delta t\, \Delta f = \frac{1}{N} \quad \text{(reciprocal grids)} \tag{A.49}$$

Since the repetition period in frequency is $f_s = 1/\Delta t$, the above expression implies that the number of samples in both domains is the same. As stated in the equation, time and frequency grids that satisfy the above property are said to be "reciprocal."

Different samples in time and in frequency correspond, respectively, to points at $t_i = i\,\Delta t$ and $f_k = k\,\Delta f$, where i and k are integer numbers. For reciprocal grids the time-frequency product in (A.37) or (A.38) reduces to

$$f_k t_i = \frac{k\,i}{N} \tag{A.50}$$

which is used in the definitions below.

Given N samples of a time domain signal $v_i = v(t_i)$, the Discrete Fourier Transform (DFT) of vector $\boldsymbol{v} = [v_0, v_1, \ldots, v_{N-1}]$ is

$$\text{DFT:} \quad V_k = \sum_{i=0}^{N-1} v_i e^{-j2\pi k\, i/N} \tag{A.51}$$

Note that adding N to the index produces exactly the same result $V_{k+N} = V_k$ confirming the periodicity mentioned above.

The Inverse Discrete Fourier Transform (IDFT) of vector $\boldsymbol{V} = [V_0, V_1, \dots, V_{N-1}]$ is

$$\text{IDFT:} \quad v_i = \frac{1}{N}\sum_{k=0}^{N-1} V_k e^{j2\pi k\, i/N} \tag{A.52}$$

where the factor $1/N$ is needed to ensure that IDFT(DFT(v)) $= v$. Again, the periodicity is apparent when N is added to the index i.

The above definitions are compatible with a numerical evaluation of (A.37) and (A.38). Consider, for example, that $v(t)$ is a bandlimited signal sampled at a rate high enough to ensure that it is confined within the principal period in both domains.[2] This means that $v(t) = 0$ for $|t| > T/2$ and $V(f) = 0$ for $|f| > f_s/2$ and the integration limits in (A.37) and (A.38) are reduced to those of the principal periods. The Fourier transform of $v(t)$ at a given frequency $V(f_k)$ can be evaluated by discretizing (A.37). Since only the N points in the principal period are nonzero, assuming reciprocal grids

$$V(f_k) = \Delta t \sum_{i=0}^{N-1} v_i e^{-j2\pi k\, i/N} = \Delta t\, V_k \tag{A.53}$$

where $v_i = v(t_i)$. On the other hand, discretizing (A.38) and substituting the above result yields

$$\Delta f \sum_{k=0}^{N-1} V(f_k) e^{j2\pi k\, i/N} = \Delta f \Delta t \sum_{k=0}^{N-1} V_k e^{j2\pi k\, i/N} \tag{A.54}$$

which is identical to v_i when (A.52) and (A.49) are considered. Equation (A.53) is consistent regarding dimensions: if, for example, $v(t)$ has units of Volt, then V_k is also in volts and $V(f_k)$ in Volts per Hertz in agreement with (A.38).

The same analysis holds for nonlimited signals in either time, frequency or both. In this case, however, aliases must be taken into account so that $v(t_i)$ and $V(f_k)$ are no longer the original sampled signals in each domain, but they are the sum of all the aliases at the corresponding grid point.

For convention, the DFT vectors are sorted by starting in the point at $t = 0$ and $f = 0$, respectively, then the positive part of the principal period (A.48), and finally, the negative part. Sometimes, particularly to present results, the positions may be shifted to reflect the principal period as written in (A.48). The reader may wish to consult a dedicated book on Discrete Fourier Transform for more details.

2 Strictly speaking, this is only an approximation since a signal cannot be both time limited and bandlimited.

A.8.1 Correlation in Time and in Frequency

Two arbitrary complex vectors $\boldsymbol{w} = [w_0, w_1, \dots, w_{N-1}]$ and $\boldsymbol{v} = [v_0, v_1, \dots, v_{N-1}]$ satisfy the following important relationship:

$$\sum_{i=0}^{N-1} w_i v_i^* = \frac{1}{N} \sum_{k=0}^{N-1} W_k V_k^* \tag{A.55}$$

where W_k and V_k are the elements of the DFTs of $\boldsymbol{w}$ and $\boldsymbol{v}$, respectively. This equation is easily demonstrated by replacing v_i^* by the conjugate of (A.52) and interchanging the order of the sums. Note that the symbols of both indices can be arbitrarily chosen (the same on both sides, for example) and that they do not need to start at 0. The above equation can be seen as the discrete version of (A.39).

A.8.2 Random Signals

Equation (A.55) is a *mathematical identity* applicable to any two arbitrary complex vectors. In particular, they can be samples of random processes without any constraint (stationarity, independent samples, ...).

The DFT of N consecutive samples of a stochastic process is used as a convenient representation in the frequency domain, although it cannot represent a nonexisting Fourier transform of the random signal. To be precise, it corresponds to the *periodic signal* that results of replicating the very same N samples in adjacent periods. The spectral lines are thus situated at specific points in the regular grid (A.49), separated $\Delta f = 1/T$ where T is the total time in which the N samples are allocated. The band containing the frequency domain signal is the principal period (A.48) going from $-1/(2\Delta T)$ to $+1/(2\Delta T)$ where ΔT is the time separation between samples. This band is repeated periodically.

The elements of the DFT represent thus spectral components at specific frequencies inside a given bandwidth. Applying the same DFT to different segments of the signal results in different values in frequency *at the same locations*. To better illustrate this point, consider N consecutive samples of a complex stationary random process $v(t)$ and define the following random variable

$$\mathrm{p} = \frac{1}{N} \sum_{N} |v(t_i)|^2 \tag{A.56}$$

where t_i represents the sampling times. This variable takes different values depending on the specific N samples used, belonging to different signal segments. According to (1.51), the mean of p is equal to the power of the process $\overline{\mathrm{p}} = \langle |v(t)|^2 \rangle$. Using now (A.55),

$$\langle |v(t)|^2 \rangle = \frac{1}{N^2} \sum_{N} \overline{|V_k|^2} \tag{A.57}$$

where V_k is the DFT component at f_k. This equation allows interpreting the term $|V_k|^2/N^2$ as "discrete power spectral density" since the sum of all terms is the mean power of the signal. A more physical interpretation can be inferred by substituting $1/N$ by $\Delta t\,\Delta f$ as stated in (A.49) and using (A.53). The contribution of the power at frequency f_k becomes

$$\overline{\frac{|V_k|^2}{N^2}} = \overline{|V(f_k)\Delta f|^2} \tag{A.58}$$

that is the average power of the total signal at frequency f_k within a bandwidth Δf, namely $V(f_k)\Delta f$. Furthermore, comparing this result with the definition of a phasor for random signals (1.152), it follows that $V(f_k)/\sqrt{\Delta f}$ is equivalent to the phasor when the frequency interval is small.

Changing the sampling rate and/or the number of samples results in different positions of the spectral lines within different frequency margins, but the validity of (A.57) is not altered. In particular, it is independent of whether the signal is sampled at the Nyquist rate to avoid aliasing or not. Equation (A.57) is consequence only of the mathematical identity (A.55), which is always valid.

A.8.3 Two-dimensional Case

Discretization can also be applied to the two-dimensional case (A.46) and (A.47). The DFT is defined as (A.51),

$$\text{DFT2:} \quad V_{k_1 k_2} = \sum_{i_1=0}^{N-1} \sum_{i_2=0}^{N-1} v_{i_1 i_2} e^{-j2\pi(k_1 i_1 + k_2 i_2)/N} \tag{A.59}$$

where $V_{k_1 k_2}$ are the elements of the matrix $\boldsymbol{V}$ formed by samples of $V(f_1, f_2)$ and $v_{i_1 i_2}$ those of matrix $\boldsymbol{v}$ with samples of $v(t_1, t_2)$. Similarly

$$\text{IDFT2:} \quad v_{i_1 i_2} = \frac{1}{N^2} \sum_{k_1=0}^{N-1} \sum_{k_2=0}^{N-1} V_{k_1 k_2} e^{j2\pi(k_1 i_1 + k_2 i_2)/N} \tag{A.60}$$

where the factor $1/N^2$ is again needed to ensure that $\text{IDFT2}(\text{DFT2}(\boldsymbol{v})) = \boldsymbol{v}$. Now, property (A.49) of the reciprocal grids, implicitly used in the above definitions, is

$$\Delta_{t_1 t_2} \Delta_{f_1 f_2} = \frac{1}{N^2} \quad \text{(reciprocal grids)} \tag{A.61}$$

where $\Delta_{t_1 t_2}$ and $\Delta_{f_1 f_2}$ are, respectively, the Jacobian determinants of the change of variables applied to (A.46) and (A.47) for discretization.

Reference

1 A. B. Carlson. *Communication Systems.* McGraw Hill, 2nd edition, 1975.

Index

Principles of Interferometric and Polarimetric Radiometry, First Edition. Ignasi Corbella.

Printed and bound by CPI Group (UK) Ltd, Croydon, CR0 4YY

17/09/2026

14964329-0004